T0332305

IFIP Advances in Information and Communication Technology

702

Editor-in-Chief

Kai Rannenberg, Goethe University Frankfurt, Germany

IFIP Advances in Information and Communication Technology

The IFIP AICT series publishes state-of-the-art results in the sciences and technologies of information and communication. The scope of the series includes: foundations of computer science; software theory and practice; education; computer applications in technology; communication systems; systems modeling and optimization; information systems; ICT and society; computer systems technology; security and protection in information processing systems; artificial intelligence; and human-computer interaction.

Edited volumes and proceedings of refereed international conferences in computer science and interdisciplinary fields are featured. These results often precede journal publication and represent the most current research.

The principal aim of the IFIP AICT series is to encourage education and the dissemination and exchange of information about all aspects of computing.

More information about this series at https://link.springer.com/bookseries/6102

Christophe Danjou · Ramy Harik ·
Felix Nyffenegger · Louis Rivest ·
Abdelaziz Bouras
Editors

Product Lifecycle Management

Leveraging Digital Twins, Circular Economy, and Knowledge Management for Sustainable Innovation

20th IFIP WG 5.1 International Conference, PLM 2023
Montreal, QC, Canada, July 9–12, 2023
Revised Selected Papers, Part II

 Springer

Editors
Christophe Danjou 🄳
Polytechnic Montreal
Montréal, QC, Canada

Ramy Harik 🄳
University of South Carolina
Columbia, SC, USA

Felix Nyffenegger 🄳
Ostschweizer Fachhochschule
Rapperswil, Switzerland

Louis Rivest 🄳
École de Technologie Supérieure
Montréal, QC, Canada

Abdelaziz Bouras 🄳
Qatar University
Doha, Qatar

ISSN 1868-4238   ISSN 1868-422X (electronic)
IFIP Advances in Information and Communication Technology
ISBN 978-3-031-62581-7   ISBN 978-3-031-62582-4 (eBook)
https://doi.org/10.1007/978-3-031-62582-4

This Springer imprint is published by the registered company Springer Nature Switzerland AG
The registered company address is: Gewerbestrasse 11, 6330 Cham, Switzerland

If disposing of this product, please recycle the paper.

Preface

Since 2003 the International Conference on Product Lifecycle Management (PLM) has brought together researchers, developers and users of Product Lifecycle Management. It aims to integrate business approaches to the collaborative creation, management and dissemination of product and process data throughout the extended enterprises that create, manufacture and operate engineered products and systems. The conference aims to involve all stakeholders of the wide concept of PLM, hoping to shape the future of this field and advance the science and practice of enterprise development.

PLM 2023 was hosted by Polytechnique Montréal, Canada, from July 9th to July 12th, 2023. The conference started on Monday, July 10th, the industrial day was held on Tuesday, July 11th and the doctoral day was a success on Sunday, July 9th. The leading organizer was Prof. Christophe Danjou.

For its 20th anniversary, PLM 2023 registered 132 participants from 16 countries who had the chance to attend two Special Sessions (Digital engineering, and Artificial intelligence in support of digital transformation), an overall set of 20 scientific sessions plus one industrial day and an amazing social program to celebrate this anniversary. In addition, PLM 2023 proposed Academic Keynotes, a PLM Pioneer presentation, Industrial Keynotes and round tables.

4 academic keynotes:
- Catherine Beaudry, Polytechnique Montréal
- Felix Nyffenegger, Eastern Switzerland University of Applied Sciences
- Louis Rivest, École de Technologie Supérieure
- Hany Moustapha, École de Technologie Supérieure

PLM Pioneer speaker:
- Francis Bernard, 1st PLM pioneer award recipient

Industrial Keynotes:
- Mohamed-Ali El Hani (Impararia) & Frédéric Gal (Bouygues Construction) – Smart modular home construction
- Wissem Maazoun (BusPas) – The Bus Stop, Reimagined
- Fethi Chebli (VPorts) – Shaping the future of advanced air mobility
- Nick Giannias (CAE) – DOME: The Digital Operations and Maintenance Environment for the offshore renewable energy lifecycle
- Jonathan Brodeur (MILA) – Adopting applied AI in manufacturing SMEs: A project lifecycle approach

Round table on "The future of PLM with AI and Digital twins":
- Moderator: Benoît Eynard, Université de Technologie de Compiègne
- Panelists: Éric Boutin (Siemens), Kevin Chagnon Beaulieu (Techso), Arnaud Guitton (Impararia), François Lamy (PTC), Florent Salako (Dassault Systems)

All submitted papers were double-blind peer-reviewed by at least two reviewers. In total, 116 contributions were submitted, of which 62 were accepted. Program chairs Christophe Danjou and Ramy Harik proposed an amazing scientific conference with presentations grouped in 20 thematic sections.

We would like to thank everyone who directly or indirectly contributed to making the PLM 2023 conference a success, particularly by participating in rich scientific discussions and meeting all of us in Montreal, Canada.

Christophe Danjou
Ramy Harik
Felix Nyffenegger
Louis Rivest
Abdelaziz Bouras

Organization

Conference Chair

Hany Moustapha — École de Technologie Supérieure, Canada

Program Committee Chairs

Christophe Danjou — Polytechnique Montréal, Canada
Ramy Harik — University of South Carolina, USA

Local Organizing Committee

Enzo Domingos — Polytechnique Montréal, Canada
Claudia François Eudoxie — Polytechnique Montréal, Canada
Lucas Jacquet — Polytechnique Montréal, Canada
Syrine Njah — Polytechnique Montréal, Canada
Loïc Parrenin — Polytechnique Montréal, Canada
Suzanne Pirié — Polytechnique Montréal, Canada
Julien Pons — Polytechnique Montréal, Canada
Zeina Tamaz — Polytechnique Montréal, Canada

Steering Committee

Chairs

Felix Nyffennegger — Eastern Switzerland University of Applied Sciences (OST), Switzerland
Louis Rivest — École de Technologie Supérieure, Canada

Members

Abdelaziz Bouras — Qatar University, Qatar
Anderson Luis Szejka — Pontifical Catholic University of Paraná, Brazil
Balan Gurumoorthy — Indian Institute of Science, India
Benoit Eynard — University of Technology of Compiègne, France
Christophe Danjou — Polytechnique Montréal, Canada
Clement Fortin — Skolkovo Institute of Science and Technology, Russia
Darli Rodrigues Vieira — Université du Québec à Trois-Rivières, Canada
Debashish Dutta — University of Michigan, USA
Frédéric Noël — Grenoble Institute of Technology, France
José Rios — Polytechnic University of Madrid, Spain
Klaus-Dieter Thoben — University of Bremen, Germany
Néjib Moalla — University of Lyon, France

Osiris Canciglieri Junior	Pontifical Catholic University of Paraná, Brazil
Paolo Chiabert	Polytechnic of Turin, Italy
Ramy Harik	University of South Carolina, USA
Romeo Bandinelli	University of Florence, Italy
Sebti Foufou	University of Sharjah, UAE
Sergio Terzi	Politecnico di Milano, Italy

Doctoral Workshop Committee

| Monica Rossi | Politecnico di Milano, Italy |
| Yacine Ouzrout | University of Lyon, France |

Scientific Committee

Abdelhak Belhi	Qatar University, Qatar
Aicha Sekhari	University of Lyon, France
Alain Bernard	Central School of Nantes, France
Alexander Smirnov	Russian Academy of Sciences, Russia
Alison McKay	University of Leeds, UK
Amaratou Saley	University of Lyon, France
Ambre Dupuis	Polytechnique Montréal, Canada
Améziane Aoussat	Arts et Métiers ParisTech, France
Anderson Luis Szejka	Pontifical Catholic University of Paraná, Brazil
Anneli Silventoinen	LUT University, Finland
Balan Gurumoorthy	Indian Institute of Science, India
Benoit Eynard	University of Technology of Compiègne, France
Brendan Sullivan	Polytechnic of Milan, Italy
Bruno Agard	Polytechnique Montréal, Canada
Camelia Dadouchi	Polytechnique Montréal, Canada
Chen Zheng	Northwestern Polytechnical University, China
Chris Mc Mahon	Technical University of Denmark, Denmark
Christophe Danjou	Polytechnique Montréal, Canada
Claudio Sassanelli	Politecnico di Bari, Italy
Clement Fortin	Skolkovo Institute of Science and Technology, Russia
Daniel Schmid	Zurich University of Applied Sciences, Switzerland
Darli Rodrigues Vieira	Université du Québec à Trois-Rivières, Canada
Debashish Dutta	University of Michigan, USA
Detlef Gerhard	Ruhr University Bochum, Germany
Dimitris Kiritsis	Federal Institute of Technology in Lausanne, Switzerland
Eduardo Zancul	University of São Paulo, Brazil
Felix Nyffenegger	Eastern Switzerland University of Applied Sciences, Switzerland
Fernando Mas	University of Seville, Spain
Frédéric Demoly	University of Technology of Belfort-Montbéliard, France
Frédéric Noël	Grenoble Institute of Technology, France

Frédéric Segonds	Arts et Métiers ParisTech, France
George Huang	University of Hong Kong, China
Gianluca D'Antonio	Polytechnic of Turin, Italy
Guilherme Brittes Benitez	Pontifical Catholic University of Paraná, Brazil
Hamidreza Pourzarei	University of Quebec, Canada
Hannele Lampela	University of Oulu, Finland
Hannu Kärkkäinen	Tampere University of Technology, Finland
Hervé Panetto	University of Lorraine, France
Ilkka Donoghue	LUT University, Finland
Johan Malmqvist	Chalmers University of Technology, Sweden
Jong Gyun Lim	Samsung Advanced Institute of Technology, South Korea
José Rios	Polytechnic University of Madrid, Spain
Julie Jupp	University of Technology Sydney, Australia
Julien Le Duigou	University of Technology of Compiègne, France
Klaus-Dieter Thoben	University of Bremen, Germany
Lionel Roucoules	Arts et Métiers ParisTech, France
Louis Rivest	University of Quebec, Canada
Marcelo Rudek	Pontifical Catholic University of Paraná, Brazil
Margherita Peruzzini	University of Modena and Reggio Emilia, Italy
Mariangela Lazoi	University of Salento, Italy
Matthieu Bricogne	University of Technology of Compiègne, France
Michael Schabacker	University of Magdeburg, Germany
Michele Marcos de Oliveira	Pontifical Catholic University of Paraná, Brazil
Mickael Gardoni	University of Quebec, Canada
Monica Rossi	Polytechnic of Milan, Italy
Néjib Moalla	University of Lyon, France
Nickolas S. Sapidis	University of Western Macedonia, Greece
Nicolas Maranzana	Arts et Métiers ParisTech, France
Nikolaos Bilalis Technical	University of Crete, Greece
Nikolay Shilov	Innopolis University, Russia
Nopasit Chakpitak	Chiang Mai University, Thailand
Osiris Canciglieri Junior	Pontifical Catholic University of Paraná, Brazil
Paolo Chiabert	Polytechnic of Turin, Italy
Paul Hong	University of Toledo, USA
Peter Hehenberger	University of Applied Sciences Upper Austria, Austria
Ramy Harik	University of South Carolina, USA
Rebeca Arista	Airbus, France
Ricardo Jardim-Goncalves	New University of Lisbon, Portugal
Rob Vingerhoeds	ISAE-SUPAERO, France
Romain Pinquié	Grenoble Institute of Technology, France
Romeo Bandinelli	University of Florence, Italy
Samira Keivanpour	Polytechnique Montréal, Canada
Sebti Foufou	University of Sharjah, UAE
Sehyun Myung	Youngsan University, South Korea
Sergio Terzi	Politecnico di Milano, Italy

Contents – Part II

Circular Economy: Characterization, Criteria and Implementation

A Technical and Systematic Characterization of Circular Strategy
Processes   3
    Gautier Vanson, Pascale Marange, and Eric Levrat

When Industry 4.0 Meets End-of-Life Aircraft Treatment: A Brief Review
and Criteria for Identifying the Core Technologies   14
    Ghita El Anbri and Samira Keivanpour

How to Foster the Circular Economy Within the Pharmaceutical Industry?
A Research Framework Proposition.   28
    Simon Massot, Duc-Nam Luu, Claus-Jürgen Maier, Nicolas Maranzana,
    and Améziane Aoussat

Optimizing Closed-Loop Supply Chain in the Electric Vehicle Battery
Industry: A Fully Fuzzy Approach   38
    Mina Kazemi Miyangaskary, Samira Keivanpour, and Amina Lamghari

Drives and Barriers for Circular Ion-Lithium Battery Economy:
A Case Study in an Automobile Manufacturer   50
    Ana Paula Louise Yamada Macicieski, Enzo Domingos,
    Juliana Hoffmann, Luciana Rosa Leite, and Carla Roberta Pereira

Application of Life Cycle Assessment for More Sustainable Plastic
Packaging - Challenges and Opportunities   61
    Simon Merschak, Christian Kneidinger, David Katzmayr,
    Johanna Casata, and Peter Hehenberger

**Interoperability Technology: Blockchain, IoT and Ontologies for
Data Exchange**

Integrating Processes, People and Data Management to Create a
Comprehensive Roadmap Toward SMEs Digitalization: an Italian
Case Study.   75
    Marco Spaltini, Federica Acerbi, Anna De Carolis, and Marco Taisch

Development of IoT Solutions According to the PLM Approach.   85
    Francesco Serio, Ahmed Awouda, Mansur Asranov, and Paolo Chiabert

Development of a Multi-plant Cross-Function Roadmapping Tool: An
Industrial Case in Food & Beverage Sector   96
  Elena Beducci, Federica Acerbi, Marco Spaltini, Anna De Carolis,
  and Marco Taisch

A Preliminary Reflection Framework of Sustainability, Smart Cities, and
Digital Transformation with Effects on Urban Planning: A Review and
Bibliometric Analysis   107
  Andreia de Castro e Silva, Elpidio Oscar Benitez Nara,
  Marcelo Carneiro Gonçalves, Izamara Cristina Palheta Dias,
  Camila Vitoria Piovesan, and Gabrielly dos Santos Domingos

Development of a Human-Centric Knowledge Management Framework
Through the Integration Between PLM and MES   119
  Giovanni Marongiu, Giulia Bruno, and Franco Lombardi

Blockchain Applications in the Food Industry: A Pilot Project
Implementation in the Ancient Grains Industry   130
  Bianca Bindi, Gloria Padovan, Giacomo Trombi, Niccolò Bartoloni,
  Virginia Fani, Marco Moriondo, Camilla Dibari, and Romeo Bandinelli

Protecting Manufacturing Supply Chains Through PLM - Blockchain
Integration and Data Model Encapsulation   140
  Abdelhak Belhi and Abdelaziz Bouras

Smart Product-Service Systems: A Review and Preliminary Approach to
Enable Flexible Development Based on Ontology-Driven Semantic
Interoperability   151
  Athon Francisco Staben de Moura Leite, Matheus Beltrame Canciglieri,
  and Osiris Canciglieri Junior

A Preliminary Discussion of Semantic Web Technologies and 3D
Feature Recognition to Support the Complex Parts Manufacturing
Quotation: An Aerospace Industry Case.   163
  Murillo Skrzek, Anderson Luis Szejka, and Fernando Mas

An Approach to Model Lifecycle Management for Supporting
Collaborative Ontology-Based Engineering.   173
  Manuel Oliva, Rebeca Arista, Domingo Morales-Palma,
  Anderson Luis Szejka, and Fernando Mas

Learning and Training: From AI to a Human-Centric Approach

Investigation of an Integrated Synthetic Dataset Generation Workflow for
Computer Vision Applications 187
    Julian Rolf, Mario Wolf, and Detlef Gerhard

Digital Technologies and Emotions: Spectrum of Worker Decision
Behavior Analysis. 197
    Ambre Dupuis, Camélia Dadouchi, and Bruno Agard

Prediction of Next Events in Business Processes: A Deep Learning
Approach. 210
    Tahani Hussein Abu Musa and Abdelaziz Bouras

Machine Learning Algorithms for Process Optimization and Quality
Prediction of Spinning in Textile Industries 221
    Hye Kyung Choi, Whan Lee, Seyed Mohammad Mehdi Sajadieh,
    Sang Do Noh, Hyun Sik Son, and Seung Bum Sim

E-Learning Content Creation for Interdisciplinary Master of Science
Program in Product Lifecycle Management (PLM) 233
    Alexandra Saliger, Yannick Juresa, Manfred Grafinger,
    and Jens C. Göbel

Enhancing Collaborative Design Through Process Feedback with
Motivational Interviewing: Can AI Play a Role?. 244
    Sabah Farshad, Yana Brovar, and Clement Fortin

Designing a Human-Centric Manufacturing System from a Skills-Based
Perspective. 254
    Marco Dautaj, Maira Callupe, Monica Rossi, and Sergio Terzi

Smart Processes: Prediction, Optimization and Digital Thread

Product Model for Lifecycle Support of Mechanical Parts 269
    Hiroyuki Hiraoka and Arata Hori

Comparative Analysis of the Sustainability of Injection Molding and
Selective Laser Sintering Technologies for Spare Part Manufacturing. 279
    Philipp Jung, Klaas Tuschen, Kristin Zagatta, and Iryna Mozgova

Hybrid Production Structures as a Solution for Flexibility and
Transformability for Longer Life Cycles of Production Systems 289
    Dorit Schumann, Marco Bleckmann, and Peter Nyhuis

A Methodology to Promote Circular Economy in Design by Additive
Manufacturing 300
    Simona Ianniello, Giulia Bruno, Paolo Chiabert, Fabrice Mantelet,
    and Frédéric Segonds

Design and Release Process for AM Parts 313
    Daniel Schmid

Investigation on Additive Manufacturing Processes Performed by
Collaborative Robot. 323
    Khurshid Aliev, Mansur Asranov, Tianhao Liu, and Paolo Chiabert

Optimization Framework for Assembly Line Design Problem with
Ergonomics Consideration in Fuzzy Environment 333
    Elham Ghorbani, Samira Keivanpour, Firdaous Sekkay,
    and Daniel Imbeau

Optimization of the Operation Management Process of a Company in the
Electronic Manufacturing Sector 344
    Marcelo Carneiro Gonçalves, Katuzi Hamasaki,
    Izamara Cristina Palheta Dias, and Elpidio Oscar Benitez Nara

Towards Zero-Defect Manufacturing in the Silicon Wafer Production
Through Calibration Measurement Process: An Italian Case 355
    Federica Acerbi, Andrea Pranzo, Cristina Sanna, Marco Spaltini,
    and Marco Taisch

Author Index 365

Contents – Part I

Technology Implementation: Augmented Reality, CPS and Digital Twin

Benefits of Digital Twin Applications Used to Study Product Design and
Development Processes 3
   Milad Attari Shendi, Vincent Thomson, Haoqi Wang, and Gaopeng Lou

A Digital Twin Framework for Industry 4.0/5.0 Technologies 14
   *Mansur Asranov, Khurshid Aliev, Paolo Chiabert,
and Jamshid Inoyatkhodjaev*

A Data Structure for Developing Data-Driven Digital Twins 25
   *Oghenemarho Orukele, Arnaud Polette, Aldo Gonzalez Lorenzo,
Jean-Luc Mari, and Jean-Philippe Pernot*

A Quality-Oriented Decision Support Framework: Cyber-Physical Systems
and Model-Based Design to Develop Design for Additive Manufacturing
Features 36
   *Claudio Sassanelli, Giovanni Paolo Borzi, Walter Quadrini,
Giuseppe De Marco, Giorgio Mossa, and Sergio Terzi*

Examining the Influence of Business Models on Technical Implementation
of Smart Services 47
   Samuel Helbling and Felix Nyffenegger

Navigating the Digital Twin: 3D Exploration for Asset Administration Shell
Content 59
   Mario Wolf, Oliver Vogt, and Detlef Gerhard

Industrial Layout Mapping by Human-Centered Approach and Computer
Vision 70
   Osmar Moreira da Silva Neto and Marcelo Rudek

A Data Management Approach for Modular Industrial Augmented Reality
Applications 80
   Jan Luca Siewert, Matthias Neges, and Detlef Gerhard

A Method to Interactive Simulations of Industrial Environments Based on
Immersive Technologies 91
   Richard Valandro, João Cláudio Nogueira, and Marcelo Rudek

Organisation: Knowledge Management, Change Management, Frameworks for Project and Service Development

How to Support Knowledge Exchange in a Multi-division Manufacturing
Firm Using a Prototype Platform?.............................. 105
  Mélick Proulx and Mickaël Gardoni

Contextualization for Generating FAIR Data: A Dynamic Model for
Documenting Research Activities 116
  *Osman Altun, Marc Hinterthaner, Khemais Barienti,
  Florian Nürnberger, Roland Lachmayer, Iryna Mozgova,
  Oliver Koepler, and Sören Auer*

Gamification as a Knowledge Management Tool..................... 127
  Pierre Miroite and Mickaël Gardoni

Drivers of Change Impacting Outcome-Based Business Models in
Industrial Production Equipment 136
  Olli Kuismanen, Karan Menon, and Hannu Kärkkäinen

Exploration of Multi-layers Networks to Elicit and Capture
Product Changes... 151
  *Vincent Cheutet, Aicha Sekhari, Chantal Cherifi, Justin Favre,
  Mohamed Douass, and Maxence Millescamps*

A Model to Predict Span Time and Effort for Product Development
Processes.. 161
  Shourav Ahmed and Vince Thomson

A Multicriteria Framework Proposition for Project Management
Approaches 171
  *Márcio Leandro do Prado, João Felipe Capioto Seelent,
  Gilberto Reynoso-Meza, and Guilherme Brittes Benitez*

Innovative Development in a University Environment Based on the Triple
Helix Concepts: A Systematic Literature Review................... 181
  *Lucas Sydorak Lessa, Michele Marcos de Oliveira,
  and Osiris Canciglieri Junior*

Understanding Service Design in the Context of Servitization............. 191
  *Ana Maria Kaiser Cardoso, Pedro da Rocha Loures Robell,
  and Guilherme Brittes Benitez*

Modelisation: CAD and Collaboration, Model-Based System Engineering and Building Information Modeling

SE Based Development Framework for Changeable Maritime Systems   203
 Brendan Sullivan and Monica Rossi

Interface Modeling for Complex Systems Design: An MBSE and PLM
System Integration Perspective   215
 Yana Brovar, Arkadii Kazanskii, Betania Tapia, and Clement Fortin

An Implementation of Integrated Approach in Product Life-Cycle
Management Tool to Ensure Requirements-In-Loop During Complex
Product Development: A Cubesat Case Study   225
 Mubeen Ur Rehman and Clement Fortin

Holistic Perspective to the Drug-Device Combination Product Development
Challenges   235
 Yaroslav Menshenin, Romain Pinquié, and Pierre Chevrier

Knowledge-Based Engineering Design Supported by a Digital Twin
Platform.   243
 Sthefan Berwanger, Henrique Diogo Silva, António Lucas Soares,
 and Cristiano Coutinho

Industrialization of Site Operations Planning and Management: A BIM-
Based Decision Support System   253
 Félix Blampain, Matthieu Bricogne, Benoît Eynard, Céline Bricogne,
 and Sébastien Pinon

On Considering a PLM Platform for Design Change Management in
Construction.   263
 Hamidreza Pourzarei, Conrad Boton, and Louis Rivest

BIM Technology Application to Propagate the Knowledge and Information
for Design Changes and Modifications Throughout the Product
Development Cycle.   277
 José Roberto Alcântara Lobo, Anderson Luis Szejka,
 and Osiris Canciglieri Junior

Towards a Multi-view and Multi-representation CAD Models System for
Computational Design of Multi-material 4D Printed Structures   287
 Hadrien Belkebir, Romaric Prod'hon, Sebti Foufou, Samuel Gomes,
 and Frédéric Demoly

A State of the Art of Collaborative CAD Solutions 298
    Hugo Locquet, Louis Rivest, and Matthieu Bricogne

KARMEN: A Knowledge Graph Based Proposal to Capture Expert
Designer Experience and Foster Expertise Transfer 309
    *Jean René Camara, Philippe Véron, Frédéric Segonds, Esma Yahia,
    Antoine Mallet, and Benjamin Deguilhem*

Author Index 323

Circular Economy: Characterization, Criteria and Implementation

A Technical and Systematic Characterization of Circular Strategy Processes

Gautier Vanson$^{(\boxtimes)}$, Pascale Marange , and Eric Levrat

Université de Lorraine, CRAN, UMR 7039, Campus Sciences, BP 70239,
54506 Vandœuvre-lès-Nancy, France
{gautier.vanson,pascale.marange,eric.levrat}@univ-lorraine.fr

Abstract. To deal with environmental, economic and social problems linked to high consumption and intense manufacturing, Europe, ecological associations propose scenarios for the implementation of the circular economy, without saying how to implement them. Several circular strategies (CS) (reuse, remanufacturing, recycling...) enable to regenerate a product and its components throughout its life cycle. Currently, these CS are not implemented in a systemic way, but in a punctual and individual way. The objective is to deepen the knowledge of the global implementation of several CS. To achieve this, the paper proposes to characterise each CS according to the requirements of the entering and leaving products. This technical characterisation allows us to analyse their complementarities and interactions. This characterisation will be illustrated by the example of remanufacturing and resynthesis. Then, an analysis of our proposal is carried out to show the contributions in order to have a holistic approach of the regeneration.

Keywords: Circular Economy · Circular strategy · Products regeneration · Holistic approach · Requirement · characterization

1 Introduction

Today, governments are urging more and more companies to adopt a circular economy (CE) through laws, or decrees. Indeed, current issues such as linear consumption through mass customization, fashion effects and obsolescence [1] are not compatible with the objectives of sustainable development (SD). The literature shows that this consumption mode is a source of environmental problems such as the increase in waste, climate change [12] and the depletion of natural resources [15]. In addition, pandemics and wars show how dependent some states are on certain primary resources [23].

The CE concept based on the pillars of SD is defined by [7]: "A CE is an industrial system that is restorative or regenerative by intention and design. It replaces the 'end-of-life' concept with restoration, shifts towards the use of renewable energy, eliminates the use of toxic chemicals, which impair reuse,

C. Danjou et al. (Eds.): PLM 2023, IFIP AICT 702, pp. 3–13, 2024.
https://doi.org/10.1007/978-3-031-62582-4_1

and aims for the elimination of waste through the superior design of materials, products, systems, and, within this, business models.". Through this definition, the consideration of a systemic vision with several regeneration cycles becomes essential to achieve the objectives of CE. The objective of our work over the last few years is to propose a holistic approach of product regeneration (i.e consider the interactions of different systems and stakeholders in a global view). In Diez 2016, we defined the regeneration paradigm as [11]: "set of actions, natural or technical, to restore a waste or its constituents to an acceptable state (functional and operational) allowing to extend its life cycle".

In the literature, many works deal with circular strategies (CS), eco-design but there are very few papers that integrate the multiple life cycles and the holistic approach. Thus, a CS standard characterization would help regeneration designers to define an adequate regeneration process for their products in a holistic approach. In order to achieve the SD objectives, a characterization considering environmental, economical and social aspects is necessary. For this first work, only the CS technical aspect is considered. Thus, this article proposes a technical and systematic characterization of CS with a formalization that allows to compare and analyze CS. The final objective is to give a better knowledge of CS to help industrial to adopt CE. This holistic vision will allow them to better apprehend the complexity of CE and the new CE laws implemented by states.

2   Circular Strategies State of the Art

In literature, the circular strategies (CS) enable to extend the product life and create loops for circular economy (CE) to recover value for a product, sub-assemblies/components or material that would otherwise be considered as waste [22]. A literature review enabled the identification of several CS: reuse, repurposing, upgrading, reconditioning, remanufacturing, resynthesis, recycling, energetic valorization and landfill. A systematic review was done for each CS. This complete work is not presented in this article, only few references are used by CS. Each selected reference is considered as the best to describe the CS in a holistic view. The objective of this section is to present the technical CS features identified in the literature. The following elements are analyzed:

- **input**: the expected level in the product breakdown structure, and the expected product quantity for carrying out a regeneration.
- **output**: the expected level in the product breakdown structure and its nature: does it have the same purpose and the same composition as the incoming product (i.e. is it identical /different to the incoming product (identical finality and composition) or is it improved (identical finality and different composition by additions/subtractions/modifications))
- **process**: the CS activities and the level of the product breakdown structure that is regenerated. This listed information of the processes will not be developed in the rest of the article but considered as the activities that enable to regenerate an input product.

Table 1. CS technical features in the literature

CS	Input	Output	Process Activities	Regeneration level
Reuse [13, 21]		The same input product	diagnostic, cleaning and requalification	
Repurposing [20, 21]		The same input product with another finality		
Upgrading [4, 18, 19]	Product	The same input product but enhanced	diagnostic, cleaning, requalification, replacement, disassembly and Re-Assembly	Product
Reconditioning [4, 14]		The same input product with few different components		
Remanufactur-ing [2, 10, 15]	Several products sub-assemblies and components	A new product composed with different components	diagnostic, cleaning, requalification, replacement, disassembly, Re-Assembly and reworking	
Resynthesis [8, 9]	Several sub-assemblies and components from different product families	A new product but now for another finality and composed with different components from different product families		Sub-assemblies / Components
Recycling [3, 5, 6, 16]	Several products material	Recycled material	diagnostic, Sort and Separate	Material
Energetic Val. [17]		Energy	Combustion or Methanization	Wastes
Landfill [24]		Wastes	Stock	No

The literature review shows that CS are developed in a silo view because researchers optimize the CS process activities rather than the life cycle of the product to be regenerated. Thus, their goal is to regenerate a product with a certain health state only once. Nowadays, CS need to be developed in a holistic vision integrating each other, explaining how to implement them, how to choose a regeneration trajectory between different possibilities and regenerate for several life cycles. In this way, the CE objectives will be achieved, i.e. not lose value (recycle a product that could be reused) and preserve the product regeneration potential (regenerate by considering others CS), it's necessary to define a regeneration process in its entirety.

Table 1 shows some differences and similarities between CS. For example, several CS have common activities which could enable to pool activities and reduce the regeneration cost infrastructures. In addition, CS as reuse, repurposing, upgrading, and reconditioning expect the same input (a product) but have a different output: the same product regenerated for reuse, the same product with a different finality for repurposing etc. These features, which are sometimes close, show that the papers analyzed in literature are not sufficient to analyze CS and to choose the optimal regeneration for a product according to its health state. Indeed, how to choose a CS between these four CS, knowing that each one expects a product to be regenerated as input? What are the technical features that enable us to say whether a regenerated product is different in its finality and composition? A formalization is necessary to deepen the features identified in the literature, to determine which are the technical elements that allow to specify a regenerated product and especially to be able to position how much one CS is different from another by the formalization. A formalization based on the product requirements's baseline is proposed. Indeed, it is according to the product's technical features, reflecting its health state, and the CS's capacity to regenerate a product, i.e. if a product can access to a CS or not, to be regenerated.

3   CS Characterization and Comparison Methodology

For regenerating an end-of-life product, a circular strategy (CS) will be chosen according to various economic, environmental, and social constraints, but mainly according to the product's health state. We must also consider the CS's capacity to regenerate the product in order to reach a desired state at the CS's output. So, the product's health state is used to categorize CS, it can be measured according to its initial product requirements.

A requirement is defined by the standard (ISO/IEC 26702:2007) as a "Statement that identifies a product or process operational, functional, or design feature or constraint, which is unambiguous, testable or measurable, and necessary for product or process acceptability". According to [25], a guide for systems engineering, there are several types of requirements: functional, performance...

It is interesting to choose the product requirements to characterize the CS because the requirements are built in the design phase and constitute a commune baseline for a product family. This baseline is then used in the manufacturing phase which uses this baseline to manufacture product instances. As it is illustrated in Fig. 1, when the products leave the factory, they are considered as new, and all their requirements are satisfied. During the use phase, the product's use will degrade the product's health state, and throughout time some requirements will no longer be satisfied, the product becomes a product to regenerate. Then, depending on the product requirements state, a CS is selected and a regenerated product is obtained. Throughout its life cycle, the product is defined by requirements sets that evolve.

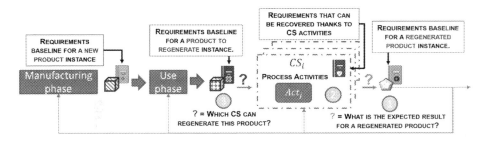

Fig. 1. Origin of the baseline requirements and the formalization method illustration.

The literature has shown that some CS can regenerate a product at multiple levels, and the requirement baselines are different according to the product life phase. Thus, the formalization of a requirement baseline $A^{l,k}$ must be defined according to the product breakdown structure level and the product life phase with:

- The l index to define the level of the product's breakdown structure. The l index can therefore take the value in the interval $[1; n]$, with n the level number in the product's breakdown structure ($l = 1$ the product level, for $l \in [2; n - 2]$ the different sub-assemblies levels, $l = n - 1$ the component level and $l = n$ the material level).
- The k index, to specify the product's life phase: if the product comes from: manufacturing phase N (New), use phase r (to regenerate) or regeneration phase R (regenerated).

Each requirements set can be decomposed according to the requirement type t: $A^{l,k} = \{A_t^{l,k}\}$ with $t = \{F(\text{Functional}), Q(\text{Performance})\}$. In this first work and to simplify, only the F and Q type are used. A requirement is defined as $Req_{j,t}^{l,k}$ with j the requirement ID. A requirement has a textual definition that can have parameters. To illustrate, a new product A leaving the factory has a requirements baseline composed of a F and a Q requirement subset: $A^{l,k} = \{A_F^{l,k}, A_Q^{l,k}\}$ with $A_F^{l,k} = \{Req_{1,F}^{1,N} : \text{"The system must provide a rotation"}\}$ and $A_Q^{l,k} = \{Req_{2,Q}^{1,N} : \text{"The nominal speed must be between 1000 and 1500 rpm"}\}$

The requirements sets of a new product $(A^{1,N})$ and regenerated product $(A^{1,R})$ product are compared according to their definition and parameters. For that, The 3 operators are defined:

- $\equiv$ indicates that the 2 requirements have the same definition and the same parameters. Ex: $Req_{1,F}^{1,N} \equiv Req_{1,F}^{1,R}$: "The system must provide a rotation"
- $=$ indicates that the 2 requirements have the same definition but with different parameters. Ex: $Req_{2,Q}^{1,N} = $ "The nominal speed must be between **1000** and **1500** rpm" compared with $Req_{2,Q}^{1,R} = $ "The nominal speed must be between **800** and **1200** rpm"
- $\neq$ indicates that the 2 requirements have different definitions. Ex: $Req_{1,F}^{1,N} = $ "The system must provide a **rotation**" compared with $Req_{1,F}^{1,R} = $ "The system must provide a **translation**"

To compare the requirements sets, augmented same operators are used with the same variables and the requirements number:

- $\equiv$ indicates that the 2 sets compared have the same requirements number, with the same definitions and parameters for each requirement.
- $=$ indicates that the 2 sets compared have the same requirements number, the same definitions but with different parameters for some requirements.
- $\neq$ indicates that the 2 sets compared have different requirements number, and/or with different definitions. In this case, the operator $\sqsubseteq$ is added to specify whether the set, although different, has the same common basic requirements of the product family. For example, the set $A_F^{1,N}$ with $A_F^{1,N} = \{Req_{1,F}^{1,N}\}$ and $A_F^{1,R} = \{Req_{1,F}^{1,R}, Req_{2,F}^{1,R}, Req_{3,F}^{1,R}\}$ so, $A_F^{1,N} \neq A_F^{1,R}$ because the product number is different but $A_F^{1,N} \sqsubseteq A_F^{1,R}$ because $Req_{1,F}^{1,N}$ and $Req_{1,F}^{1,R}$ have the same definition.

To categorize CS, the following methodology uses the previous formalization:

- **Step 1**: The starting point of the analysis is to measure the gap between the regenerated product expected and the original product (new). So, the first step consists in comparing both requirements sets, respectively $A_t^{l,R}$ and $A_t^{l,N}$ for each requirement type t.
- **Step 2**: Consists in identifying the requirement type t that could be regenerated by the CS activities.
- **Step 3**: Consists in comparing the requirements sets of a product to regenerated $A_t^{l,r}$ to the regenerated product expected $A_t^{l,R}$ for each requirement type t.

To illustrate this methodology, remanufacturing and resynthesis processes are categorized:

Step 1: the requirements set of a remanufactured product $A^{1,R}$ and a resynthesized product $B^{1,R}$ are compared with an original product $A^{1,N}$.

- For remanufacturing, the F requirements are identical to the original product (new) ($A_F^{1,R} \equiv A_F^{1,N}$) and the Q requirements have different parameters (worse performance than the original product) ($A_Q^{1,R} = A_Q^{1,N}$).
- For resynthesis, the requirements number and definitions are different because the product finality has changed (For $t = \{F, Q\}$: $B_t^{1,R} \neq A_t^{1,N}$).

Step 2: Both CS regenerate components and sub-assemblies and are able to recover F and Q requirements because they have the same activities. The difference is that resynthesis regenerates sub-assemblies and components from several other different products.

Step 3: Both CS receive some components and sub-assemblies, which may have requirements not satisfied, but with the CS activities satisfaction will be recovered for F and Q requirements (For $t = \{F, Q\}$: $A_t^{l,r} \neq A_t^{l,R}$ and $A_t^{l,r} \neq B_t^{l,R}$).

The following section analyzes the CS categorization and positions the CS.

4   Discussion

The characterization made it possible to clarify the differences and complementarities between circular strategies (CS). Indeed, by comparing the sets of F and Q requirements of a CS output product, a more relevant positioning is proposed. In addition, the CS characterization confirms that all CS do not enable a complete regeneration of all the product levels in a sustainable way. The next subsection shows the positioning.

4.1   Gap Between Regenerated Product Compared to an Original Product

The CS characterization showed that no CS can achieve the equivalent of an original product but with inferior performance. So the more a regenerated product has different requirements from the original product, the more it is distant.

Therefore, the CS of this regenerated product is more distant from the manufacturing process of the original (new) product. In addition, the more a regenerated product is close to an original product, the higher its health state, and the more likely it would go through a regeneration cycle again. Thus, the shortest gap for a product between an original and regenerated product will be preferred when choosing a CS.

Reuse, reconditioning, remanufacturing and recycling allow to get closer to the original requirements with the difference that the Q requirements have parameters representing their performance intervals lower than the original ones $(A_Q^{l^R} = A_Q^{l^N})$. Thus, to decide which of these four CS used, it is necessary to look at the regeneration activities. It is more advantageous to reuse a product because it requires fewer activities and the product level (l) is the highest (1). However, to be reused, the incoming product must be at product level and all its functional requirements must be satisfied.

For upgrading, the objective is to improve the product by adding, deleting or improving the requirements. Thus, an updated product has different requirements set than the original one, but with the initial functional requirements common to the original product $(A_F^{1^N} \sqsubseteq A_F^{1^R})$.

Other CS such as repurposing and resynthesis completely change the product's finality, which further differentiates the requirements of these regenerated products from the original product $(B_F^{1^R} \neq A_F^{1^N})$. Finally, energy recovery and landfill are even further away, as the output is energy or waste.

Figure 2 illustrates the above explanations by positioning CS according to the comparison gap between a regenerated product and an equivalent new product. In addition, CS are positioned according to what the CS activities can regenerate.

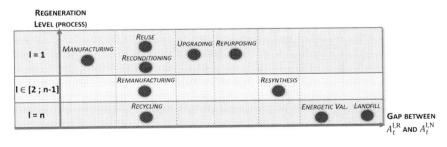

Fig. 2. CS positioning according to the comparison gap between a regenerated product with an equivalent new product and the regeneration level process.

4.2 An Analysis of Holistic Approach

To implement a complete circular economy (CE) it is necessary to regenerate a product at all its levels. From a sustainable development point of view it is obvious that favoring short regeneration loops with few activities with the highest product level like reuse is more advantageous. On the contrary, recycling a product that still has functional and reusable sub-assemblies is a significant loss.

The analysis of all circular strategies (CS) input shows that it is possible that a product does not fulfill any condition to access a CS. For example, if there is a product with a requirement set that does not have the right requirements satisfied to access product-level CS, or sub-assembly-level CS then the product will be regenerated for its material whereas probably several sub-assemblies and components are still functional. A need for a holistic approach in the design phase becomes important in order to plan the different regeneration and avoid skipping regeneration levels.

Thanks to the CS characterization and the expected input requirements (image of the product health state to regenerate) it is possible to position CS by specifying supply chains for the product flow to be regenerated. For example, to avoid switching from reuse to recycling without skipping intermediate CS. Four groupings are proposed:

- **Product from use phase**: The product can be sent to reuse, upgrading and reconditioning.
- **Product, sub-assemblies and components from reconditioning**: The product can be sent to repurposing and sub-assemblies/components to remanufacturing or resynthesis.
- **Sub-assemblies and components from remanufacturing**: The sub-assemblies and components are either sent to other supplier regeneration cycles that can regenerate them or they are sent to recycling.
- **Material from recycling**: The material considered as waste is used as an energy source (energetic valorization) or landfilled if this is not possible.

Figure 3 illustrates these explanations by positioning CS in a holistic approach. To simplify the figure, external cycles of different products that supply

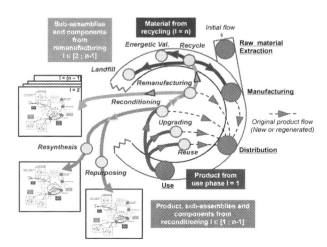

Fig. 3. CS positioning synthesis to avoid value loss and preserve the regeneration potential of a product throughout its life cycle.

some CS with elements to regenerate have not been added. In addition, fig.3 could be instantiated for each element disassembled. Indeed, sub-assemblies are potentially the finished product of a sub-supplier.

By defining the access conditions of each CS in adequacy with the CS regeneration capacities and the expected result, it is possible to constrain the regeneration paths to avoid losing value and preserve the regeneration potential of a product throughout its life cycle.

5   Conclusion

According to the need to adopt a circular economy (CE) and the problems highlighted in literature, circular strategies (CS) are implemented in a punctual/individual way and the processes are mostly artisanal, this paper proposes to use a holistic vision to define the regeneration process which enables to support the CE. However, in this assumption, it would be necessary to analyze CS to study their differences and complementarities. So technical CS features from review are identified in Sect. 2 and Sect. 3 proposes a formalization to complete the CS characterisation to specify the CS more precisely, to be able in Sect. 4, to position CS according to their technical features. Finally, a possible technical regeneration path is proposed to avoid losing value and preserving the regeneration potential of a product throughout its life cycle. So, this first characterization work raised several questions and perspectives:

This first technical and systematic characterization of CS is a beginning. Indeed, in order to get a CS holistic approach, complementary characterizations from an environmental, economic and social point of view should complete this work. By integrating these elements into the CS categorization it will be possible to choose the optimal CS for a product to be regenerated, with a multi-criteria decision-making system. This system will need to be fed with complementary data (product, market and CS features, ...) [26].

This paper promote the need of a holistic approach to define a regeneration process and implement multiple CS. We therefore propose in the design phase to develop, in addition to the product, a regeneration system that will support the product family throughout the product life cycle. Thus, a product, its manufacturing system and its regeneration system must be co-designed to consider their interactions. So, the CS characterization in Sect. 3 can be used to determine the outputs, activities, and input conditions of the CS to build the regeneration system. Through this co-engineering, it would therefore be possible to avoid skipping regeneration loops and implement a realistic CE.

References

1. Giles Slade. Made to break: Technology and obsolescence in America. Harvard University Press (2006)
2. Östlin, J., Sundin, E., Björkman, M.: Product life-cycle implications for remanufacturing strategies. J. Clean. Prod. **17**(11), 999–1009 (2009)

3. Al-Salem, S.M., Lettieri, P., Baeyens, J.: Recycling and recovery routes of plastic solid waste (PSW): a review. Waste Manag. **29**(10), 2625–2643 (2009)
4. Ijomah, W.L., Danis, M.: 8 - Refurbishment and reuse of WEEE. In: Waste Electrical and Electronic Equipment (WEEE) Handbook, pp. 145–162. Woodhead Publishing (2012)
5. Ravi, V.: Evaluating overall quality of recycling of e-waste from end-of-life computers. J. Clean. Prod. **20**(1), 145–151 (2012)
6. Binnemans, K., et al.: Recycling of rare earths: a critical review. J. Clean. Prod. **51**, 1–22 (2013)
7. Ellen Mac Arthur Foundation. Towards the circular economy. In: Ellen Mac Arthur Foundation (2013)
8. Sane, C., Tucker, C.S.: Product Resynthesis as a reverse logistics strategy for an optimal closed-loop supply chain. In: Volume 4: 18th Design for Manufacturing and the Life Cycle Conference; 2013 ASME/IEEE International Conference on Mechatronic and Embedded Systems and Applications (2013)
9. Kang, S.W., et al.: Product Resynthesis: knowledge discovery of the value of end-of-life assemblies and subassemblies. J. Mech. Des. **136**(1) (2013)
10. Abbey, J.D., et al.: Remanufactured products in closed-loop supply chains for consumer goods. Prod. Oper. Manag. **24**(3), 488–503 (2015)
11. Diez, L., et al.: Maintenance as a cornerstone for the application of regeneration paradigm in systems lifecycle. In: Complex Systems Design & Management: Proceedings of the Sixth International Conference on Complex Systems Design & Management, pp. 185–197 (2016)
12. Ghisellini, P., Cialani, C., Ulgiati, S.: A review on circular economy: the expected transition to a balanced interplay of environmental and economic systems. J. Clean. Prod. **114**, 11–32 (2016)
13. Cooper, D., Gutowski, T.: The environmental impacts of reuse: a review. J. Ind. Ecol. **21**(1), 38–56 (2017)
14. Gaur, J., Amini, M., Rao, A.K.: Closed-loop supply chain configuration for new and reconditioned products: an integrated optimization model. Omega **66**, 212–223 (2017). New Research Frontiers in Sustainability
15. Hazen, B.T., Mollenkopf, D.A., Wang, Y.: Remanufacturing for the circular economy: an examination of consumer switching behavior. Bus. Strateg. Environ. **26**(4), 451–464 (2017)
16. Kumar, A., Holuszko, M., Espinosa, D.C.R.: E-waste: an overview on generation, collection, legislation and recycling practices. Resour. Conserv. Recycl. **122**, 32–42 (2017)
17. Machin, E.B., Pedroso, D.T., de Carvalho, J.A.: Energetic valorization of waste tires. Renew. Sustain. Energy Rev. **68**, 306–315 (2017)
18. Pialot, O., Millet, D., Bisiaux, J.: "Upgradable PSS": Clarifying a new concept of sustainable consumption/production based on upgradablility. J. Clean. Prod. **141**, 538–550 (2017)
19. Khan, M.A., et al.: Review on upgradability - a product lifetime extension strategy in the context of product service systems. J. Clean. Prod. **204**, 1154–1168 (2018)
20. Scott, K.A., Todd Weaver, S.: The intersection of sustainable consumption and anticonsumption: repurposing to extend product life spans. J. Publ. Pol. Mark. **37**(2), 291–305 (2018)
21. Bauer, T., Zwolinski, P., Nasr, N., Mandil, G.: Characterization of circular strategies to better design circular industrial systems. J. Remanufacturing **10**(3), 161–176 (2020)

22. Morseletto, P.: Targets for a circular economy. Resour. Conserv. Recycl. **153**, 104553 (2020)
23. Roy, S.: Economic impact of COVID-19 pandemic. A preprint, pp. 1–29 (2020)
24. Zhang, Y., et al.: Impact assessment of odor nuisance, health risk and variation originating from the landfill surface. Waste Manage. **126**, 771–780 (2021)
25. SEBok. Guide to the Systems Engineering Body of Knowledge (2022). https://sebokwiki.org/wiki/Stakeholder_Requirements_Definition
26. Vanson, G., Marangé, P., Levrat, E.: End-of-life decision making in circular economy using generalized colored stochastic petri nets. Auton. Intell. Syst. **2**(1), 1–18 (2022)

When Industry 4.0 Meets End-of-Life Aircraft Treatment: A Brief Review and Criteria for Identifying the Core Technologies

Ghita El Anbri[✉] and Samira Keivanpour

Department of Mathematics and Industrial Engineering, Polytechnique Montréal 2500 Chem. de Polytechnique, Montréal, QC H3T 1J4, Canada
{ghita.el-anbri,samira.keivanpour}@polymtl.ca

Abstract. Only a few studies discuss the applications of industry 4.0 in the context of end-of-life aircraft treatment. This paper aims to discuss the implications of these technologies in end-of-life aircraft processes. The selected technologies including block-chain, the internet of things, digital twins, big data, artificial intelligence, augmented/virtual reality, collaborative robots are discussed. In addition, the criteria to evaluate the different technologies and a conceptual model for selecting the core technologies will be presented.

Keywords: Industry 4.0 · End-of-life aircraft · Analytical Network Process (ANP)

1 Introduction

The aerospace industry is more and more challenged by the growing disposal of end-of-life (EoL) aircraft [54]. Indeed, more than 15,000 aircraft are projected to reach the end of their life in two decades [2]. Despite the growth of EoL aircraft, the status quo remains the disposal in deserts [10, 41]. There are several alternatives to EoL aircraft treatment and thus several initiatives are taking place. AFRA (Aircraft Fleet Recycling Association) focuses on decommissioning aircraft in the most sustainable way [1]. The European Commission introduced the PAMELA project which stands for the Process for Advanced Management of EoL aircraft and aims to dispose of aircraft in the most sustainable and safe way possible [12]. The performance of EoL management can be increased with the help of industry 4.0 and its tool. Industry 4.0 in the EoL context is a new field of study that started in the mid-2010s [8]. Only a limited amount of literature addresses EoL products and industry 4.0 implications [38]. This paper aims to address the applications of industry 4.0 technologies in the EoL aircraft context and introduce a conceptual framework for evaluating the relationship and performance of the core technologies. This study's contributions are as follows:

- Identifying the implications of Industry 4.0 in EoL aircraft context considering the operational context

- Developing a multi-criteria decision framework for identifying the key criteria for evaluating the industry 4.0 technologies

The rest of the paper is organized as follows; the literature on EoL aircraft treatment processes and subprocess is discussed. After that, a brief literature review is provided on industry 4.0 technologies that are used or can be used in the EoL context. Then, a synthesis is provided for classifying the relation between aircraft processes and industry 4.0 technologies and a preliminary conceptual model to choose the core technologies is presented. Finally, the conclusion is presented with some remarks and the future research direction.

2 End-of-Life Aircraft

2.1 Aircraft Decommissioning Processes

There is no specific standard for an efficient decommissioning process of EoL aircraft [27]. Authors discuss four steps for the EoL aircraft treatments: decontamination, disassembly, dismantling and material recovery or landfill. IATA presented the "Best Industry Practices for Aircraft Decommissioning" (BIPAD) [17]. First, the aircraft go through decontamination of hazardous material (HAZMAT) and other non-desirable substances. After, in the disassembling step, valuable components such as the engine, avionics, and landing gears will be extracted (Figure 1). To reuse those valuable components, it is necessary to recertify them and keep their airworthiness status through reverse logistics (RL) phases. After the disassembling, the dismantling step is processed for recycling needs. The final phase is to sort the components and material left into two categories, the recyclable, and the non-recyclable which will go to the scrap and landfilled. The aircraft is a complex product that can attain 2.3 million components as the Boeing 787 [52]. As

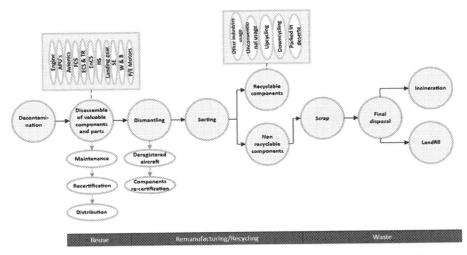

Fig. 1. EoL aircraft decommissioning processes (Based on information in [17])

a result, the challenging tasks of disassembly and dismantling require smart sorting and advanced technologies to prioritize the disassembly and dismantling process.

2.2 Proposed Approaches for Addressing EoL Aircraft Problem

Several models and guidelines have been proposed for the management of EoL aircraft. Some of the literature attempts to tackle the problem at its source and focuses on the very first stage of product development: the design phase. In this phase, there is a lack of thinking about the end of the product's life [43]. Sabaghi and al. used multi-criteria decision-making (MCDM) to assess the EoL aircraft disassembly process [43]. They suggest a difficulty disassembly indicator to measure the level of complexity of disassembly tasks. Five parameters have been chosen through literature and experts' opinions and they combined the DOE-TOPSIS method to run this MCDM problem. The first parameter is the accessibility of the parts that need to be disassembled. The second is the 'mating face' complexity of two components merged or more. The third is the 'tools' used to disassemble, and the fourth parameter is the 'connection type' that influenced the complexity of the disassembly. The final parameter is the 'quantity and variety of connections. Ribeiro et al. [41], introduced a framework in the preliminary design for the treatment of EoL aircraft and addresses economic and environmental factors. The framework aims to integrate the 3R approach (reuse, recycle and remanufacture) at the design phase. They focused on the EoL phase during the conceptual design phase, as this is an important stage for the designer, who needs feedback from EoL phase. The framework uses requirements and aircraft information such as the sizing of components and detailed design to develop the EoL part. Then, an estimation of performance parameters is analyzed, and a Life Cycle Assessment (LCA) is completed with simulation to find the optimum final aircraft design. Eco-design thinking can be used to optimize the EoL process by reducing the environmental impact in a cost-effective manner [27]. Eco-design used with LCA can design a better product by considering environmental impacts during the life cycle (LC) with less cost and low social impacts [45].

Other studies have proposed application of the LCA approach to evaluate the impact of decommissioning, dismantling, or disassembling an EoL aircraft to optimize its sequences with the triple bottom line objectives (economic, environmental, and social) [15, 40, 41, 47, 56]. Howes et al. [15] also used LCA assessment to determine the consequences of decom- missioning on the environment. Domingues and al. [11] used the LCA in decision-making by combining it with MCDM methods to analyze the environmental impacts of different vehicles. LCA evaluates the environmental impacts of vehicles' alternatives and the MCDM methods classified the alternatives based on the different indicators: 'abiotic depletion, acidification, eutrophication, fuel consumption, global warming, photochemical oxidation, NOx, CO, particulate matter, and ozone layer depletion' [11]. Zahedi et al. [53] proposed a model to evaluate and optimize disassembly for EoL aircraft processing and to minimize the envi- ronmental impact. They developed the 'disassembly difficulty indicator' that measures the complexity of a disassembly sequence in a semi-destructive EoL aircraft context. This model also guides designers in making future-generation planes with more EoL-oriented incentives. Application of Industry 4.0 technology in the LC of complex products is a fresh research theme. More empirical studies are required to assess the challenges and discuss the application perspectives. In this research, EoL aircraft treatment will be considered for examining and evaluating the impact of advanced technologies for improving the complex product recovery network.

3   End-of-Life 4.0

3.1   Industry 4.0 and EoL Aircraft Treatment

The 4th industrial revolution, known as Industry 4.0, gave access to several disruptive technologies that can improve the management of EoL products [19]. Its technologies help preserve the value of a product. Several of those technologies can be used to preserve the value of a product through its life cycle and supply chain, such as blockchain, Internet of Things (IoT) and Big Data. [22]. The EoL treatment in the context of industry 4.0 can be referred to as 'EoL 4.0'. As opposed to manufacturing inputs, recycled inputs can be damaged by use, bent, twisted, deformed, or even fatigued [38]. Advanced technologies help predict the remaining useful life (RUL) and fatigue [51], track and exchange data in real-time [38] that can reinforce partnerships with the different stakeholders involved in the reverse supply chain and increase the quality of resources. Furthermore, the EoL 4.0 context is interesting in western countries because it diminishes labor shortage [38, 50]. The following subsections discuss the application technologies under the umbrella of Industry 4.0 in EoL product treatment.

Application of Blockchain and Internet of Things
Blockchain could address a certain challenge regarding EoL aircraft. Keivanpour [20] proposed to use blockchain for the EoL aircraft recovery part market for its traceability features and smart contract possibilities. Blockchain can also provide better data security, better visi- bility in the supply chain through traceability and improved information sharing. This technology can also be used for its safety guarantees and auditable easy outcomes [48]. Aleshi and al. [3] used blockchain technology to propose a Secure Aircraft Maintenance Record (SAMR) to address the need to provide greater integrity and transparency of aircraft maintenance records. The radio-frequency identification (RFID) system combined with blockchain technology contributes to the complex credentials and documents that need to be collected for the certification of an aircraft. It will improve the efficiency of the process and the reliability of these documents all along the supply chain [44].

The IoT can also be applied in the supply chain to enable traceability features. Maintenance traceability is critical in the aerospace sector as more than 20,000 aircraft components will be replaced throughout their LC [48]. The authors highlighted the lack of standardized records and traceability for the aircraft use of materials, maintenance. In response to this, [44] proposed to combine the RFID tag system with IoT technology in Airbus' supply chain. This could be achieved through blockchain processes that simplify information access and increase data security. However, the combination of IoT and RFID has some limitations, such as energy consumption, data synchronization, intercommunication difficulties and safety [19]. Wang and Li designed a model that uses blockchain and IoT to create a framework for monitoring the supply chain to ensure that airworthiness traceability is met for the entire LC of an aircraft [48].

Application of Digital Twin

Digital twin (DT) technology provides a digital copy of a physical object with real-time data transfer information. It can help to simulate and optimize dismantling or disassembling processes with better efficiency through data optimization [19]. It was initially introduced by NASA and shifted from spatial simulation to aircraft [31, 49]. DT application in the EoL stage is the least studied topic in the life cycle of a product with only 4.9% [25]. This technology can fulfil different roles: it can diagnose a problem or predict a scenario in a disassembly process of a component via simulation [25]. These different roles can be addressed by providing real-time data in a decentralized manner. Indeed, DT integrates the proper information at the correct time in the product's life cycle. This leads to the enhancement of material flow and helps to adopt the circular economy by allowing transparency and improved collaboration between the different stakeholders. With the availability of information in real time, the different actors can anticipate and foresee possible problems and act accordingly in the value chain. Hence, collaboration is simplified in this way [36]. Moreover, digital technologies help the transition to the circular economy model by making data accessible through transparency. This enables reverse material flows [34] and can be called a 'digital reverse logistic twin' [46]. DT enables reverse flow by using artificial intelligence to predict the data behavior of the physical and digital world combined with the use of a mathematical model to simulate the critical parameters for a quantitative decision model [18, 46]. The digital mirror model enables simulation with high reliability. Wang and Wang [49] proposed using DT in recovery, recycling, and remanufacturing processes for waste from electric and electronic equipment (WEEE). They presented the WEEE cyber-physical system based on DT technology and through product life cycle. This system is based on a cloud system and includes:

– Design phase: an LCA is made to assess environmental impact and simulation for design for recycling, disassembly, remanufacturing, and archiving when needed.
– Manufacturing and remanufacturing phases: CAx model
– Consumer: attached tag of tracking/identification technology (e.g., RFID, QR code) so they can update information for repair, replacement, or upgrade through an application. These technologies are static and are linked to a DT that becomes unique at this phase. This DT supports different recycling, and remanufacturing EoL activities. It is personalized and procures a competitive advantage by its adaptability to scenarios, simulations, and clients' requirements.
– EoL collecting and recycling: collecting is generated with a computer-aided logistic system (CALS) and is driven based on the most recent system status.
– Remanufacturing phase: DT is based on the product history data, and a plan is made for optimized recycling processes.

DT technology can be integrated with other industry 4.0 technologies such as RFID, artificial intelligence, or different cyber-physical systems. International standards are designed to support interoperability in these diverse systems [49].

Application of Big Data and Artificial Intelligence

Data plays an essential role in the aviation industry. To optimize EoL processes, a significant volume of data is required throughout the life cycle of an aircraft. Specifically, data mining can provide a better understanding and decision-making in the different stages during the treatment of an EoL aircraft [19]. Big data analysis (BDA) enables 'smart decision-making' as the data permit to predict maintenance and other product behavior [57]. A big set of data permits the use of algorithms. Mascle and Balasoiu [28] use the wave propagation algorithm to propose an optimal disassembly analysis and sequence execution. Ding and al. Have used data technology for the remanufacturing of EoL goods. Indeed, they used this technology to predict and optimize the cost for the remanufacturing step and to propose a cost model [9]. [55] created an architecture with BDA for cleaner maintenance of complex products.

Artificial intelligence (AI) and machine learning (ML) are broadly utilized for the remaining useful life (RUL) [51] and in the RL for disassembly, remanufacturing, and recycling [22]. Amin and Kumar [4] used AI to propose a model for forecasting the residual operating life of a turbofan aircraft engine. They used NASA's commercial-Modular Aero-Propulsion System data sets. First, they used sensors to measure different raw data such as temperature, pressure, speeds, fuel flow and coolant bleed. After, the authors used the min-max normalization method within the $[-1,1]$ range and used a two-dimension matrix with sensor data and time sequences. These data were used with the deep learning method of Convolutional Neural Network that flattered outputs for the next step. In the next step, the long-short-term memory layer or the recurrent neural network layers were used and compared for the remaining useful life prediction with a single neuron attached at the end to provide the result. After this step, a performance evaluation is made with a root mean squared error and simulations are performed [4]. AI allows for the introduction of more intelligent technologies such as augmented and virtual reality (AR/VR), as discussed further below, through its human-like intelligence and discernment. ML can be used in the disassembly process. Grochowski and Tang [13] aimed to apply ML to be cost-efficient through the disassembly processes. There are two principal challenges for the disassembly sequence: first, the modulization of the disassembly process can be complex because it is necessary to know what activity needed to be performed at any moment during the disassembly. For this matter, the authors proposed to use the disassembly Petri net (DPN) that represents a non-deterministic decision-making model during disassembly. However, DPN cannot be used alone because it doesn't acknowledge which actions are the most appropriate to take. To respond to this challenge, the hybrid Bayesian network (HBN) is combined with DPN. It determines the uncertain parameters that characterize the disassembly process through an acyclic graph that represents quantitative and qualitative information related to the disassembly. They applied it for a computer disassembly to optimize the sequence [13]. Yan and al. [51] combined DL technology and device electro- cardiogram (DECG) for RUL estimation and maintenance. On one hand, DECG permits industrial application maintenance to monitor the product cycle and provides predictive data in the product lines improving operator efficiency and non-planned downtime [51]. On the other hand, DL technology has the benefit of extracting features automatically which simplifies prediction for RUL.

Application of Virtual Reality and Artificial Reality
Virtual reality (VR) and augmented reality (AR) are two technologies of industry 4.0. VR enables the 'virtual disassembly' and 'virtual maintenance' possibilities with simulations and optimization of the physical processes [37]. Qiu and al. [37] proposed an interactive model based on VR that analyses constraints, interaction, and methods of operator assembly and disassembly through 'virtual human' (VH). AR can be defined as the overlaying of virtual signs, icons, or imagery on physical reality [6]. Introduced by Ronald Azuma [5], it can benefit the aviation industry. Ceruti and al. [6] investigated the use of AR and additive manufacturing for the maintenance process. They proposed to use AR to better assist operators in a maintenance process to achieve better efficiency and reduce the possibility of errors. Mo and al. [29] present virtual disassembly through the virtual environment Motive3D. They proposed virtual disassembly analyzer (VDA). VDA uses CAD models' inputs, then proceed to generate a disassembly sequence and assess it with the data deployment tool (DDT). Together, DDT and VDA constitute the server Motive3D that enables the visualization of the disassembly process in 3D. Motive3D shows gesture of operators with special data gloves that shows rotation, movement, translation, grasp, release, mouse movement or selection.

Application of Collaborative Robots
Collaborative robots (CR) or collaborative human-robot (H-R) are programmed to assess, plan, and perform tasks while considering human constraints and the limitations of the environment [21]. Manual disassembly is not optimal from an economic point of view. On the other hand, the use of regular robots is not reliable enough for complex products [35] such as EoL aircraft parts. Parsa and Saadat [35] recommended that the CR can assess the sequence to be disassembled based on remanufacturing potential. The operator will be helpful in difficult tasks considering the required flexibility. Hjorth and al. [14] conduct a literature review on CR for the non-destructive disassembly process of EoL objects. They've highlighted some gaps in the field as the strategies put in place with H-R collaboration are mainly the avoidance of contact with the operator or the lack of a combination of the verbal and the non-verbal communication method between the worker and the CR. Ottogalli and al. Assessed with VR technology the CR in the assembly line of an aircraft because most of the task is executed manually [33]. They developed a framework to assess the process of semi-automation tasks, the return on investment and the ergonomics of the workers. Liu and al. [24] proposed to use of an algorithm to optimize the robotic disassembly sequence with a collaborative H-R in the remanufacturing process.

3.2 Summary of Industry 4.0 Technologies in EoL Aircraft Problem

Table 1 presents 7 technologies of industry 4.0 and their implications in EoL aircraft, how those technologies can be used, the related literature and the industry 5.0 perspectives. Industry 5.0 is a new movement that introduces the efficiency and precision of industry 4.0 technologies while integrating the subjectivity and intelligence of humans [23]. Leng and al. [23] discussed that industry 5.0 is driven by three key characteristics: 'Human centricity', sustainability, and resiliency. 'Human centricity' is the principle that humans bring higher levels of tolerance to the systems.

Table 1. Industry 4.0 technologies and EoL aircraft process

Technologies	Advantages	Processes intervention	How it is used	Liter- ature	Industry 5.0 application
Blockchain	Traceability Visibility Security Information Sharing/accessibility	Maintenance Recovery Remanufacture Recycling Re-using	a. Tracking of aircraft parts b. Maintenance records of aircraft available in real-time c. Smart contract possibilities d. Data sharing and monitoring for airworthiness status e. Secure the data	[3] [20] [44] [48]	**Sustainability**: clients expect the supplier to deliver sustainability-related information on products. The utilization of blockchain tech- nology as a solution for the absence of traceability in the supply chain. [23] **Resilience**: product life cycle data [23]
IoT	Traceability Accessibility Prediction of comportments	Maintenance Standardized Records Sales, and storage	a. Maintenance records of aircraft available in real-time b. Standardization of maintenance record c. Monitoring for airworthiness status	[19] [23] [44] [48]	**Resilience**: internet of everything (IoE) offers new capabilities such as: (a) allowing predictive maintenance operations performed to avoid issues instead of planning once a defect occurs. (b) Combining IoT sensors with DT technology to collect real-world data [26]
Digital Twins	Simulation/optimization process Accessibility Decentralized Transparency	Simulation Recycle Recovery Remanufacture	a. Real-time simulation for optimization of dismantling or disassembling process for an aircraft b. Diagnostic operational problem c. Can predict scenario	[19] [25] [31] [36] [49]	**Human centricity and resilience**: Society 5.0: merging physical environment with cyberspace Metaverse: immersion for H-R/CR [23]

(continued)

Table 1. (*continued*)

Technologies	Advantages	Processes intervention	How it is used	Liter- ature	Industry 5.0 application
Big Data	Simulation/optimization process Prediction of comportment Reliability/Accuracy	Predictive maintenance, Recycling Re-using Disassembly Remanufacture	a. Predict maintenance b. Predict comportments and tendency c. Predict and optimizes for remanufacturing d. Green maintenance	[9] [19] [28] [55] [57]	Society 5.0: Data-driven society **Resilience characteristic**: mass-personalization [23] BD: Faster decision-making [26]
AI	Simulation/optimization process Prediction of comportment	Disassembly Remanufacture Recycling	Predict residual life of an aircraft parts	[4] [22]	**Resilience:** 'Better task allocation' [23] 'Better handling of reitative jobs' [23] Manufacturing activities' hyper customization [26]
AR / VR	Reliability Error Minimization Ergonomy	Maintenance Simulation Disassembly Remanufacture	a. More efficient process fordisassembly of parts b. Maintenance process optimization	[5] [6] [37]	**Resilience:** extended reality (combination of VR/AR): is used for remote assistance, assembly line monitoring, training, and maintenance [26]
Collaborative robot	Optimization process Productivity Cost prevention Ergonomy	Disassembly Remanufacture	Disassembly of semi-automation process for safety purposes, to be more cost-effective, and productive	[14] [21] [24] [33] [35]	'**Human centricity' and resilience:** Operator 5.0 / Cobots (collaborative robots): system/self-resiliency, enhancing flexibility and dexterity, analyzing and optimize human work [23, 26]

3.3 Evaluation of Technologies

Blockchain permits traceability, reliability, and security but can be complex in the adoption process. IoT also allows for visibility and traceability, although its energy consumption might be a challenge. DT can be combined with AI or used alone to simulate scenarios and optimizes them by mimicking the physical world into a digital world with real-time scenarios. AR/VR helps analyze ergonomically and minimizes errors in certain processes such as disassembly or remanufacturing. CR can optimize operations and add values, but it can be a challenge to integrate into a disassembly chain and to be trustable by operators. Addressing the challenges and opportunities of these technologies in EoL aircraft problem to maximize its value and minimize the negative impact on sustainability is a complex task. To determine which of the technologies of Industry 4.0 will be valuable in the EoL sustainable decommissioning of an aircraft, multi-objective decision-making (MCMD) approach can be applied. One of the most applied MCDM methodologies is the analytics hierarchical process (AHP) which was first introduced by Saaty [42]. This method provides a classification of different alternatives considering qualitative and quantitative information and evaluates the consistency of the results and criteria. The gap in this method is the lack of consideration of the relation between criteria, so Saaty proposed a variation: the analytic network process (ANP) method. The ANP provides a network structure to analyses the alternatives [30]. This method has been used to assess industry 4.0 technologies. Chang, Chang, and Lu [7] used the MCDM approach of ANP and the technology-organization-environment framework (TOE) to assess in small and medium enterprises the utilization of technology 4.0. Ravi and al. Combined ANP and the balanced scorecard approach (BS) to classify EoL computers into the reverse logistic context [39]. Ordooadi proposed to consider quality, flexibility, cost, and impact on human resource when considering the adoption of technologies [32]. As already discussed, the importance of traceability features and maintenance reports and requirements is critical in the aerospace industry. Real-time data exchange, security, reliability, certification information, substitutability, energy consumption, environmental impact, social impact, and economic impact, are key criteria to consider when evaluating and classifying the technology alternative in an EoL context. Those airworthiness information needs to be considered when evaluating technologies from industry 4.0.

As shown in Figure 2, this paper presents ANP framework of criteria to evaluate the best options among industry 4.0 technology for EoL aircraft treatment. Four categories of criteria have been considered in the model through the literature: economic, environmental, and social, for the triple bottom line and the technological performance (level 1). Sub criteria related to the aerospace industry are considered in the level 2. In level 3 the preliminary indicators are presented and will be used when evaluating the different alternatives (level 4).

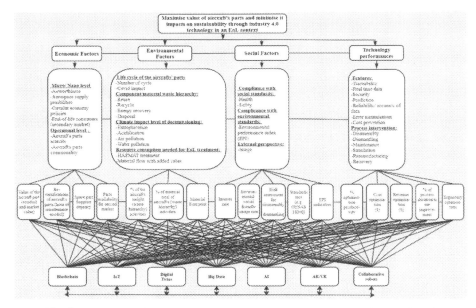

Fig. 2. ANP model for the technology of industry 4.0 to decommission an aircraft

4 Conclusion

Valorization of EoL aircraft is an important research field considering the 20 000 air-crafts that will be decommissioned in the next 20 years. The aerospace industry focuses on maximizing the value of those EoL airplanes that are often just parked in desert. For successful operation of reused and remanufactured parts, the reliability of the components, ease of dis- assembling, inspection, cleaning, and maintenance data should be considered. Industry 4.0 tools provide several opportunities with analyzing real-time data and facilitate the decision-making process The blockchain technology, the Internet of Things, digital twins, big data, artificial intelligence, augmented reality/virtual reality, and collaborative robots are dis- cussed. An MCDM based framework is proposed to assess and classify those technologies with the analytical network process. As the future research, secondary data from EoL aircraft treatment of a pilot project will be used to assess the impacts of Industry 4.0 in the different sub-processes of EoL aircraft treatment based on waste hierarchy approach. Moreover, the preliminary ANP model in this paper will be completed, readiness of the alternatives will be added as a key factor and applied in a real case study.

Acknowledgement. This paper is supported by the Natural Sciences and Engineering Research Council of Canada (NSERC) and the 'parrainage, integration, émergence and démarrage' (P.I.E.D.) program.

References

1. AFRA: AFRA Mission. Récupéré sur AFRA association (2006). https://afraassociation.org/about-us/afra-mission/
2. AFRA: Home (2022). https://afraassociation.org/
3. Aleshi, A., Seker, R., Babiceanu, R.F.: Blockchain model for enhancing aircraft maintenance records security. In: International Symposium on Technologies for Homeland Security (HST), pp. 1–7. IEEE, Woburn (2019)
4. Amin, U., Kumar, K.D.: Remaining useful life prediction of aircraft engines using hybrid model based on artificial intelligence techniques. In: IEEE International Conference on Prognostics and Health Management (ICPHM), Detroit, pp. 1–10 (2021)
5. Azuma, R.T.: Survey of augmented reality. Teleoperators Virtual Environ. **6**, 355–385 (1997)
6. Ceruti, A., et al.: Maintenance in aeronautics in an Industry 4.0 context: the role of augmented reality and additive manufacturing. J. Comput. Des. Eng., 516–526 (2019)
7. Chang, S.-C., Chang, H.-H., Lu, M.-T.: Evaluating industry 4.0 technology application in SMEs: using a hybrid MCDM approach. Mathematics, 1–21 (2021)
8. de Oliveira Junior, F. S., et al.: A practical approach to support end-of-life commercial aircraft parking, market relocation, retirement and decommissioning strategic decisions. Int. J. Prod. Res., 5144–5163 (2020)
9. Ding, Z., et al.: A big data cost prediction method for remanufacturing End-of-Life products. In: Proceedings of the 51st CIRP Conference on Manufacturing Systems, pp. 1362–1367 (2018)
10. Dolganova, L., Bach, V., Rodl, A., Kaltschmitt, M., Finkbeiner, M.: Assessment of critical resource use in aircraft manufacturing. Circular Econ. Sustain. (2022). https://link.springer.com/article/https://doi.org/10.1007/s43615-022-00157-x
11. Domingues, A.R., et al.: Applying multi-criteria decision analysis to the life-cycle assessment of vehicles. J. Clean. Prod. **107**, 749–759 (2015)
12. European Commission: Process for Advanced Management of End of life Aircraft. Récupéré sur PAMELA on LIFE Public Database (2005). https://webgate.ec.europa.eu/life/publicWebsite/index.cfm?fuseaction=search.dspPage&n_proj_id=2859
13. Grochowski, D.E., Tang, Y.: A machine learning approach for optimal disassembly planning. Int. J. Comput. Integr. Manuf. **22**, 374–383 (2009)
14. Hjorth, S., Chrysostomou, D.: Human–robot collaboration in industrial environments: a literature review on non-destructive disassembly. Robot. Comput. Integr. Manuf. **73**, 1–18 (2022)
15. Howe, S., Kolios, A.J., Brennan, F.: Environmental life cycle assessment of commercial passenger jet airliners. Transp. Res. Part D Transport Environ. **19**, 34–41 (2013)
16. EU-Industry 5.0. https://research-and-innovation.ec.europa.eu/research-area/industrial-research-and-innovation/industry-50_en
17. IATA: Best Industry Practices for Aircraft Decommissioning (BIPAD), Switzerland (2018)
18. Ivanov, D., Dolgui, A.: A digital supply chain twin for managing the disruption risks and resilience in the era of Industry 4.0. Prod. Planning Control **32**, 775–788 (2021). https://doi.org/10.1080/09537287.2020.1768450
19. Keivanpour, S.: End of life management of complex products in an Industry 4.0 driven and customer-centric paradigm: a research agenda. In: MOSIM2020, Agadir.7 (2020)
20. Keivanpour, S.: Designing a system architecture for the management of the recovered parts from End of Life aircraft using fuzzy decision-making and blockchain (2022)
21. Kragic, D., et al.: Interactive, collaborative robots: challenges and opportunities. In: Proceedings of the 27th International Joint Conference on Artificial Intelligence (IJCAI 2018), Stockholm, pp. 1–8 (2018)

22. Krstic, M., et al.: Evaluation of the smart reverse logistics development scenarios using a novel MCDM model. Cleaner Environ. Syst. **7** (2022)
23. Leng, J., et al.: Industry 5.0: prospect and retrospect. J. Manuf. Syst. **65**, 279–295 (2022)
24. Liu, J., et al.: Collaborative optimization of robotic disassembly sequence planning and robotic disassembly line balancing problem using improved discrete Bees algorithm in remanufacturing. Rob. Comput. Integr. Manuf. **61**, 1–18 (2020)
25. Lo, C., et al.: A review of digital twin in product design and development. Adv. Eng. Inform. **48**, 1–15 (2021)
26. Maddikunta, P.K., et al.: Industry 5.0: A survey on enabling technologies and potential applications. J. Ind. Inf. Integr. (2022)
27. Mascle, C., et al.: Process for advanced management and technologies of aircraft EoL. In: Proceedings of the 12th Global Conference on Sustainable Manufacturing, vol. 26, pp. 299–304 (2015). Procedia CIRP, Johor Bahru
28. Mascle, C., Balasoiu, B.-A.: Algorithmic selection of a disassembly sequence of a component by a wave propagation method. Rob. Comput. Integr. Manuf. **19**, 439–448 (2003)
29. Mo, J., et al.: Virtual disassembly. Int. J. CAD/CAM **2**, 29–37 (2002)
30. Munier, N., Eloy, H.: Uses and limitations of the AHP method - a non-mathematical and rational analysis. Management for Professionals (2021)
31. NASA: Draft Modeling, Simulation, Information Technology & Processing Roadmap (2010)
32. Ordoobadi, S.: Application of ANP methodology in evaluation of advanced technologies. J. Manuf. Technol. **23**, 229–252 (2011)
33. Ottogalli, K., et al.: Virtual reality simulation of human-robot coexistence for an air-craft final assembly line: process evaluation and ergonomics assessment. Int. J. Comput. Integr. Manuf. **34**, 975–995 (2021)
34. Pagoropoulos, A., et al.: The emergent role of digital technologies in the circular economy: a review. In: Proceedings of the 9th CIRP IPSS Conference: Circular Perspectives on Product/Service-Systems, pp. 19–24 (2019). Procedia CIRP, Copenhagen
35. Parsa, S., Saadat, M.: Human-robot collaboration disassembly planning for end-of-life product disassembly process. Robot. Comput. Integr. Manuf. **71**, 1–15 (2021)
36. Preut, A., et al.: Digital twins for the circular economy. Sustainability **13**, 1–15 (2021)
37. Qiu, S., et al.: Virtual human modeling for interactive assembly and disassembly operation in virtual reality environment. Int. J. Adv. Manuf. Technol. **69**, 2355–2372 (2013)
38. Rahman, M.A., Perry, N., Müller, J.M., Kim, J., Bertrand, L.: End-of-Life in industry 4.0: Ignored as before. Resour. Conserv. Recycl. **154**, 2 (2020)
39. Ravi, V., et al.: Analyzing alternatives in reverse logistics for end-of-life computers: ANP and balanced scorecard approach. Comput. Ind. Eng. **48**, 327–356 (2005)
40. Ribeiro, J.S., de oliveira Gomes, J.: Proposed framework for End-of-Life aircraft recycling. In: Proceedings of the 12th Global Conference on Sustainable Manufacturing, vol. 26, pp. 311–316 (2015). Procedia CIRP, Johor Bahru
41. Ribeiro, J., De Oliveira Gomes, J.: A framework to integrate the End-of-Life air-craft in preliminary design. Procedia CIRP **15**, 508–514 (2014)
42. Saaty, R.W.: The analytic hierarchy process - what it is and how it is used. Math. Model.. Model. **9**, 161–176 (1987)
43. Sabaghi, M., Mascle, C., Baptiste, P.: Evaluation of products at design phase for an efficient disassembly at End-of-Life. J. Clean. Prod. **116**, 177–186 (2016)
44. Santonino III, M.D., Koursaris, C.M., Williams, M.J.: Modernizing the supply chain of airbus by integrating RFID and modernizing the supply chain of airbus by integrating RFID and blockchain processes blockchain processes. Int. J. Aviat. Aeronaut. Aerosp. **5** (2018)
45. Su, D., Casamayor, J., Xu, X.: An integrated approach for eco-design and its application in LED lighting product development. Sustainability **13**, 488 (2021)

46. Sun, X., et al.: Towards the smart and sustainable transformation of reverse logistics 4.0: a conceptualization and research agenda. Environ. Sci. Pollut. Res. **29**, 69275–69293 (2022)
47. Veeramanikandan, R., et al.: Life cycle assessment of an aircraft component: a case study. Int. J. Ind. Syst. Eng. **27**(4), 485–499 (2017)
48. Wang, C., Li, B.: Research on traceability model of aircraft equipment based on blockchain technology. In: Proceedings of the 1st International Conference on Civil Aviation Safety and Information Technology (ICCASIT), pp. 88–94. IEEE, Kunming (2019)
49. Wang, X.V., Wang, L.: Digital twin-based WEEE recycling, recovery and remanufacturing in the background of Industry 4.0. Int. J. Prod. **57**, 3892–3902 (2019)
50. World Economic Forum: Labour shortages have risen across OECD countries, here's how to plug the gaps. Récupéré sur World Economic Forum - Workforce and employment, 2 December 2022. https://www.weforum.org/agenda/2022/12/labour-shortages-rise-across-oecd-countries/
51. Yan, H., et al.: Industrial big data analytics for prediction of remaining useful life based on deep learning. IEEE **6**, 17190–17197 (2018)
52. ZACKS: 10 Facts About Boeing's Revolutionary 787 Dreamliner. Récupéré sur NASDAQ, 31 March 2017. https://www.nasdaq.com/articles/10-facts-about-boeings-revolutionary-787-dreamliner-2017-03-31
53. Zahedi, H., Mascle, C., Baptiste, P.: A quantitative evaluation model to measure the disassembly difficulty; application of the semi-destructive methods in aviation End-of- Life. Int. J. Prod. Res. **54**(12), 3736–3748 (2016)
54. Zahedi, H., Mascle, C., Baptiste, P.: Advanced airframe disassembly alternatives: an attempt to increase the afterlife value. Procedia CIRP **40**, 168–173 (2016). http://www.sciencedirect.com/science/article/pii/S2212827116001086
55. Zhang, Y., et al.: A big data analytics architecture for cleaner manufacturing and maintenance processes of complex products. J. Clean. Prod. **142**, 626–641 (2017)
56. Zhao, X., et al.: Disposal and recycle economic assessment for aircraft and engine End of Life solution evaluation. Appl. Sci. **10**, 1–24 (2020)
57. Zheng, P., et al.: Smart manufacturing systems for Industry 4.0: Conceptual framework, scenarios, and future perspectives. Front. Mech. Eng. **13**, 137–150 (2018)

How to Foster the Circular Economy Within the Pharmaceutical Industry? A Research Framework Proposition

Simon Massot[1,2(✉)], Duc-Nam Luu[1], Claus-Jürgen Maier[3], Nicolas Maranzana[1], and Améziane Aoussat[1]

[1] Arts et Métiers Institute of Technology, LCPI, HESAM Université,
151 Boulevard de L'Hôpital, 75013 Paris, France
`simon.massot@ensam.eu`
[2] Sanofi, 82 Avenue Raspail, 94250 Gentilly, France
[3] Sanofi, Industriepark Höchst, 65926 Frankfurt am Main, Germany

Abstract. Pharmaceutical industries play an integral role in our society as their primary purpose is to provide products that enhance human health and well-being. Due to its intrinsic activities, the impact of the healthcare sector on the environment is consequent. However, the preservation of our environment is a prerequisite to ensure human health resilience. The circular economy (CE) is a model of production and consumption aimed to preserve the environment. The CE principles, such as "Reduce, Reuse, Recycle" [1, 2] are required in an expanding world where resources are limited [3], and the generation of waste increases each year. Considering the size of the pharmaceutical sector in the global economy and the fact that less than 10% of our economy is circular [4], it is essential to implement the concept of circularity to limit the tremendous resource consumption. Nevertheless, most of the work that has been done in this area focuses on a few specific steps of drug and medical device development processes [5, 6], creating room for implementation opportunities. The goal of this paper is to provide a mapping of the research that has been done in the field of CE, related to the pharmaceutical sector. As circularity is driven by design, this paper focuses on the whole value chain of pharmaceutical products, from raw material extraction to the end of life. The findings of this study are categorized regarding product life-cycle steps. This allows the identification of current research trends and opportunities for future research and potential circular solutions.

Keywords: Circular Economy · Pharmaceutical Industry · Design Process · Lifecycle

1 Introduction

The industrial revolutions and the development of human society have led to important consequences for our environment, such as the depletion of natural resource. The negative impacts of our society have been growing exponentially for many years and are

© IFIP International Federation for Information Processing 2024
Published by Springer Nature Switzerland AG 2024
C. Danjou et al. (Eds.): PLM 2023, IFIP AICT 702, pp. 28–37, 2024.
https://doi.org/10.1007/978-3-031-62582-4_3

considered as the biggest challenge of the 21st century [7]. The concept of planetary boundaries, established in 2009, defines environmental limits in which humanity can operate safely and depicts the emergency of the situation [8]. Pharmaceutical industries play an integral role in our modern societies due to their intrinsic purpose which is developing medicines and medical devices to prevent, heal, cure, and diagnose diseases to maintain the population's health and well-being. However, the production of drugs is responsible for important pollution and negative environmental impacts.

The Circular Economy (CE) is a model of production and consumption that aims to implement environmental sustainability and its principles into organizations. Integrating circularity into an organization means continually reusing products, materials, and resources whenever possible [9]. In opposition to a linear business model "take-make-use-dispose", the CE's purpose is to create loops in the product life cycle, to retain the value of the resources. Today, only 7,2% of our world is considered circular, and the majority of organizations work in a linear economy [4]. The CE is a promising concept that could help pharmaceutical companies in achieving sustainability goals and improve their environmental performances. Nevertheless, since the purpose of circularity is to reduce the environmental footprint of products, circularity must be taken into account across the entire product life cycle using a multicriteria approach, such as the eco-design one.

This paper proposes a mapping to categorize the existing literature on the CE within the pharmaceutical industry regarding the several life cycle steps of medicines. The first parts of this paper introduce the CE, and the pharmaceutical industry. Then, the literature analyzed is presented and, a mapping is proposed which allows the identification of the main research axis for CE in the healthcare sector, and future potential opportunities.

In conclusion, the goal of this paper is to answer to the following research question: *How can the CE research for the pharmaceutical sector be classified to foster further research and opportunities?*

2 Circular Economy

2.1 Definition and Principles

The CE is a concept born in the middle of the 1990s which integrates different principles and concepts. For many authors, the CE is frequently depicted as an easier way to allow the operationalization of sustainable development into businesses [1]. Therefore, the CE is a trending concept as the number of publications increased significantly in the last few years. Numerous organizations have proposed their definition and most of them are based on the same principles. Kirchherr et al., conducted a study of 114 definitions of the CE, from 2005 to 2017 [1]. A systematic analysis of the definitions allowed the authors to demonstrate that even if most of the definitions are different, some common points can be highlighted. First, the fact that in opposition to a linear economy "take-make-use-dispose", the CE aims to have better management of resources throughout the life cycle of systems and to retain the value of products and materials. Second, the CE principles aim to generate loops in the product life cycle. There are three types of loops: short, medium, and long, and they do not present the same benefits or consequences for the environment. Indeed, the shorter the loop is, the better to keep the value of the material/product. Hence, it is more interesting to reuse a product than to recycle it.

The following definition from the Ellen MacArthur Foundation (EMF) was established in 2013 and is well recognized among peers and in the industry [2]: *"A systems solution framework that tackles global challenges like climate change, biodiversity loss, waste, and pollution. It is based on three principles, driven by design: eliminate waste and pollution, circulate products and materials (at their highest value), and regenerate nature [...]"*.

Transitioning to a CE will be systemic, deep, and transformative. It is not only recommended but it is a crucial part of the transformation of industries to improve their environmental performances [10]. Even the EU made a proposal for a new Eco-design for Sustainable Products Regulation, published on 30 March 2022, that is the cornerstone of the European Commission's approach to more environmentally sustainable and circular products [11].

2.2 Review of Existing CE Frameworks

Since the apparition of the CE and its numerous definitions and interpretations, several frameworks have been proposed and aim to help organizations structure the integration of the CE within their business models (Table 1).

Table 1. Example of the most common CE frameworks

3R [1]	Most common conceptualization of the "how-to" of the CE Several adaptations proposed in the literature (4R, 6R, etc.) •Reduce, the consumption of natural resources/materials •Reuse, the discarded product by another consumer •Recycle, process materials or products to obtain the same or lower quality
9R [5]	One of the adaptations of the 3R framework Additional principles: refuse, rethink, repair, refurbish, remanufacture, repurpose. Emphasizes the differentiation between short, medium, and long loops to help the identification of specific levers for practitioners
Butterfly diagram [2]	It represents continuous flows of materials in a CE. Those flows are divided into biological and technical cycles The cycles are divided into several loops, the inner loops refer to short loops, and the larger ones to long loops
ReSOLVE [12]	Translation of the CE principles into six actions: Regenerate, Share, Optimize, Loop, Virtualize, and Exchange Each action represents a major circular business opportunity. Actions can be considered individually but a compounding effect can appear if many of them are addressed

Many CE frameworks exist in the literature as a basis for CE actions, such as the ones proposed by the circle economy organization. However, the frameworks presented previously represent some of the most important frameworks and their associated concepts. The use of such CE frameworks within the pharmaceutical industry is an opportunity to reduce the environmental burden generated by this sector.

3 The Pharmaceutical Industry

This section aims to give an overview of some specificities of the pharmaceutical industry which help to understand why a focus on this specific sector is proposed, and why it is important to foster CE studies.

3.1 Specificities of the Pharmaceutical Industry

The pharmaceutical industry is a highly regulated industry, which needs to comply with numerous strict standards and regulations. As the purpose of this industry is to provide the population with medicines to protect and preserve their health, the developed products need to be aligned with high-quality and safety standards. Those standards might differ from one country to another because many regulations are established at a national level. These regulations are considered in a singular development process and potential environmental levers were identified as described by Luu et al., 2022 [13].

3.2 Environmental Impact of the Pharmaceutical Sector

The pharmaceutical industry plays an integral role in our modern societies. However, as with any human activity, pharmaceutical companies are responsible for important environmental impacts. Several studies focused on this topic and assessed the impact of this sector on the environment regarding different aspects such as climate change, biodiversity, natural resources, and so on. According to Eckelman who has led numerous studies on the impact of the healthcare sector, the pharmaceutical industry contributes to approximately 4.5% of worldwide global greenhouse gas emissions [14]. Not only do those emissions pollute our environment, but they are also responsible for the apparition of diseases into the population [15].

Due to their intrinsic nature, Active Pharmaceutical Ingredients (API) and their metabolites have tremendous impacts on our ecosystems. A large-scale study has been conducted across the world, analyzing the presence of 61 API in more than 1000 sampling sites and more than 250 rivers, and has shown that the presence of those API is important and increase each year [16]. The global population growth coupled with the pollution of our environment contributes to an increasing number of diseases internationally, which can have a direct effect on the environment.[17].

In recent years, the impact of the healthcare sector on the environment increased due to the covid crisis. According to Wuyts, the waste generation increased by 65% at the peak of the crisis [18]. The use of single-use technologies, already common in the pharmaceutical industry and deleterious for the environment, played an important role during the crisis because it helped to accelerate the vaccine production processes implementation [19].

4   Circular Economy and Pharmaceutical Industry: A Review

In this part, the methodology followed to conduct the literature review is presented, then the classification of the papers studied is proposed to better understand how the existing literature is structured.

4.1   Methodology

To analyze the existing scientific literature on CE within the pharmaceutical industry, a semi-systematic literature review was performed [20]. Most of the research was done by using google scholar or science direct and numerous scientific papers and publications were analyzed. Specific keywords corresponding to the areas of CE and pharmaceutical industry/drug development were used. The articles corresponding to the criteria were selected.

4.2   Classification of the Literature

As it was mentioned previously, the number of publications on the CE has been increasing through the years. The analysis of the literature on the CE within the healthcare sector has also proven the same tendency, this topic is on the rise and many stakeholders of this sector are willing to consider this new model in their practices.

The complexity of this research can be explained by the broadness of the scope regarding the CE field and its potential applications. In recent years, many papers including specific terms such as "circular economy" and "pharmaceutical industry" were published. In their study, Ang et al. identified 182 publications linked to circularity in the pharmaceutical industry [5]. As it is well explained in their article, each paper corresponds to CE principles from the 9R framework. An analysis of the cited publications allowed us to identify that in many papers, the terms "pharmaceutical industry, healthcare sector, medical industry" are not mentioned. The publications refer to CE principles and propose diverse ways to reach circularity, but they do not always focus on the pharmaceutical industry. Nevertheless, even though some articles do not specify the aforementioned keywords, those papers are still relevant because they propose circular actions for specific processes, waste treatment, etc., which could directly be implemented into the pharmaceutical industry processes. Therefore, many articles cannot be directly linked to the pharmaceutical sector because their purpose is to improve different manufacturing processes in general, for different types of industries. Yet, it is still understandable to categorize those papers into pharmaceutical circularity. This enlightens the complexity linked to the categorization of this literature. It is not an easy task to determine which publication concerns the pharmaceutical industry or not, due to the several development processes (chemicals, biologicals) and the number of circularity possibilities within this sector.

Moreover, the papers' study permitted the identification of additional classification parameters: fundamental versus applied research, and scientific literature versus standards and regulations. On a higher level, the semi-systematic literature review conducted highlighted the fact that two types of papers stand out: papers focusing on strategy and management, with action plans, public awareness improvement propositions, etc.; papers focusing on the operational level, with solutions for waste management, or solutions regarding alternative chemistry/ process/ functions.

This literature review allowed a better understanding of the existing literature on the CE in general, and to understand how it takes place in the pharmaceutical industry. This topic has gained interest in recent years, not only in the research field but also in many industries and for national /international authorities. The current trends emphasize the future growing evolution of this topic and the necessity for pharmaceutical stakeholders to implement this concept into their goals and practices.

5 Proposal of Mapping for CE into the Healthcare Sector

5.1 Mapping

The previous parts of this paper emphasize the crucial role of the life cycle in the improvement of the environmental performance of products or services. Therefore, the classification of the papers analyzed throughout the literature review was done regarding the life cycle steps of a product. However, it was decided not to select the generic life cycle stages proposed by the eco-design approach. The diagram from Luu, adapted from Keoleian and Menerey, 1994 [21], was chosen for the realization of the mapping, it represents the life cycle stages selected to categorize each paper (Fig. 1).

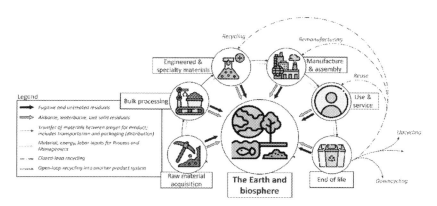

Fig. 1. Generic life cycle of a product integrating short, medium, and long CE loops

As we can see, six different steps compose this life cycle and all of them refer to the elementary level called "The earth and biosphere". According to Keoleian and Menerey, this level helps us to understand that each product (and life cycle stage) requires resources from the earth and biosphere to be created and generates waste that accumulates in the same place. Surrounding this elementary level, we can find the main life cycle steps of a product. At first, there is the raw material extraction, then, the bulk processing and engineered & specialty materials, afterwards there is manufacturing & assembly, then, use & service, and finally the end of life. This diagram is interesting because it allows the identification of numerous loops, and thus, the identification of different CE frameworks. As an example, the link between end-of-life and the earth and biosphere can be considered as the biological cycle from the EMF butterfly diagram.

The selection of those life cycle stages instead of others was based on several assumptions. The transport/distribution steps are represented by the arrows connecting the life cycle stages. This does not mean that no circular actions can be implemented regarding this specific stage, but more importantly that it can be considered as an underlying part of the other stages studied. The main differences with this diagram are the additional steps in between the raw material acquisition and the manufacturing & assembly. Those steps are respectively named bulk processing and engineered & specialty materials. The CE is based on different principles such as "Reduce, Reuse, Recycle" which require a focus on resources and materials [22]. Because they are the main drivers in CE, it was considered that adding more stages regarding these elements was essential. The literature review conducted, and analyses of the papers ended up with the categorization of each paper regarding the aforementioned life cycle steps. The following diagram depicts the number of times the topics are studied in the different papers (Fig. 2).

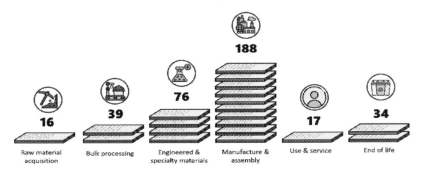

Fig. 2. Breakdown of CE-related topics in the pharmaceutical industry mentioned in the literature, regarding product life cycle steps

A little bit less than 300 publications were analyzed resulting in 370 topics mentioned. There are publications that focus on several topics. The 182 publications cited in the framework proposed by Ang et al., to implement the CE into the pharma industry were assessed, and more than one hundred additional articles published in the last four years, found on google scholar were selected. This chart depicts the current trends in terms of CE within the pharmaceutical industry. It is easily identifiable that most of the papers published focus on the manufacture & assembly step. The second part with the highest number of publications is "engineered & specialty materials", closely followed by the bulk processing part. Then, there is the end of life, with less than 40 papers, and finally, the raw material acquisition and the use & service parts, with fewer than 20 publications.

To achieve CE goals and principles, each life cycle step of the product/service has to be considered with the aim of reducing the corresponding environmental impacts. It is not new that every life cycle stage generates impacts that need to be tackled. Nevertheless, the mapping proposed highlights some differences between the number of publications available for all the steps. Hence, we can question the significant difference in the number of publications available for each stage.

Results Interpretation. There are only a few publications that address the issues of raw material acquisition and use & service. Due to its main principles, it can be understood that the raw material stage is not the first focus because it is more important to reuse or recycle resources already in the loops than to extract new ones from the earth. The European Commission is part of many of those papers, as an example, they published a chart for raw materials resilience. Papers with a larger scope, focusing on several life cycle steps including raw materials and use/service, were proposed in the last few years, such as the new packaging and plastic waste regulation proposal or the CE action plan. The European Federation of Pharmaceutical industries and Associations (EFPIA) released a white paper on CE in the pharmaceutical sector tackling several issues such as the use phase and public awareness. Despite this type of paper, only a very low number of scientific papers are available in the literature regarding the scope of our study.

Referring to the waste management part, the scarce number of articles available can be surprising, mostly when thousands of articles tackling the management of waste can be found in the literature. Nevertheless, this can be explained by the fact that only a few papers consider the waste resulting from the use phase of a product (e.g.: expired medicines brought back to the pharmacy by the patient). Papers considering the issue of waste management in the pharmaceutical industry mainly focus on manufacturing waste. In this study, it has been considered that this kind of waste management is part of the manufacturing process. This argument that explains the huge quantity of publications dealing with the manufacturing stage.

The articles' analysis did not permit a deep review of those papers, and thus can have limited our understanding. In consequence, the differentiation between papers classified in bulk processing or engineered materials cannot be completely accurate. Therefore, we considered that those parts can be gathered. Consequently, together they represent the second highest focus. In this category, many papers study the transformation of waste from other sectors to resources for the pharmaceutical industry (e.g.: the transformation of pig mucosa into low-molecular weight heparin).

Finally, the most consequent part, manufacture & assembly, has been the main focus for science during the last years. Numerous papers discuss new molecular entities or new synthesis processes to improve the yields and decrease the environmental burden of the drug development process. Regulatory constraints may also contribute to the large number of publications on this topic.

5.2 Limits and Opportunities

This mapping is interesting because it emphasizes the current trends in the pharmaceutical circularity field. What we can conclude from it is the significant number of opportunities linked to the implementation of CE in the pharmaceutical sector. It is possible to identify where the major issues for the healthcare sector are and where further research should be conducted. As an example, the toxicity of the medicines on the environment is of high importance and the link between ecotoxicology and circularity needs to be further investigated, thus potentially increasing the number of publications for the end-of-life step. The growing use of biotechnologies for the development of drugs may also bring circular solutions regarding the raw materials acquisition. Those technologies based on biological systems and living organisms could represent opportunities for the production of drugs from renewable raw materials or intermediates, but the global environmental footprint should be considered.

The elements mentioned above can also reflect the limits of this paper. The mapping proposed can be considered as a preliminary mapping as the literature review conducted did not allow the analysis of all the existing literature on this topic. As it was presented, the keywords used for the research, or the chosen criteria, did not provide a complete view of the available papers. Nonetheless, this is the first approach to this topic and this preliminary review allows the identification of a trend in the examined publications. This literature review, depicted as preliminary, needs to be further investigated. The following elements are ideas that could help to conduct a deep and detailed literature review. First, it is important to add additional keywords in the research, such as "drug, medicine, medical, ecotoxicology, biotechnology, etc.". Second, the analysis of each selected article has to be thorough. Here, for most of the articles, only the abstracts were read, thus limiting the ability to clearly identify specific details of each paper article. Finally, the trending topics identified can help to foster research in those fields of research.

6   Conclusion and Future Work

The CE is a recognized approach that helps to tackle the biggest environmental challenges from the 21st century, moving from a linear model to a circular one. The number of publications available on this topic is increasing each year making it one of the most trending topics when talking about environmental sustainability. Due to its processes and the nature of its products, the pharmaceutical industry has significant impacts on the environment. Therefore, it is crucial that the CE can be implemented in the pharmaceutical industry especially since circular principles are not fully embedded yet.

This paper defines a mapping of the publications available in the literature about the CE within the pharmaceutical industry. To reach this objective, we identified life cycle steps in accordance with the CE that could represent the stakes of circular actions. The analyzed publications were sorted into each life cycle stage of the product. The mapping proposed emphasized the high number of papers in two stages of the product life cycle. This classification not only allows the identification of potential solutions to implement the CE into the pharmaceutical industry but also permits the identification of opportunities for future research.

This study acknowledges the broadness of the scope of the CE when it comes to circular principles implementation. However, this preliminary review presents the basis for further study in the field of circularity for pharmaceutical industries and highlights a certain level of uncertainty with the needs for a deeper literature review.

References

1. Kirchherr, J., Reike, D., Hekkert, M.: Conceptualizing the circular economy: an analysis of 114 definitions. Resour. Conserv. Recycl. **127**, 221–232 (2017)
2. Ellen MacArthur Foundation: Towards the Circular Economy: An Economic and Business Rationale for an Accelerated Transition, vol. 1 (2013)
3. Panza, L., et al.: Open product development to support circular economy through a material lifecycle management. Int. J. Prod. Lifecycle Manage. **14**(2/3) (2022)
4. Circle Economy: The Circularity Gap Report 2023, Amsterdam (2023)
5. Ang, K.L., et al.: Sustainability framework for pharmaceutical manufacturing (PM): a review of research landscape and implementation barriers for circular economy transition. J. Clean. Prod. **280**(2), 124264 (2020)
6. Ranjbari, M., et al.: Mapping healthcare waste management research: past evolution, current challenges, and future perspectives towards a circular economy transition. J. Hazard. Mater. **422**, 126724 (2022)
7. UNEP: 21 Issues for the 21st century: results of the UNEP foresight process on emerging environmental issues (2017)
8. Rockström, J., et al.: Planetary boundaries: exploring the safe operating space for humanity. Ecol. Soc. **14**, 3–32 (2009)
9. Bocken, N.M.P., et al.: Product design and business model strategies for a circular economy. J. Ind. Prod. Eng. **33**, 308–320 (2016)
10. EU: A new CE action plan for a cleaner and more competitive Europe (2020)
11. EC: Proposal for a regulation establishing a framework for setting ecodesign requirements for sustainable products and repealing directive 2009/125/EC (2022)
12. McKinsey's Center for Business and the Environment, Ellen MacArthur Foundation: Growth within: a circular economy vision for a competitive Europe (2015)
13. Luu, D.-N., et al.: Eco-design and medicine: Opportunities to implement eco-design in the pharmaceutical R&D process. J. Clean. Prod. **365**, 132785 (2022)
14. Eckelman, M.J., et al.: Health care pollution and public health damage in the United States: an update. Health Aff. **39**, 2071–2079 (2020)
15. Ogden, N., Gachon, P.: Climate change and infectious diseases: what can we expect? Can. Commun. Dis. Rep. **45**, 76–80 (2019)
16. Wilkinson, J.L., et al.: Pharmaceutical pollution of the world's rivers. Proc. Nat. Acad. Sci. **119**, e2113947119 (2022)
17. Pimentel, D., et al.: Ecology of increasing diseases: population growth and environmental degradation. Hum. Ecol. **35**, 653–668 (2007)
18. Wuyts, W., et al.: Circular economy as a COVID-19 cure? Resour. Conserv. Recycl. **162**, 105016 (2020)
19. Luu, D.-N., et al.: Recycling of post-use bioprocessing plastic containers - mechanical recycling technical feasibility. Sustainability **14**, 15557 (2022)
20. Snyder, H.: Literature review as a research methodology: an overview and guidelines. J. Bus. Res. **104**, 333–339 (2019)
21. Keoleian, G.A., Menerey, D.: Sustainable development by design: review of life cycle design and related approaches. Air Waste **44**, 645–668 (1994)
22. Saidani, M.: Monitoring the circular economy with circularity indicators, applications to the heavy vehicle industry. Ph.D. thesis, Paris-Saclay (2018)

Optimizing Closed-Loop Supply Chain in the Electric Vehicle Battery Industry: A Fully Fuzzy Approach

Mina Kazemi Miyangaskary[1]([✉]), Samira Keivanpour[1], and Amina Lamghari[2]

[1] Department of Mathematical and Industrial Engineering, Polytechnique Montreal, Montreal, QC, Canada
mina.kazemi-miyangaskary@polymtl.ca

[2] Department of Management, University of Quebec, Trois-Rivières, QC, Canada

Abstract. Increasing vehicle emissions are major causes of global warming which is the most serious threat to human life. To alleviate this process, the Net-zero regulations enforce car manufacturers and encourage the population to shift from gasoline- to Electric vehicles. Although EV usage is unprecedently amplified, existing uncertainty in the supply chain of batteries of electric vehicles (BEVs) endangers EV's future market. For example, the scarcity of battery minerals, and the vagueness of supply chain parameters like costs. Reverse logistics in the BEVs supply chain can cope with the shortage of raw materials, and fuzzy theory is a promising approach to handle the vagueness. This study aims to put forward a fully fuzzy multi-Objective mathematical model by considering the uncertainty to optimize the BEVs closed-loop supply chain according to sustainable development principles in Canada. To do so, three objective functions are developed. Two objective functions maximize the profits of all supply chain players and service levels. The last one minimizes environmental impacts. Eventually, the model obtains the optimal amount of material flow, as decision variables, between all components of the supply chain.

Keywords: Battery Electric Vehicle · Sustainable Closed-Loop Supply Chain · Fully Fuzzy Multi-Objective Programming · Echelon Utilization

1 Introduction

Road transport generated 43.3% of GHG emissions in 2019, in Quebec (Ministère de l'Environnement et de la Lutte contre les changements climatiques (MELCC), 2021). Emitted GHG contributes to global warming that poses significant threats to human societies. For example, it increases the frequency of extreme weather like heatwaves and droughts that can disrupt food supplies, affect critical infrastructure, and cause widespread displacement. To avert the global warming implications, governments take various actions, from establishing binding actions to increasing public awareness to reduce vehicle emissions (Yarahmadi, Morency and Trepanier, 2023). For example,

C. Danjou et al. (Eds.): PLM 2023, IFIP AICT 702, pp. 38–49, 2024.
https://doi.org/10.1007/978-3-031-62582-4_4

according to the Canadian Net-Zero Emissions Accountability Act, Canada committed to decreasing its GHG emissions to zero by 2050 (Association, 2021). Over the last decade, due to zero-emission regulations, vehicle industries have shifted from developing fossil fuel engines to electric vehicles. Statistics imply that electric cars will play a significant role in the future of transportation. One of the essential parts of EVs is their Lithium-ion battery. The battery of an EV (BEV) composes almost 50% of the cost of an EV (Gu et al., 2018) and designates the EVs range. It is predicted that the EV demand will surge; lithium-ion battery demand will increase.

Figure 1 depicts the prediction of the universal market size of EVs (a) (Nogueira, Sousa and Alves, 2022) and BEV production by 2030 (b) (Shine, 2022). Obviously, both the EV markets and BEV production will notably increase by 2030.

Achilles' Heel of BEVs is its supply chain because, first, a few countries possess most of the minerals. For instance, the Democratic Republic of Congo produces more than 50% of cobalt globally, and China reserves over 66% of natural graphite (Igogo, Sandor, Mayyas and Engel-Cox, 2019). The scarcity of critical minerals is another issue. For example, the International Energy Agency (IEA) announced lithium shortage would be a global challenge by 2025 (Tracy, 2022).

Although battery minerals storage is restricted, focusing on a closed-loop supply chain could significantly compensate for this problem. A closed-loop supply chain comprises two parts: forward flow to generate products from raw materials and reverse flow to collect and reuse product waste. Scientific resources demonstrate that almost 90% of battery electric vehicles can be recycled. The reverse logistics can be composed of different actors, involving remanufacturing, refurbishing, echelon utilization, and recycling.

Reviewing the literature revealed that there are two critical gaps. First, according to the authors' best knowledge, previous studies used crisp mathematical models to optimize the BEVs supply chain (Zhang, Tian and Han, 2022). Although owing to the deficiency of the developed model in handling uncertainties related to the supply chain components, their results cannot be reliable. At the same time, uncertainties can limit the developed model in the real world. For example, parameters like the demand for BEVs and raw materials, order quantity, costs, and return rate of BEVs are vague. Moreover, their values are not certain because they depend on other issues. For example, transportation costs fluctuate because of political factors and fuel prices, and BEV demand is a function of various factors like EV demand. Therefore, to optimize the network parameters and variables should be fuzzy.

Second, most conducted research is needed to model the BEVs supply chain comprehensively. For example, Gonzales-Calienes, Yu and Bensebaa, (2022) utilized Geographic Information System to optimize the reverse flow, but the forward flow needs to be addressed. On the other hand, Li, Dababneh and Zhao, (2018) optimized the supply chain without contributing to the echelon utilization and market. A complete closed-loop supply chain of BEVs embarks by mine, refinery, BEV factory, EV manufacture, EV retailer, and EV market. It is followed by a collection and sorting center, recycling center, echelon utilization, echelon market, and disposal center.

To fill the gap, it is necessary to employ methods to consider all supply chain components, at the same time, take ambiguity, uncertainty, and indecision into account.

To do that, first, this study aims to consider sustainable development criteria to put forward a multi-objective model.

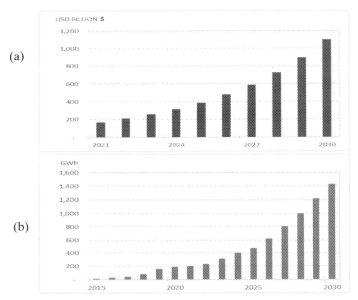

Fig. 1. The universal trend of the EV market (a) (Nogueira, Sousa and Alves, 2022) and BEV production (b) (Shine, 2022) by 2030

In addition, the developed model considers all BEV supply chain players. Furthermore, the proposed model is fully fuzzy, meaning all parameters and variables are fuzzy. Finally, it is worth mentioning that fuzzy techniques, like fuzzy mathematical programming, are potent approaches to dealing with environmental uncertainty and bridging the gaps between real-world conditions and the developed model.

The remainder of this paper is structured as follows. In the next section, previous studies that explored the problem are described. Section 3 discusses what steps should be taken to achieve the desired objective. Finally, the contribution of the paper is discussed, and the results are reviewed.

2 Literature Review

To further clarify our contributions, we summarize and arrange Table 1 to highlight the research gap concisely and clearly between our work and relevant studies. As demonstrated, the number of supply chain components in the present study is more comprehensive than previous efforts. In forward flow, mine, BEV manufacture, EV factory, EV retailer, and EV market. In reverse flow, players are collection and sorting centers, echelon utilization, recycling, disposal, and echelon market.

In addition, a previous practice rarely developed the supply chain according to sustainable development criteria. Lastly, this study develops a fully fuzzy multi-objective model contrary to almost all former deterministic optimization models.

Table 1. Comparison between the proposed model and conducted research in the BEV supply chain

Author (year)	Supply Chain components[1]	Criteria			Technique	Model type	Problem type	
		Ec	En	So			Certain	Uncertain
(Li, Dababneh and Zhao, 2018)	M, B, E, Re, A, C, Y, RM, D	*			Artificial intelligence	PSO[2]	*	
(Gu *et al.*, 2018)	M, B, Eu, A, Y, H	*			Game theory	Nash equilibrium	*	
(Zhang, Tian and Han, 2022)	E, Re, A, Y, D, H	*	*		Game theory	Stackelberg model	*	
(Gonzales-Calienes, Yu and Bensebaa, 2022)	C, Y, Eu	*	*		GIS	shortest path	*	
(Huster *et al.*, 2022)	B, E, A, Y, RM		*		Simulation	Discrete event	*	
(Pamucar, Torkayesh and Biswas, 2022)	Ranking of recycling centers	*	*	*	MCDM	Fuzzy WASPS		*
(Scheller *et al.*, 2020)	B, Y, Eu, RM	*			Simulation & Artificial intelligence	AIMMS & GUROBI	*	
(Zhang, Chen and Tian, 2023)	E, Re, A, Y, D, H	*	*		Game theory	Stackelberg model	*	
Purposed model	**M, B, E, Re, A, C, Y, Eu, D, H**	*	*	*	**Mathematical model**	**FFMOM[3]**		*

1. M: mine; B: BEV manufacture; E: EV manufacture; Re: retailer; A: EV market; C: collection center; Eu: echelon utilization; Y: recycling; D: disposal center; H: echelon market; RM: remanufacture
2. Ec: Economical; En: Environmental; So: Social;
3. Particle Swarm Optimization
4. Fully Fuzzy Multi-Objective Model

3   Model Description and Formulation

Figure 2 illustrates the designed research methodology. The methodology is developed in two phases, phases1, and 2 consisting of three and one steps.

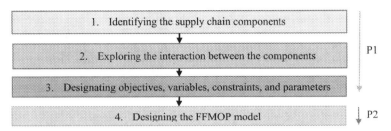

Fig. 2. Steps of the research methodology

The main objective of phase 1 is business understanding related to BEVs. Indeed, through three steps, all components are identified, their interactions are explored, and finally, objective functions, decision variables, and parameters are defined. Eventually, in phase 2 the FFMOP model is developed.

Figure 3 depicts the structure of the BEV closed-loop supply chain. The configuration includes two flows, forward and reverses flow showing in the order in black and green dot arrow. In forward flow, minerals are turned into the battery through the refining process and BEV manufacturing. Next, batteries are installed in EVs and are sold in the market using retailers. With time, the life of batteries is decreased; when they lose 20% of their life, batteries are replaced (Lai et al., 2021).

So, the used batteries are collected in the collection centers, and via a qualification assessment, their high quality is shipped to echelon markets for reuse, and low-quality ones are transported to recycling centers. More than 90% of a battery is recycled and is send to refining centers for more processing. The main difference between the designed supply chain network and previous ones is reverse flow. Indeed, not only all reverse flow components are taken into account but also their relationships are investigated. Then the model designates the optimal values of all variables.

Next step, objective functions are developed. In this study, three objective functions are designed to cover all aspects of sustainable development.

The following defines assumptions, notations (sets, decision variables and parameters) used in our proposed model.

3.1 Model Assumptions

- The capacity of all facilities is determined and limited.
- The BEVs closed-loop supply chain is pull system. In pull supply chain, production is based on a real demand and inventory cost outweighs the benefit of stocking products (Koo, 2020). So, there is no inventory cost.
- Quantity discounts are not considered in the purchase.
- Parameters and variables in the model like BEV demand, EV demand, costs, BEV return rate, energy consumption, facilities capacity, and BEV and mineral batteries order amount etc. are fuzzy.
- Shortage of products to supply customer's demand is allowed and incur a cost.

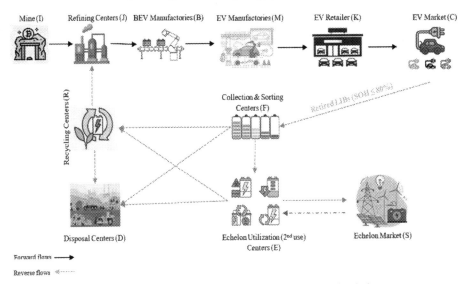

Fig. 3. Thematic configuration of the BEVs supply chain

3.2 Notations

The following notation is used to formulate the problem. Also, "~" denotes that the parameter or variable is fuzzy.

Sets

- I: Set of Suppliers (Mines), index by i.
- J: Set of Refining centers, index by j.
- B: Set of BEV Manufactories, index by b.
- M: Set of EV Manufactories, index by m.
- K: Set of EV Retailer, index by k.
- C: Set of EV Market, index by c.
- F: Set of Collection & Sorting Centers, index by f.
- S: Set of Echelon Markets, index by s.
- R: Set of Recycling Centers, index by r.
- D: Set of Disposal Centers, index by d.
- E: Set of Echelon Utilization Centers, index by e.
- G: Set of nodes, index by g.

Parameters

- $\widetilde{PS}_{bm}$: Unit sale price of EV batteries from BEV manufactories b to EV manufactories m (Dollar)

- $\widetilde{PS}_{mk}$: Unit sale price of EVs from EV manufactories m to EV retailer k (Dollar)
- $\widetilde{PS}_{kc}$: Unit sale price of EVs from EV retailer k to EV market c (Dollar)
- $\widetilde{P}_{ib}$: Unit purchase cost of raw material from mine i to BEV manufactories b (Dollar)
- $\widetilde{P}_{rb}$:Unit purchase cost of raw material from recycling center r to BEV manufactories b (Dollar)
- $\widetilde{D}\ b$: The demand for BEV (Ton)
- $\widetilde{D}\ EV$: The demand for EV (Ton)
- $\widetilde{M}$: Required raw materials to produce an EV battery unit (Ton)
- $\widetilde{C}_g$: The capacity of facilities g ∈ I,J,B,M,K,F,S,R,D,E (Ton)
- $\widetilde{Pr}_g$: Unit Processing cost in node g ∈ I,B,M,F,R,D,E (Dollar)
- $\widetilde{T}_{gh}$: Unit Transport cost from node g to node h (Dollar)
- $\widetilde{Sh}\ b$: Unit Shortage cost of EV batteries in BEV manufactories (Dollar)
- $\widetilde{Sh}\ m$: Unit Shortage cost of EV in EV manufactories (Dollar)
- $\widetilde{Sh}\ k$: Unit Shortage cost of EV in EV retailer (Dollar)
- $\widetilde{R}$: Return percentage of retired batteries from EV markets
- $\widetilde{Rs}$: Return percentage of second retired batteries from echelon markets
- $\widetilde{PS}_{es}$: Unit sale price of retired batteries from echelon utilization centers e to echelon markets s (Dollar)
- $\widetilde{PS}_{rb}$: Unit sale price of raw material from recycling centers r to BEV manufactories b (Dollar)
- $\widetilde{P}_{cf}$: Unit purchase cost of retired batteries from EV markets c to Collection & Sorting Centers f (Dollar)
- $\widetilde{Ec}_g$: Unit of energy consumed in processing in node g ∈ I,J,B,M,K,C,F,S,R,D,E (Dollar)

Decision Variables

- $\widetilde{QR}_{ib}$: Quantity of raw material purchased by BEV manufactories b from mines I (Ton)
- $\widetilde{QR}_{rb}$: Quantity of raw material purchased by BEV manufactories b from recycling centers r (Ton)
- $\widetilde{QB}_{bm}$: Quantity of EV batteries transported from BEV manufactories b to EV manufactories m (Ton)
- $\widetilde{QB}_{mk}$: Quantity of EV transported from EV manufactories m to EV retailer k (Ton)
- $\widetilde{QB}_{kc}$: Quantity of EV transported from EV retailers k to EV market c (Ton)
- $\widetilde{QB}_{fr}$: Quantity of retired batteries transported from collection centers f to recycling centers r (Ton)
- $\widetilde{QB}_{fe}$: Quantity of retired batteries transported from collection centers f to echelon utilization centers e (Ton)

- $\tilde{QB}_{fd}$: Quantity of retired batteries transported from collection centers f to disposal centers d (Ton)
- $\tilde{QB}_{es}$: Quantity of retired batteries transported from echelon utilization centers e to echelon markets s (Ton)
- $\tilde{QB}_{er}$: Quantity of retired batteries transported from echelon utilization centers e to recycling centers r (Ton)
- $\tilde{QB}_{ed}$: Quantity of retired batteries transported from echelon utilization centers e to disposal centers d (Ton)
- $\tilde{QC}_{rd}$: Quantity of scraps transported from recycling centers r to disposal centers d (Ton)

3.3 Mathematical Model

The following presents the mathematical model of the fully fuzzy multi-objective problem in this study:

$$\mathbf{MaxF_1}\left(\tilde{\mathbf{X}}\right) = \left(T\tilde{R}F \oplus T\,\tilde{R}\,R\right) \ominus \left(T\tilde{C}R \oplus T\tilde{C}P \oplus T\tilde{C}T \oplus T\tilde{C}S\right) \quad (1)$$

$$\mathbf{Min\ F_2}(\tilde{X}) = \left[\sum_b \tilde{EC}_b \otimes \sum_m \tilde{QB}_{bm}\right] \oplus \left[\sum_i \tilde{EC}_i \otimes \sum_b \tilde{QR}_{ib}\right] \oplus \left[\sum_m \tilde{EC}_m \otimes \sum_k \tilde{QB}_{mk}\right]$$

$$\oplus \left[\sum_f \tilde{EC}_f \otimes (\tilde{R} \otimes \sum_c \tilde{QB}_{kc})\right] \oplus \left[\sum_e \tilde{EC}_e \otimes \left(\sum_r \tilde{QB}_{er} \oplus \sum_s \tilde{QB}_{es} \oplus \sum_d \tilde{QB}_{ed}\right)\right]$$

$$\oplus \left[\sum_r \tilde{EC}_r \otimes \left(\sum_b \tilde{QR}_{rb} \oplus \sum_d \tilde{QC}_{rd}\right)\right]$$

$$\oplus \left[\sum_d \tilde{EC}_d \otimes \left(\sum_f \tilde{QB}_{fd} \oplus \sum_e \tilde{QB}_{ed} \oplus \sum_r \tilde{QC}_{rd}\right)\right] \quad (2)$$

$$\mathbf{MaxF_3}(\tilde{X}) = \left(\frac{\sum_c \sum_k \tilde{QB}_{kc}}{\tilde{DEV}}\right) \oplus \left(\frac{\sum_k \sum_m \tilde{QB}_{mk}}{\tilde{DEV}}\right) \oplus \left(\frac{\sum_b \sum_m \tilde{QB}_{bm}}{\tilde{Db}}\right) \quad (3)$$

$$\mathbf{St}: \quad \sum_b \tilde{QR}_{ib} \le \tilde{C}_i \quad \forall\, i \in I \quad (4)$$

$$\sum_i \tilde{QR}_{ib} \oplus \sum_r \tilde{QR}_{rb} \le \tilde{C}_b \quad \forall b \in B \quad (5)$$

$$\sum_b \tilde{QB}_{bm} \le \tilde{C}_m \quad \forall m \in M \quad (6)$$

$$\tilde{R} \otimes \sum_c \tilde{QB}_{kc} \le \sum_f \tilde{C}_f \quad (7)$$

$$\sum_f \tilde{QB}_{fe} \oplus \left(\tilde{Rs} \otimes \sum_s \tilde{QB}_{es}\right) \le \tilde{C}_e \quad \forall e \in E \quad (8)$$

$$\sum_e \tilde{QB}_{es} \le \tilde{C}_s \quad \forall\, s \in S \quad (9)$$

$$\sum_f \widetilde{QB}_{fr} \oplus \sum_e \widetilde{QB}_{er} \leq \tilde{C}_r \quad \forall r \in R \tag{10}$$

$$\tilde{R} \otimes \sum_k \sum_c \widetilde{QB}_{KC} = \sum_r \widetilde{QB}_{fr} \oplus \sum_d \widetilde{QB}_{fd} \oplus \sum_e \widetilde{QB}_{fe} \quad \forall f \in F \tag{11}$$

$$\sum_f \widetilde{Q}_{fe} \oplus \left(\tilde{R}e \otimes \sum_s \widetilde{Q}_{es} \right) = \sum_r \widetilde{QB}_{er} \oplus \sum_d \widetilde{QB}_{ed} \oplus \sum_s \widetilde{QB}_{es} \quad \forall e \in E \tag{12}$$

$$\sum_i \widetilde{QR}_{ib} \oplus \sum_r \widetilde{QR}_{rb} = \tilde{M} \otimes \sum_m \widetilde{QB}_{bm} \quad \forall b \in B \tag{13}$$

$$\sum_k \sum_c \widetilde{QB}_{KC} \leq \tilde{D}EV \tag{14}$$

$$\sum_k \sum_m \widetilde{QB}_{Km} \leq \tilde{D}EV \tag{15}$$

$$\sum_b \sum_m \widetilde{QB}_{bm} \leq \widetilde{Db} \tag{16}$$

$$\text{All of variables} \geq \tilde{0} \tag{17}$$

The first objective function maximizes the profits of all supply chain components. In fact, this function considers the economic aspect of sustainable development. To do that, total income is subtracted from total cost. The income involves summation of income of forward (T R̃ F) and reverse flow (T R̃ R).

$$\mathbf{T \tilde{R} F} = \left(\sum_b \sum_m \widetilde{PS}_{bm} \otimes \tilde{Q}B_{bm} \right) \oplus \left(\sum_m \sum_k \widetilde{PS}_{mk} \otimes \tilde{Q}B_{mk} \right) \oplus \left(\sum_k \sum_c \widetilde{PS}_{kc} \otimes \tilde{Q}B_{kc} \right) \tag{18}$$

$$\mathbf{T \tilde{R} R} = \sum_e \sum_s \widetilde{PS}_{es} \otimes \tilde{Q}B_{es} \tag{19}$$

The cost is calculated from summation of cost of purchase of raw material $(T\tilde{C}R)$, processing $(T\tilde{C}P)$, transportation $(T\tilde{C}T)$, and shortage $(T\tilde{C}S)$.

$$\mathbf{T \tilde{C} R} = \left(\sum_i \sum_b \tilde{P}_{ib} \otimes \tilde{Q}R_{ib} \right) \oplus \left(\sum_r \sum_b \tilde{P}_{rb} \otimes \tilde{Q}R_{rb} \right) \tag{20}$$

$$T\check{C}P = \left(\sum_i \tilde{Pr}_i \otimes \sum_b \widetilde{QR}_{ib}\right) \oplus \left(\sum_b \tilde{Pr}_b \otimes \sum_m \widetilde{QB}_{bm}\right) \oplus \left(\sum_m \tilde{Pr}_m \otimes \sum_k \widetilde{QB}_{mk}\right)$$

$$\oplus \left(\sum_k \tilde{Pr}_k \otimes \sum_c \widetilde{QB}_{kc}\right) \oplus \left(\sum_f \tilde{Pr}_f \otimes \left(\tilde{R} \otimes \sum_c \widetilde{QB}_{kc}\right)\right)$$

$$\oplus \left(\sum_e \tilde{Pr}_e \otimes \left(\sum_f \widetilde{QB}_{fe} \oplus \sum_s \tilde{R}s \otimes \widetilde{QB}_{es}\right)\right) \tag{21}$$

$$\oplus \left(\sum_r \tilde{Pr}_r \otimes \left(\sum_b \widetilde{QR}_{rb} \oplus \sum_d \widetilde{QC}_{rd}\right)\right)$$

$$\oplus \left(\sum_d \tilde{Pr}_d \otimes \left(\sum_f \widetilde{QB}_{fd} \oplus \sum_e \widetilde{QB}_{ed} \oplus \sum_r \widetilde{QC}_{rd}\right)\right)$$

$$T\check{C}S = \left(\tilde{Sh}b \otimes \left(\tilde{D}b - \sum_b \sum_m \widetilde{QB}_{bm}\right)\right) \oplus \left(\tilde{Sh}m \otimes \left(\tilde{D}EV - \sum_m \sum_k \widetilde{QB}_{mk}\right)\right)$$

$$\oplus \left(\tilde{Sh}k \otimes \left(\tilde{D}EV - \sum_k \sum_c \widetilde{QB}_{kc}\right)\right)$$

$$\tag{22}$$

$$T\check{C}T = \left(\sum_i \sum_b \tilde{T}_{ib} \otimes \widetilde{QR}_{ib}\right) \oplus \left(\sum_r \sum_b \tilde{T}_{rb} \otimes \widetilde{QR}_{rb}\right) \oplus \left(\sum_r \sum_b \tilde{T}_{rb} \otimes \widetilde{Qc}_{rd}\right)$$

$$\oplus \left(\sum_b \sum_m \tilde{T}_{bm} \otimes \widetilde{QB}_{bm}\right) \oplus \left(\sum_m \sum_k \tilde{T}_{mk} \otimes \widetilde{QB}_{mk}\right)$$

$$\oplus \left(\sum_f \sum_e \tilde{T}_{fe} \otimes \widetilde{QB}_{fe}\right) \oplus \left(\sum_f \sum_r \tilde{T}_{fr} \otimes \widetilde{QB}_{fr}\right)$$

$$\oplus \left(\sum_f \sum_d \tilde{T}_{fd} \otimes \widetilde{QB}_{fd}\right) \oplus \left(\sum_e \sum_s \tilde{T}_{es} \otimes \widetilde{QB}_{es}\right) \oplus \left(\sum_e \sum_r \tilde{T}_{er} \otimes \widetilde{QB}_{er}\right)$$

$$\oplus \left(\sum_e \sum_d \tilde{T}_{ed} \otimes \widetilde{QB}_{ed}\right)$$

$$\tag{23}$$

Function 2 minimizes the negative impacts of the supply chain on the environment by reducing energy consumption in processing the facilities.

Function 3 maximizes the social impacts of sustainable development via increasing the service level. The service level defined as is a ratio between satisfied demand to total demand. Constraint 10 shows facilities' capacities.

Constraints 11 and 12 in the order are related to collection centers and echelon utilization centers and show that the total amount of their input produces is equal to the outputs. Constraint 13 indicates that the quantity of battery production determines the amount of purchase of raw materials. Constrains 14 shows that the amount of EVs that retailers supply to the market should be equal to or less than market demand. Constrains 15 persists that the quantity of the EVs that manufacturers supply to retailers should be equal to or less than k retailer demands. And constraint 16 demonstrates that the amount of battery demand should always be more significant than battery production. Finally, constrain (17) all variables are non-negative.

3.4 Solutions Approaches

There are various techniques to solve FFMOP problems. The present study employs the technique (Sharma and Aggarwal, 2018) to solve the proposed FFMOP model because its computation complexity is less than other techniques, solves problems with triangular and trapezoidal fuzzy numbers, and can be used for other LR flat fuzzy numbers. According to this method, the proposed FFMOP model is converted into a CLP model (for more information, see (Sharma and Aggarwal, 2018)); then, by solving the CLP model, the optimal values of the decision variable are obtained.

The main advantage of the proposed model is that the optimal values of decision variables and objective functions are calculated as fuzzy values. It means decision makers and practitioners can consider the real-world uncertainties in their decision-making processes. For example, when the outputs are shown via triangular fuzzy numbers the optimal values are three values lower boundary, upper boundary, and mean, while previous studies determined only one value as an optimal result.

4 Conclusion

This research initially contributes to considering uncertainty in optimizing the closed-loop supply chain of EV batteries. In addition, designing the developed model was based on the principles of sustainable development.

The three objective models were designed. To make the generated model more realistic fuzzy logic was utilized, and a fully fuzzy multi-objective was developed. The model includes three objective functions. The first maximize profits, while the second and third functions minimize negative environmental impacts of the supply chain and maximize social benefits, respectively.

Performing the model on real datasets provides opportunities for scholars and practitioners to investigate the outcomes of various scenarios and improve electric battery supply chain management. In addition, Applying the model in a real case study in Canada is the agenda of the authors.

References

Association, C.R.E.: Powering Canada's Journey to Net-Zero CanREA's 2050 Vision (2021)
Gonzales-Calienes, G., Yu, B., Bensebaa, F.: Development of a reverse logistics modeling for end-of-life lithium-ion batteries and its impact on recycling viability—a case study to support End-of-Life electric vehicle battery strategy in Canada. Sustainability (Switz.) **14**(22) (2022). https://doi.org/10.3390/su142215321

Gu, X., et al.: Developing pricing strategy to optimize total profits in an electric vehicle battery closed loop supply chain. J. Clean. Prod. **203**, 376–385 (2018)

Huster, S., et al.: A simulation model for assessing the potential of remanufacturing electric vehicle batteries as spare parts. J. Clean. Prod. **363**, 132225 (2022)

Igogo, T., et al.: Supply chain of raw materials used in the manufacturing of light-duty vehicle lithium-ion batteries (2019)

Lai, X., et al.: Turning waste into wealth: a systematic review on echelon utilization and material recycling of retired lithium-ion batteries. Energ. Storage Mater. **40**, 96–123 (2021). https://doi.org/10.1016/j.ensm.2021.05.010

Li, L., Dababneh, F., Zhao, J.: Cost-effective supply chain for electric vehicle battery remanufacturing. Appl. Energy **226**, 277–286 (2018)

Ministère de l'Environnement et de la Lutte contre les changements climatiques (MELCC): Ges 1990–2019: Inventaire Québécois Des Émissions De Gaz À Effet De Serre En 2019 Et Leur Évolution Depuis 1990 (2021)

Nogueira, T., Sousa, E., Alves, G.R.: Electric vehicles growth until 2030: impact on the distribution network power. Energy Rep. **8**, 145–152 (2022)

Pamucar, D., Torkayesh, A.E., Biswas, S.: Supplier selection in healthcare supply chain management during the COVID-19 pandemic: a novel fuzzy rough decision-making approach. Ann. Oper. Res. **10**, 2–43 (2022)

Scheller, C., et al.: Decentralized planning of lithium-ion battery production and recycling. Procedia CIRP **90**, 700–704 (2020)

Shine, I.: The world needs 2 billion electric vehicles to get to net zero. But is there enough lithium to make all the batteries? (2022)

Tracy, B.S.: Grants for Charging and Fueling Infrastructure; Section 40201, Earth Mapping Resources Initiative; Section 40207, Battery Processing and Manufacturing; Section 40208, Electric Drive Vehicle Battery Recycling and Second-Life Applications Program; Section 4 (2022)

Yarahmadi, A., Morency, C., Trepanier, M.: New data-driven approach to generate typologies of road segments. Transportmetrica A Transport Sci. (2023). https://doi.org/10.1080/23249935.2022.2163206

Zhang, C., Chen, Y.X., Tian, Y.X.: Collection and recycling decisions for electric vehicle end-of-life power batteries in the context of carbon emissions reduction. Comput. Ind. Eng. **175**(195), 108869 (2023). https://doi.org/10.1016/j.cie.2022.108869

Zhang, C., Tian, Y.X., Han, M.H.: Recycling mode selection and carbon emission reduction decisions for a multi-channel closed-loop supply chain of electric vehicle power battery under cap-and-trade policy. J. Clean. Prod. **375**, 134060 (2022)

Koo, J.: Push system vs. pull system: adopting a hybrid approach to MRP (2020). https://tulip.co/blog/what-is-a-push-system-vs-a-pull-system/

Sharma, U., Aggarwal, S.: Solving fully fuzzy multi-objective linear programming problem using nearest interval approximation of fuzzy number and interval programming. Int. J. Fuzzy Syst. **20**(2), 488–499 (2018). https://doi.org/10.1007/s40815-017-0336-8

Drives and Barriers for Circular Ion-Lithium Battery Economy: A Case Study in an Automobile Manufacturer

Ana Paula Louise Yamada Macicieski[1] (ID), Enzo Domingos[2]([✉]) (ID), Juliana Hoffmann[1], Luciana Rosa Leite[1] (ID), and Carla Roberta Pereira[3] (ID)

[1] Santa Catarina State University, Joinville, SC, Brazil
[2] Polytechnique Montréal, Montréal, QC, Canada
enzogdomingos@gmail.com
[3] The Open University Business School, Milton Keynes, UK

Abstract. The growing concern for sustainability and the development of new technologies have made the electric vehicle one of the solutions and alternatives for global mobility. However, the increase in sales of these vehicles also impacts the amount of lithium-ion batteries produced, which encompass rare mineral extraction processes. Therefore, the concept of circular economy has been explored and applied by organizations, aiming to close the product life cycle, reduce the demand for resources and improve the supply chain. Through a single case study, the objective of this paper is to identify the drivers and barriers in the implementation of the circular economy for lithium-ion batteries used in electric cars. To this end, semi-structured interviews were conducted with three employees of a car manufacturer located in Brazil. From the content analysis, thirteen drivers and sixteen empirical barriers were identified. The unavailability of national technology and qualified suppliers for the implementation of the second and third use of batteries were identified as the main barriers. Regarding the drivers, objectives and strategies of the organization and regulations (governmental and environmental) were highlighted as the main boosters. We believe such results can help those organizations that intend to apply the circularity model to electric vehicle batteries to anticipate drivers and barriers and hence explore the opportunities presented in this study.

Keywords: Circular Economy · Electric Vehicles · Lithium-Ion Batteries

1 Introduction

The Industrial Revolution was one of the points of transformation of contemporary society, which altered the way of manufacturing enabling mass production and defined the model of economy, which remains to the present day. This model is defined as linear, in which materials are extracted, transformed and discarded. However, this model is reaching its physical limits, because natural resources are finite and the accumulation and generation of waste is inevitable [1]. Therefore, circular economy (CE) emerges as an alternative to the linear economy model, being designed to close the life cycle of a product, through the 9R's (in particular: Reduction, Reuse and Recycling) thus making production and consumption more sustainable and keeping the value and usefulness of products and resources [2, 28].

At the same time, companies are seeking the sustainability of their business, having as one of the objectives to achieve competitive advantages in the market, through the development of new products and technologies, while seeking to reduce environmental impacts and the emission of polluting gases. One of these innovations is the electric vehicles or EVs. According to the International Energy Agency [3], worldwide sales of EVs doubled in 2021, reaching a record of 6.6 million new vehicles, which is equivalent.

to approximately 9% of sales global levels. Other authors indicate that the trend in the coming years is the growth of the production of EVs, because they are a great alternative for conventional vehicles to combustion, which depend on fossil fuels and have higher rates of carbon emissions [4–6].

However, one of the concerns arising from this growth is the availability of the raw materials used in batteries of the EVs, which are usually lithium-ion type, because the elements required for production such as lithium, cobalt, nickel, manganese and graphite are on the European Union's list of critical raw materials [7]. Lithium is expected to increase demand by six times by 2030 [3]. Thus, the low availability of these resources and the high demand induce increased prices, impacting the production costs of the EVs, since the battery is one of the main components and corresponds to approximately 40% of the total cost of the vehicle [6, 8]. Furthermore, the supply of critical materials for the production of EVs is also susceptible to risks due to unequal geographical distribution and almost monopolistic control of necessary resources [5].

Taking into account the current economic scenario, studies by the Ellen MacArthur Foundation [9] and the European Commission [10] indicate that the circularity model assists in economic development, generating jobs and reducing a country's social inequality, as well as increasing resource efficiency and contributing to innovation.

Drawing on this context, the present study aims to identify the drivers and barriers to the implementation of circular economy for lithium-ion batteries used in electric cars. To do so, a single case study was conducted in a rental car manufacturer in Southern of Brazil. We believe the outcome of this study can bring initial contributions to organizations in the transition to the circular economy, especially the ones in the automotive sector.

2 Theoretical Rationale

2.1 Circular Economy

According to Kirchherr, Reike and Hekkert [2], CE is a new field of study, as about 73% of the definitions are from the last 5 years. They also indicate that many studies on CE are conducted by non-academic figures, defined as gray literature [11, 12, 27]. According to the Ellen MacArthur Foundation [9], CE is "an economy that is restorative or regenerative by intent and design."

More recently, researchers and professionals have identified and highlighted the various existing drivers [e.g. 13, 14] for the implementation of the CE, in order to overcome many barriers present in this process [e.g. 15, 16, 27]. Comparing with innovation, Jesus and Mendonça [15] indicate that drivers and barriers can be divided into "harder" (technical and economic type) and "softer" (regulatory and cultural aspects) – Table 1. In addition, the authors define drivers as factors that allow and encourage the transition to CE, while barriers are impediments or bottlenecks that obstruct this change. In general, there is not only a driver or barrier that has greater prominence, but a mixture of factors that facilitate and limit, deriving from particular conditions where the organization is inserted.

Table 1. Typology and definition of drivers and barriers for CE

		Drivers	Barriers
"Harder" factors	Technician	Availability of technologies that facilitate resource optimization, remanufacturing and regeneration of by-products such as insum for other processes, development of convenient sharing solutions and with a superior consumer experience	Inadequate technology, delay between design and dissemination, lack of technical support and training
	Economical/ Financial/ Marketing	Increased demand for resources and consequent pressure of resource exhaustion, increased cost of resources and volatility of supply, leading to incentives for cost reduction and stability solutions)	Large capital requirements, significant transaction costs, high upfront costs, asymmetric information, uncertain return and profit
"Softer" factors	Institutional/Regulatory	Increasing environmental legislation, environmental standards and waste management guidelines	Misaligned incentives, lack of a favorable legal system, poor institutional structure
	Social/Cultural	Social awareness, environmental literacy and changing consumer preferences	Rigidity of consumer behavior and business routines

2.2   Electric Vehicles

Unlike combustion vehicles that have only one internal combustion engine, the Electric vehicles (EVs) use an electric propulsion system, consisting of an electric motor that uses the chemical energy stored in rechargeable battery. This energy is converted into electric energy to power the engine, being transformed into mechanical energy, enabling the movement of the vehicle [17]. Therefore, the fuel of electric vehicles is electricity, which can be obtained in different ways [18]: Connecting directly to the external source of electricity, for plugs or overhead cables; using the electromagnetic induction system; from the reaction of hydrogen and oxygen with water in a fuel cell; or by means of mechanical braking energy (regenerative braking, in which energy is obtained by braking the vehicle).

Batteries are considered the main component and the most critical technology of an EV, as they have a key role in the context of electric mobility and the highest cost in the value chain [4]. In general, the battery assembly encompasses four elements: the cooling system, the battery cell, the packaging and the battery management system [19]. Several studies [e.g. 20, 21] demonstrate that lithium-ion batteries can be remanufactured at a cost equivalent to 60% of a new battery, in addition to reducing carbon emissions and resource consumption. Additionally, other authors [22, 23] indicate a range of possibilities of redirection of batteries, mainly in the use in energy storage systems, allowing the increase of the service life of almost 10 years.

3   Method

The first stage of this research involved the identification of the gap and the development of the theoretical rationale. Next step comprised the case study conduction through data collection from semi-structured interviews, followed by analysis and interpretation of the data. Finally, the empirical elements of the case study will be compared with the knowledge gained from the literature, in order to discuss the results and draw the conclusions and final considerations of the study.

Yin [24] points out that the choice of single case study is appropriate under several circumstances, such as having a case that can "represent a significant contribution to the formation of knowledge and theory, confirming, challenging, or expanding the theory". Therefore, the choice of the single case study is justified, because the selected "case" is a national innovation, which enables the sustainable growth of electromobility, both locally and globally.

Thus, the object of the case study is a manufacturer of premium automobiles and motorcycles, recognized for developing initiatives focused on circular economy, sustainability and reducing emissions carbon dioxide. The company has two industrial units in Brazil – one responsible for the manufacturing of motorcycles, and the other for vehicles. In addition, the company also has a financial services unit located in São Paulo, which offers financing options and consortia. This study focus on the manufacturing of vehicles, located in the southern region of Brazil. This unit has developed the model of circularity of lithium-ion batteries, used in electric vehicles sold in the country, which are imported from other factories of the brand.. The infrastructure for the activities of assembly, bodywork/welding, painting and logistics was built in 2014, and it has currently around 600 direct employees being responsible for the manufacturing of four models to combustion.

The semi-structured interviews were designed considering not only aspects about the elements which lead to the adoption of lithium-ion batteries circular economy, but also the barriers to its implementation in the company. In order to present the theme and schedule the interviews, the employees were contacted by e-mail or in person. The three employees (plant director, manufacturing manager, and outsourced technician, identified respectively as interviewee A, B and C) selected for the interviews had great influence or active participation in the implementation of the circular economy model of lithium-ion batteries. The interviews took place in person in October 2022, and all of them were recorded to allow further transcription and analysis. In addition, other internal and external communication materials and sustainability reports of the company were also used as secondary data, providing support for the analysis.

4   Findings

Aware of the need for more sustainable manufacturing of components that are inserted in electric vehicles, such as the electric motor, high-voltage storage and battery cells, the organization in study initiated efforts towards this aim. Thus, based on the shared responsibility with its suppliers, distributors and concessionaires, the organization ensures the proper management of components. One of the business initiatives offered to all EVs' owners is the free collection of batteries that are at the end of the life cycle. Figure 1 presents the circular economy model, which was designed by the company's environmental area, for lithium-ion batteries of one of the models of EVs.

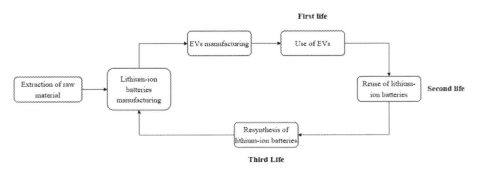

Fig. 1. Circular Economy model for EVs lithium-ion batteries

In this proposed model, the process begins with the extraction of the raw materials needed for the production of lithium-ion batteries, which will be later included in the manufacture of the EVs. After manufacturing, electrical models are made available for sale and use, which are called as the "first life" of the battery. The end-of-cycle phase of it occurs when the storage capacity reaches about 80% of the original value [25].

When this certain level is reached, the batteries must be collected by the dealers for replacement and then forwarded to their second use or "second life", in which are redirected to store electricity from solar panels. This system allows you to create a fast charging station for EVs, which can operate both connected and disconnected from the public power grid, enabling an infrastructure of sustainable recharge in regions where there is no access to the public electricity grid.

The "third life" of the lithium-ion battery is when there is no possibility of recharging solutions. Now the materials are extracted for resynthesis, that is, for the production of new batteries. This process aims to recycle rare chemical elements through the hydrometallurgy process, which is more sustainable when compared to pyrometallurgy (the most applied technique nowadays).

4.1 Identification of Drivers and Barriers

Drivers. Thirteen drivers (or factors that promote the transition to CE) were identified in the content analysis of the interviews and classified by type (internal/external) and factors (technical, institutional/regulatory, economic/financial/marketing and social/cultural).

Interviewee A is responsible for the strategic management of the industrial plant and has been working in the organization for 15 years. They mention two internal drivers (Driver D1 and D2) and two external drivers (D3 and D4), focusing mainly on the technical factor, emphasizing that there was no technology available at the time, so innovation was paramount to allow the implementation of the model, which would help promote the brand. According to them, *"the main impulse was the issue of recycling the battery"* (D1), and the use of the circular economy emerged as a possibility to avoid the direct recycling of the battery (D2), since this process was not yet fully understood at the time of implementation when the project started (2014). As external factors, this interviewee suggests that the project of circular economy was promoted because it was an innovation at the time (D3). As a matter of fact, the interviewee mentions that *"the name of the company was also benefited, because then there were several stakeholders investing money in the products [...]"* (D4).

The innovation issue was also mentioned by interviewee B (9 years at the company) when they mention that the company has an innovation acceleration program (D8). As they argue, *"The program was conducted to propose the automotive lithium-ion battery circular model to a partner company, which worked only with computer lithium batteries. Thus, the company's acceleration program fostered innovation in partner companies, enabling the circularity model to work"*. Interviewee B has the responsibility of managing the environmental area of the plant, and besides the innovation program, he comments on two other internal drivers, one of a more regulatory nature (D5) and another linked to the technical factor (D6). As D5 this interviewee mentions the fact that the organization has always been a pioneer in the adoption of sustainability strategies (D5), and that *"within this strategic plan culture, the employee has the responsibility, which today they call circularity, to have their products from the design thought to the end of life, the proper disposal. The perspective of how to achieve circularity before discarding the product and how to develop the technique for this model"* (A7).

Also highlighted by interviewee B, an external driver identified (D7): the National Solid Waste Policy (PNRS), a national law enacted in 2010, which requires the proper disposal of solid waste generated by organizations, was received with ease by the company. According to interviewee B, *"In a reverse logistics program, the National Solid Waste Policy, which involves some items of post-use of the vehicle, it was noticed that only one point of the law had not yet been solved and was in progress. In this case, the destination of high voltage batteries from electric vehicles."* This point was solved in the circular economy model proposed by the company, from the aforementioned avalanches as company's sustainability strategy, together with the internal innovation and acceleration of ideas program were the main drivers for the implementation of CE lithium-ion batteries, used in EVs.

Contrary to what was exposed by previous interviewees, the outsourced employee (interviewee C) presents more external drivers (D9, D10 and D11) than internal (D12 and D13), most of which are related to institutional/regulatory factors. The interviewee emphasizes the importance of the Paris Agreement, ISO 14001 and PNRS, in addition to sustainability as a strategic objective of the brand, to ensure the application of circularity in products manufactured by the company. According to this interviewee, the fact that the company is European also contributed to boost the battery circularity project, *"The circular economy comes mainly from the European industries, especially regarding the Paris Agreement. When the climate agreement comes into force, they say: we have to reduce the emission of CO2, otherwise the planet will heat up, and so on. Then a series of actions start to be taken both by the government and by the companies to try to achieve this reduction"*. Thus, they point out that both the concern with sustainability as a strategic focus of the company and the implementation of the circular economy are responses to external demands.

It is also important to mention that interviewee C was the only one to highlight a social and cultural factor as a driver, indicating that the responsibility as an environmental engineer is to seek better alternatives to the processes carried out in order to avoid negative environmental impacts.

Barriers. The same categorization of the drivers (by type and factor) was used for the barriers. According to interviewee A, the greatest challenges faced for the adoption of CE for lithium-ion batteries of EVs are classified as internal and technical barriers (B1 and B2), since it was necessary to develop knowledge and new technologies to enable the design of the second and third use of batteries. They mention that such barriers range from *"the challenge of knowing how the power bank technology, for example, would work using lithium-ion batteries in the second life"*, to more operational issues such as *"the length of the cable, some cars have a plug in the front, some cars have a plug in the back, and this changes the design of the product, and therefore the circularity process"*. In addition, the interviewee points out that the greatest external challenge (B3) is to expand and insert this model in the market, in their words, *"how it would be to really sell this model of circular economy to whoever is interested"*.

Eight other challenges are presented by interviewee B, which covered all four classification factors (technical, social/cultural, institutional/regulatory, economic/financial/marketing). The external technical factor presented the highest number of barriers (B4, B5 and B6), since there were no qualified national suppliers to develop the recycling and redirection solution of batteries, according to the CE model elaborated by the company. For example, he mentions the search for "*developing a national technology*" so that the lithium-ion batteries do not need to return to Europe for the recycling process. This would reduce costs and risks in the supply chain. Thus, according to interviewee B, "*since 2018, we have been looking for a logistics operator that could provide a solution for the disposal of these batteries, among other reverse logistics items, and since May 2020 we have been looking for recycling solutions at the end of battery use, because this logistics operator was not doing it, and was not finding it*".

In addition, another prominent factor was related to the external social/cultural environment (B7 and B8), due to the demand to raise awareness among concessionaires and consumers about the importance of collection and referral lithium-ion batteries for the company. Other difficulties related to institutional/regulatory (B9 and B10) and economic/financial/marketing (B11) factors reveal the importance of communicating strategic planning and concessionaires and the after-sales team, in order to ensure the continuity of the model.

Interviewee C mentions five barriers, two of which are internal social/cultural (B12 and B13) indicating that lack of awareness and resistance to change occur during the implementation process of innovation, as well as in the application of the circularity of the VEs batteries. In this sense, the interview mentions that "*the political effort is very big, you have to talk to a lot of people*" to make the circularity project happen - from the consumers to the operators of the model. In addition, barriers related to the economic/financial/marketing factor of the external (B14) and internal (B15) types show the need to present financial or (such as the markets), both for partner companies and for the board, in order to enable the implementation of the CE. Another barrier indicated was internal institutional/regulatory (B16), since to convince senior management it was necessary to demonstrate the importance of the circularity system. According to the interviewee, this barrier occurs because there is difficulty in monetizing the return to the business, "*the return is sometimes intangible*". For example, the return on aligning the brand with the market expectations is intangible for the production director.

5 Conclusion

This study drew upon the theoretical results of Agyeman *et al.* [26] and Jesus and Mendonça [15], in which the barrier and driver were classified by type (internal/external) and factors (technical, institutional/regulatory, economic/financial/marketing and social/cultural). In summary, 14 out of 16 empirical barriers may be associated with barrier factors presented in the literature. The interviewees pointed out that the unavailability of technology and national suppliers trained for the application of the second and third use of batteries were the most critical barriers faced. In addition, some difficulties related to business partners were also mentioned since the development of technology and the obtaining of batteries depend on the support of dealerships and specialized companies.

However, no corresponding barriers were found in the literature for elements of communication and financial return of EC. For instance, the barrier related to the customer communication challenge, which is critical to ensure the return/delivery of lithium-ion batteries to the company. In the same vein, there is the intangibility of the return of the circularity model for EVs lithium-ion batteries can hinder the adoption of CE. These points can be considered opportunities for the development of battery supply chains considered in this work.

Two drivers for the circular economy of lithium-ion batteries of electric vehicles showed greater prominence in the case studied, namely: objectives and strategies of the organization and regulations (governmental and environmental impacts). On the other hand, the barriers with the highest number of mentions refer to the unavailability of technology and national partners to enable the application of the second and third use of batteries lithium ion.

In practice, the strategies proposed by the company were developed in partnership with an educational institution and four private companies, in addition to a public educational institution. This CE model allows to close the life cycle of the lithium-ion battery (used in one of the electric models of the brand) and enable the expansion of electric mobility as well as the reduction of carbon emissions and raw material extractions. Furthermore, results can support organizations that intend to implement the circular economy model for lithium-ion batteries, whether they are vehicle manufacturers or other stakeholders, to anticipate and prepare for the barriers faced during the process. The results of this work can help organizations that intend to implement the circular economy model for lithium-ion batteries, whether they are vehicle manufacturers or other interested parties, to anticipate the levers and barriers faced during the process. In addition, such elements present themselves as opportunities not yet explored, which can be used for future applications.

It is noteworthy that like every study, the present study has some limitations. One of them is related to the technical procedure chosen; because it is a single case study, it is not possible to generalize the results obtained, i.e., the drivers and barriers empirically identified may be different in other contexts. Therefore, for future studies, it is recommended to use a case study with multiple sources of analysis, to contemplate suppliers and consumers, since the present study focused on the micro approach, under the vision of only one company.

References

1. Ellen MacArthur Foundation: A circular economy in Brazil: an initial exploration (2017)
2. Kirchherr, J., Reike, D., Hekkert, M.: Conceptualizing the circular economy: an analysis of 114 definitions. Resour. Conserv. Recycl. **127**, 221–232 (2017). https://doi.org/10.1016/j.res conrec.2017.09.005
3. International Energy Agency: Global EV Outlook 2022 - Securing supplies for an electric future (2022)
4. Bermúdez -Rodríguez, T., Consoni, F.L.: An approach to the dynamics of scientific and technological development of lithium-ion batteries for electric vehicles. Rev. Bras. Ino- vação. **19**, e0200014 (2020). https://doi.org/10.20396/rbi.v19i0.8658394
5. Deng, S., et al.: Planning a circular economy system for electric vehicles using network simulation. J. Manuf. Syst. **63**, 95–106 (2022). https://doi.org/10.1016/j.jmsy.2022.03.003
6. Du, S., Gao, F., Nie, Z., Liu, Y., Sun, B., Gong, X.: Life cycle assessment of recycled NiCoMn ternary cathode materials prepared by hydrometallurgical technology for power batteries in China. J. Clean. Prod. **340**, 130798 (2022)

7. Lewicka, E., Guzik, K., Galos, K.: On the possibilities of critical raw materials production from the EU's primary sources. Resources **10**, 50 (2021). https://doi.org/10.3390/re-source s10050050

8. Rajaeifar, M.A., Heidrich, O., Ghadimi, P., Raugei, M., Wu, Y.: Sustainable supply and value chains of electric vehicle batteries. Resour. Conserv. Recycl. **161**, 104905 (2020). https://doi. org/10.1016/j.resconrec.2020.104905

9. Ellen MacArthur Foundation: Towards a circular economy: business rationale for an accelerated transition (2015)

10. Humphris-Bach, A., Essig, C., Morton, G., Harding, L.: EU Resource Efficiency Scoreboard 2015 (2015)

11. Geissdoerfer, M., Savaget, P., Bocken, N.M.P., Hultink, E.J.: The circular economy – a new sustainability paradigm? J. Clean. Prod. **143**, 757–768 (2017). https://doi.org/10.1016/j.jcl epro.2016.12.048

12. Ghisellini, P., Cialani, C., Ulgiati, S.: A review on circular economy: the expected transition to a balanced interplay of environmental and economic systems. J. Clean. Prod. **114**, 11–32 (2016). https://doi.org/10.1016/j.jclepro.2015.09.007

13. Govindan, K., Hasanagic, M.: A systematic review on drivers, barriers, and practices towards circular economy: a supply chain perspective. Int. J. Prod. Res. **56**, 278–311 (2018). https:// doi.org/10.1080/00207543.2017.1402141

14. Lieder, M., Rashid, A.: Towards circular economy implementation: a comprehensive review in context of manufacturing industry. J. Clean. Prod. **115**, 36–51 (2016). https://doi.org/10. 1016/j.jclepro.2015.12.042

15. Jesus, A., Mendonça, S.: Lost in transition? Drivers and barriers in the eco-innovation road to the circular economy. Ecol. Econ. **145**, 75–89 (2018). https://doi.org/10.1016/j.ecolecon. 2017.08.001

16. Kirchherr, J., et al.: Barriers to the circular economy: evidence from the European Union (EU). Ecol. Econ. **150**, 264–272 (2018). https://doi.org/10.1016/j.ecolecon.2018.04.028

17. Theotonio, S.B.: Electric and Hybrid Vehicles - Patent Overview in Brazil, Rio de Janeiro (2018)

18. Delgado, F., Costa, J.E.G., Febraro, J., da Silva, T.B.: Electric Vehicles (2017)

19. de Souza, L.L.P., Lora, E.E.S., Palacio, J.C.E., Rocha, M.H., Renó, M.L.G., Venturini, O.J.: Comparative environmental life cycle assessment of conventional vehicles with different fuel options, plug-in hybrid and electric vehicles for a sustainable transportation system in Brazil. J. Clean. Prod. **203**, 444–468 (2018). https://doi.org/10.1016/j.jclepro.2018.08.236

20. Foster, M., Isely, P., Standridge, C.R., Hasan, M.M.: Feasibility assessment of remanufac- turing, repurposing, and recycling of end of vehicle application lithium-ion batteries. J. Ind. Eng. Manag. **7**, 698–715 (2014). https://doi.org/10.3926/jiem.939

21. Ramoni, M., Zhang, H.-C.: Remanufacturing processes of electric vehicle battery. In: Pro- ceedings of the 2012 IEEE International Symposium on Sustainable Systems and Technology (ISST), p. 1. IEEE, Boston (2012). https://doi.org/10.1109/ISSST.2012.6228014

22. Akram, M.N., Abdul-Kader, W.: Electric vehicle battery state changes and reverse logistics considerations. Int. J. Sustain. Eng. **14**, 390–403 (2021). https://doi.org/10.1080/19397038. 2020.1856968

23. Haruna, H., Itoh, S., Horiba, T., Seki, E., Kohno, K.: Large-format lithium-ion batteries for electric power storage. J. Power. Sources **196**, 7002–7005 (2011). https://doi.org/10.1016/j. jpowsour.2010.10.045

24. Yin, R.K.: Case Study: Planning and Methods. Bookman, Porto Alegre (2015)

25. Alamerew, Y.A., Brissaud, D.: Modelling reverse supply chain through system dynamics for realizing the transition towards the circular economy: a case study on electric vehicle batteries. J. Clean. Prod. **254**, 120025 (2020). https://doi.org/10.1016/j.jclepro.2020.120025

26. Agyemang, M., Kusi-Sarpong, S., Khan, S.A., Mani, V., Rehman, S.T., Kusi-Sarpong, H.: Drivers and barriers to circular economy implementation. Manag. Decis. **57**, 971–994 (2019). https://doi.org/10.1108/MD-11-2018-1178
27. Rosa, P., Sassanelli, C., Terzi, S.: Circular business models versus circular benefits: an assessment in the waste from electrical and electronic equipments sector. J. Clean. Prod. **231**, 940–952 (2019). ISSN 0959-6526, https://doi.org/10.1016/j.jclepro.2019.05.310
28. European Commission, Directorate-General for Research and Innovation, Schempp, C., Hirsch, P.: Categorisation system for the circular economy – a sector-agnostic categorisation system for activities substantially contributing to the circular economy, Publications Office (2020). https://data.europa.eu/doi/https://doi.org/10.2777/172128

Application of Life Cycle Assessment for More Sustainable Plastic Packaging - Challenges and Opportunities

Simon Merschak$^{(\boxtimes)}$ ⓘ, Christian Kneidinger ⓘ, David Katzmayr,
Johanna Casata, and Peter Hehenberger ⓘ

University of Applied Sciences Upper Austria, Wels Campus, Stelzhamerstr. 23,
4600 Wels, Austria
`simon.merschak@fh-wels.at`

Abstract. The European plastics industry is in transition to meet its 2050 net zero and circularity targets. In 2020, the overall European recycling rate for post-consumer plastics packaging reached 46%. The European Union set a target for recycling 50% of plastic packaging by 2025 and 55% by 2030. These targets can only be achieved by increasing the use of recycled materials in packaging. Therefore, it is required to design new plastic packaging products, production processes and recycling processes with a focus on easy recyclability and reduced environmental footprint. This publication points out possible benefits, opportunities and challenges of using the well-known Life Cycle Assessment (LCA) methodology for plastic packaging products. First steps towards the development of reduced LCA-based models, which can be used to support the development of plastic packaging products, are presented. One example is a generic recycling process model, which can be used for different packaging materials and provides the basis for further analysis. Another example is the development of an assessment template for current plastic recycling processes. The publication concludes with the identification of open research questions.

Keywords: Plastic Packaging · Life Cycle Assessment · Plastic Recycling

1 Introduction

The legislative requirements for plastic recycling in the European Union are currently undergoing serious changes. According to Directive (EU) 2018/852 of the European Parliament, which is the centerpiece of EU legislation on packaging and packaging waste, until end of the year 2025 a minimum of 50 % by weight of all plastic packaging waste must be recycled. Until end of 2030 a recycling rate of 55% is required by law [1]. This goal shall be achieved, among other actions, through a higher share of recycled material in packaging material production.

© IFIP International Federation for Information Processing 2024
Published by Springer Nature Switzerland AG 2024
C. Danjou et al. (Eds.): PLM 2023, IFIP AICT 702, pp. 61–71, 2024.
https://doi.org/10.1007/978-3-031-62582-4_6

The plastics industry is therefore challenged to develop new products and production processes as well as optimized recycling technologies. To increase the share of recycled material in packaging, on the one hand, the recyclate must be of sufficient quality. On the other hand new product packaging must be designed in a way that recyclate can be used for its production and that these new products can be recycled efficiently themselves. This means that topics like "Design for Recycling" and "Design from Recycling" [2] will become more important in the design process of new packaging materials. To evaluate if a new product design is advantageous regarding sustainability over the previous design, the whole product lifecycle has to be considered. This includes manufacturing, use phase and the end-of-life phase. The assessment should be done at an early stage of the product development, because then changes of the product design are still possible at low costs. A possible methodology for the assessment of environmental impacts over their entire life cycle it the well known life cycle assessment methodology (LCA). In this work, possible benefits and challenges of an early LCA of plastic packaging are identified and first steps towards an comprehensive modified LCA approach for plastic packaging are presented. It therefore provides initial answers on the overarching research question of how plastic recycling can become more environmentally friendly and how LCA can support this.

2   Background Information

This section provides an overview of related topics and gives background information which helps to understand the intended field of application of the later presented methods and tools.

2.1   Plastic Packaging

Plastic packaging plays an important role in protecting, preserving, storing and transporting goods. Different materials and compounds can be used to realize an optimal packaging. In 2021, 39.1% of European plastics demand was destined for packaging materials and over 50% of all European goods are packaged in plastics. Main materials which are used for packaging are polyetylene (PE-LD, PE-HD), polypropylene (PP), polystyrol (PS) and polyethylene terephthalate (PET) [3]. Multi-layer film packaging, which is often used for food packaging to enhance the shelf life of the products, consists of two or more materials with different properties combined in a single layered structure. This provides huge challenges for the recycling process [4]. The current linear value chain for plastic packaging has to be transformed to a circular value chain. Several attempts for this transition can be found in literature. In [5], a way to support the creation of circular value chains through the implementation of an information sharing system is proposed.

2.2   Plastic Recycling Methods

There are different ways to recycle plastic waste. Currently the most common approach of plastic recycling is mechanical recycling. Alternative approaches like chemical recycling, dissolution recycling and organic recycling are also available. *Mechanical recycling* is simple, inexpensive and has a low demand on energy and resources compared to chemical recycling. For mechanical recycling, the plastic waste is processed into secondary raw material without significantly changing the material's chemical structure. Most types of thermoplastics can be mechanically recycled with little or no impact on quality.

Chemical recycling is an umbrella term for a set of technologies (pyrolysis, gasification, hydro-cracking, depolymerisation) that change the chemical structure of plastic waste. Chemical, thermal, or catalytic processes break long hydrocarbon chains of plastics into shorter fractions or monomers. These shorter molecules can than be used as feedstock for chemical reactions to produce new recycled plastics.

Dissolution recycling is a purification process. A selected polymer in the plastic waste is selectively dissolved in a solvent. So it can be separated from the waste and recovered in a pure form without changing its chemical structure.

Organic recycling is a controlled microbiological treatment of biodegradable plastics waste under aerobic conditions (composting) or anaerobic conditions (biogasification). It applies to specific polymers and does not produce plastic material, which can be directly reprocessed [6].

2.3 Life Cycle Assessment (LCA)

The LCA methodology is standardized in its basics with ISO 14040 [7] and in detail with ISO 14044. According to ISO 14040, LCA deals with the environmental aspects and potential impact of the whole product lifecycle. This includes the raw material acquisition, production processes, product use phase and the end-of-life. The general environmental impact assessment requires the consideration of resource use, human health, and ecological consequences. In ISO 14044 the LCA methodology is divided into 4 different phases: goal and scope definition, inventory analysis, impact assessment and interpretation.

3 Benefits of LCA in Plastic Packaging

The well known LCA methodology can be used in different stages of the plastic packaging lifecycle to reduce the ecological footprint. On the one hand, it can be used in the design process of new plastic packaging to reduce the environmental impact over the entire lifecycle. LCA results can support design decisions and can also lead to a better recycability of the product at its end-of-life stage. On the other hand LCA can also be used to analyze existing recycling processes, to identify emission hot-spots and to optimize the process steps.

The LCA methodology can be used in the development process of new packaging designs to quantify the influence of design decisions on the environmental footprint. For the comparison of different packaging designs, the whole lifecycle has to be considered. This includes the production of the packaging material, the transport emissions in the use phase and the end-of-life phase which contains the recycling process emissions. Early available LCA results can help to select the packaging design with the lowest environmental impacts. Environmental hot-spots can be identified and possibly avoided. Either through adopted material selection, optimized product geometry or through changes of the manufacturing processes and manufacturing technology. Design decisions often have contrary impacts on the different lifecycle phases. For example, material and weight reduction by use of composite materials can have a positive effect on the raw material and transport emissions, but can have negative impacts on the recycability in the end-of-life phase. Another example is shelf life of food products. The use of multi layer foils in food packaging can increase the shelf life of food products and reduce food waste. Depending on the food product, this can significantly reduce the environmental footprint. But multi layer foils are difficult to recycle. Therefore developers need reliable information as basis for design decisions and the whole product lifecycle has to be considered to find an optimal solution.

LCA results can also be beneficial for the optimization of existing recycling processes and recyclate quality. A possible benefit is the identification of environmental hot-spots in the recycling process. The LCA methodology is capable of analyzing different kinds of environmental impacts such as the carbon footprint, water use, particle emissions and other emissions. Process steps with high emissions can be identified and optimized. Furthermore different recycling technologies can be compared and an optimized recycling process chain can be implemented. This leads to reduced emissions and better quality of the recyclates.

4   Challenges and Solutions for LCA in Plastic Packaging

There are several challenges for the optimization of the plastic packaging life-cycle and for the use of LCA as supportive methodology. In this work, three challenges and their solution, which support the application of LCA in plastic packaging recycling, are presented. The integration of the discussed methods and tools is shown in Fig. 1. In order to optimize the environmental footprint of the whole plastic packaging lifecycle, as a first step, the current situation must be analyzed. Therefore a standardized assessment template for the recycling process of different polymers can be helpful. Further, a generic recycling process model is required to speed up the modeling process for LCA. It should be possible to tailor this generic model to a detailed recycling process model of all common plastic packaging materials. Finally an in-dept analysis of hot-spot recycling process steps must be performed to gain knowledge on how these processes can be optimized. As a first step, energy intensive process steps like the extrusion process are analyzed in detail. In this section, possible solutions to overcome the mentioned challenges are presented.

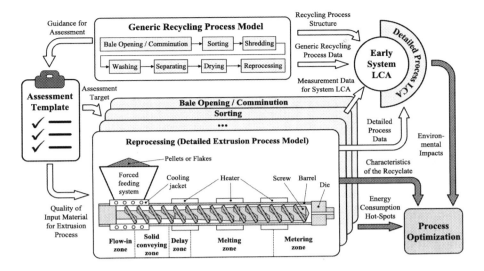

Fig. 1. Integration of the proposed models and tools.

4.1 Assessment Template for Existing Recycling Processes

In order to calculate the environmental impacts of an existing plastic recycling process, comprehensive information and process data is required. For the collection of the required data, an assessment template was developed. It supports the assessment of energy and material inputs, recycling process setup and recyclate quality and ensures that all relevant information is considered. The challenge was to identify relevant indicators from the literature, which are characteristic for plastic recycling process emissions and recyclate quality. A comprehensive literature study, considering publications of European environmental agencies [8,9], scientific books [10] and European standards for plastic recycling (EN 15342:2008 to EN 15348:2014), revealed a variety of relevant indicators. Based on this identified indicators, an assessment template was set up and structured in Microsoft Excel. For a better overview, all indicators were assigned to one of the three main categories which are preparation for recycling, recycling processes and secondary material. In the first two segments of the template, system in- and outputs are described while the third segment is used to evaluate the quality of the recyclate obtained. The main categories of indicators used for the assessment template are:

1. **Preparation for recycling**
   (a) Method of gathering
   (b) Transportation (mode, distance,...)
   (c) Inputs of sorting and cleaning systems (electricity, water,...)
2. **Recycling processes**
   (a) Inputs of recycling processes (electricity, water...)
   (b) Emission of greenhouse gases (CO_2, CH_4,...)

(c) Additives (light stabilizers, plasticizers,...)

(d) Output per day (tons)

3. **Characteristics of secondary material**

   (a) Optical (color, transparency,...)

   (b) Mechanical (density, elasticity, grain size,...)

   (c) Thermal (softening temperature, mass reduction,...)

   (d) Chemical (alkalinity, residual moisture,...)

   (e) Rheological (melt flow rate, dry flow rate, intrinsic viscosity,...)

The developed assessment template ensures that all relevant data for the LCA of recycling processes is collected and documented in a standardized way. This data can then be used for the calculation of the environmental impacts. The template is responsive regarding the type of material and expandable in all its categories and single indicators. Further work to implement LCA calculations based on the inserted data directly in the assessment template is required. Furthermore, different use cases will be assessed with the template to test its applicability. Based on best- or worst-case examples, assessment score-ranges for more or less effective recycling processes should also be identified in the subsequent research progress.

4.2   Generic Recycling Process Model

For the mechanical recycling process of plastic packaging, a sequence of different processes steps is required. Seven main steps, which are described below, have to be considered for mechanical plastic packaging recycling. To speed up the LCA of the whole plastic recycling process and to ensure that no process step is missed, a generic model of the recycling process is helpful.

1. **Bale Opening/Comminuting** - The plastic waste is delivered to the recycling facility in form of bales, which must be opened and shredded before further processing can take place. Therefore, commonly a single shaft shredder is used. [11]

2. **Sorting** - The collected plastic waste consists of a variation of different plastic types and other materials, such as metal or wood. It is sorted using different technologies to achieve the highest possible purity of the recyclate. [12] There are different types of sorting technologies that can be used in different combinations. In a first step, non-plastic impurities in the waste stream are typically removed by manual sorting. Waste screens and drum separators segregate materials based on their size. Air separators and ballistic separators can than be used to remove light materials like plastic films or paper. Magnetic separators and eddy current separators remove metal objects. Different types of plastic can be identified by use of near infrared sensors (NIR) and the separation is then obtained by a strong jet of air. [11]

3. **Shredding** - In this step, the size of the plastic parts is reduced. Through the reduced volume, in the following process steps, a larger amount of plastic can be processed at the same time and the density of the material is increased. [12]

4. **Washing** - The washing of the shredded plastic waste is a step which is not always required for all sorts of plastics. Sometimes the plastic flakes are processed directly after shredding without washing. [12] There are different ways to clean the plastic flakes. Friction washers, where the cleaning process is based on intensive mechanical friction, are used to remove glued-on labels. In dry friction cleaners, impurities are removed by friction without water through high concentration of material in rapidly rotating throwing blades and separated by a surrounding screen basket. [11]

5. **Separating** - A further separation takes place to remove as many impurities as possible. This separation is applied to protect the subsequent machines from damage and to separate different polymers. Sink-float separation is often used in plastic recycling processes. For this process, water is used as separation liquid. Some polymers, such as PET, PVC and PS, sink to the bottom and others, such as PE, PP and EPS, float on the surface. [11] So a separation of different polymer types can be achieved.

6. **Drying** - After the separation a drying process is conducted. This process can be very energy intensive. The washed particles are dried to a certain moisture level, which is about 0.1 percent by mass. [12]

7. **Reprocessing** - After washing, the plastic flakes are reprocessed. Cutting mills can be used to produce fine plastic flakes which can than be used as raw material for the subsequent extrusion process.

Based on the above-mentioned recycling steps, a generic model of the whole recycling process, including average values for energy, water and additive consumption of each process step, was developed. This generic model contains relevant information which is required for the LCA of different plastic packaging recycling processes. The aim of such a generic recycling process model is to speed up the modeling process for LCA and to ensure that no process step is missed.

4.3   Detailed Extrusion Process Model

The extrusion process is an important part of the recycling process of plastic packaging. It is an energy-intensive process that has influence on the quality of the recycled material. Therefore it is worth to have a closer look on this recycling process step. Most often, single screw extruders are used for recycling, as they are very reliable and offer a great price-performance ratio [13]. When the polymeric material has entered the single screw extruder, the material is conveyed, compressed, heated, melted, pumped, and homogenized. Additional degassing or mixing steps are possible. After the material has left the extruder, it is filtered and later the material is cooled and granulated. Models of the process are needed to optimize the quality of the recycled material and to minimize environmental impact by reducing wear and optimizing energy efficiency. To model the function of a single-screw extruder, the process is divided into several functional zones which are displayed in Fig. 1. The material enters the extruder in the flow-in zone. Forced feeding systems are usually used in recycling processes. In the solids conveying zone, the material is conveyed, compacted, and compressed. Here the tribological conditions strongly influence the behavior. To achieve good solids conveying behavior, the friction between the polymer and the barrel should be high, and the friction between the polymer and the screw should be low. The internal and external coefficient of friction are important parameters for the description of this zone [14,15]. In the delay zone, the material then starts to melt. But the biggest proportion is melted at the barrel in the melting zone where specific melting mechanisms can be observed. The screw design has significant impact on the melting behavior in the extruder and on the throughput. When the material is completely melted, the melting zone ends and the metering zone begins. Here the melt is conveyed and homogenized. Additional degassing zones can be applied to remove volatile substances. Furthermore, distributive mixing elements can be applied to distribute particles and dispersive mixing elements can be applied to break up agglomerates [14]. The material leaves the extruder through a die, where the melted plastic is brought into the desired shape.

The extrusion process itself requires at least as much energy as needed to heat and melt the polymeric material. Furthermore, there are losses of the motor drive and gearbox, forced and natural cooling, and the energy needed for auxiliary devices. [16] Most of the existing models to describe single functional zones were developed for pure virgin polymeric materials. When using recycling polymers, some special issues occur. The process is sensitive to the material itself and also to the shape of the raw material. Differing raw material shapes can lead to a strongly changing behavior in the feeding zone [17], in the solids conveying zone [15,17] and in the delay and melting zone [18]. Mixtures of materials lead to strongly changing behavior, especially in the delay and melting zone [18]. Even small proportions of other materials can significantly decrease the melting capability of a single screw extruder. At the same time, the polymeric material itself may be harmed by increased energy input. Possibly additives must be mixed in to achieve a good quality of the extrudate. To improve the LCA modeling, detailed models describing the recycling process itself and also methods to describe the quality and the applicability of the recycled polymer are needed. In a following step, these models are created.

4.4 Integration and Open Research Questions

The combination of the previously explained methods and tools supports the LCA of plastic recycling processes. As shown in Fig. 1, each part delivers specific information. The generic recycling process model guides the user of the assessment template through the assessment process and helps to model the whole recycling process for LCA. It also provides average data for energy consumption, if data of specific process steps is missing. The assessment template is used to collect recycling process data for LCA of existing processes and provides information about the input material quality for the detailed extrusion process model. Finally, the extrusion process model delivers detailed energy consumption data for LCA. Additional outputs of the extrusion process model are the characteristics of the recyclate and energy consumption hot-spots of the extrusion process. Together with the environmental impact results from LCA this information can be used for further process optimization.

The current results are first steps towards more sustainable plastic packaging and there are still some open research questions. One of the most important questions is how all required data for LCA can be acquired? The data collection process is typically very time consuming and sometimes data can not be obtained. Especially in the development process, a fast assessment of the environmental impact of design decisions is required. It would be beneficial to implement a knowledge-base which contains average values for required data and additional information which can be used for an early emission hot-spot evaluation. This would speed up the assessment process. Therefore, a detailed assessment and documentation of common plastic manufacturing and recycling processes regarding their environmental emissions is necessary. Furthermore, existing design guidelines for environmental friendly plastic packaging design should be analyzed with a focus on the end-of-life phase. Another question is, how to deal with uncertain data? Especially in the product development phase it is often not clear, which technologies and machines will be used for manufacturing and recycling. So generic emission factors must be used for LCA in this phase. It would be valuable to know the uncertainty of these factors. Environmental hot-spot processes should also be identified to provide guidance for effective optimization steps. Finally, all these topics must be included in existing development processes of plastic packaging.

5 Conclusion

The use of LCA in the development process of new plastic packaging and for the optimization of plastic recycling processes provides opportunities for the reduction of environmental impacts. First steps towards the use of LCA for plastic packaging are shown in this publication and several open research topics were identified. An important step in conducting LCA is the collection of comprehensive data. Therefore an assessment template was developed which supports the data acquisition and delivers information about material quality and recycling processes. Furthermore, first steps towards a generic recycling process model were made. This generic model contains relevant information for a fast modeling of plastic recycling processes and provides the basis for subsequent LCA. A deeper investigation of the extrusion process, which is an important step of the plastic recycling process, was also performed. Finally, further challenges and open research questions were identified.

Acknowledgements. This research was carried out in the course of the project "NaKuRe" and was financed by research funds granted by the government of Upper Austria in the course of the program "FTI-Struktur", project number: "Wi-2021-305611/21-Au/NaKuRe"

References

1. European Parliament: Directive (EU) 2018/ of the European Parliament and of the Council of 30 May 2018 amending Directive 94/62/EC on packaging and packaging waste (2018)
2. Martínez Leal J., Pompidou S., Charbuillet C., Perry N.: Design for and from Recycling: A Circular Ecodesign Approach to Improve the Circular Economy. Sustainability. **12**(23) (2020)
3. Plastics Europe: Plastics – the Facts 2022. https://plasticseurope.org/knowledge-hub/plastics-the-facts-2022/ (2022). (Accessed 11 February 2023)
4. Soares C., Ek M.,Östmark E., Gällstedt M., Karlsson S.:Recycling of multi-material multilayer plastic packaging: Current trends and future scenarios. Res. Conservation Recycling **176**(1) (2022)
5. Rosa P., Terzi S.: Supporting the development of circular value chains in the automotive sector through an information sharing system: the TREASURE project. In: Noël, F., Nyffenegger, F., Rivest, L., Bouras, A. (eds.) Product Lifecycle Management. PLM in Transition Times: The Place of Humans and Transformative Technologies. PLM 2022. IFIP Advances in Information and Communication Technology, vol 667. Springer, Cham. (2023). https://doi.org/10.1007/978-3-031-25182-5_8
6. Shamsuyeva M., Endres H. J.: Plastics in the context of the circular economy and sustainable plastics recycling: comprehensive review on research development, standardization and market. Composites Part C: Open Access **6**(1) (2021)
7. DIN EN ISO 14040: 2009-11, Environmental management - Life cycle assessment - Principles and framework (ISO 14040:2006); German and English version EN ISO 14040:2006
8. Knappe F., Reinhardt J., Kauertz B.: Technische Potenzialanalyse zur Steigerung des Kunststoffrecyclings und des Rezyklateinsatzes. TEXTE 92/2021. Umweltbundesamt (2021)
9. Neubauer C., Stoifl B., Tesar M., Thaler P.: Sortierung und Recycling von Kunststoffabfällen in Österreich: Status 2019. Umweltbundesamt GmbH (2021)
10. Maier, R., Schiller, M.: Handbuch Kunststoff-Additive. Carl Hanser, Munich (2016)
11. Letcher, T.: Plastic waste and recycling - Environmental Impact, Societal Issues, Prevention and Solutions, 1st edn. Elsevier, London (2020)
12. Worrell, E., Reuter, M.: Handbook of recycling: State-of-the-art Practitioners, Analysts, and Scientists, 1st edn. Elsevier, Waltham (2014)
13. Kneidinger, C.: Solids Conveying, Melting and Melt Conveying in Single Screw Extruders and Extrusion Dies. Dissertation, JKU (2021)
14. Gogos, C. G., Tadmor, Z.: Principles of polymer processing. 2nd edn. Wiley-Interscience (2014)
15. Kneidinger, C., Längauer, M., Zitzenbacher, G., Schuschnigg, S., Miethlinger, J.: Modeling and estimation of the pressure and temperature dependent bulk density of polymers. Int. Polym. Proc. **35**(1), 70–82 (2020)
16. Abeykoon, C., McMillan, A., Nguyen, B. K.: Energy efficiency in extrusion-related polymer processing: a review of state of the art and potential efficiency improvements. Renewable Sustainable Energy Rev. **147**. (2021)

17. Thieleke, P., Bonten, C.: Enhanced processing of regrind as recycling material in single-screw extruders. Polymers **13**(10) (2021)
18. Kneidinger, C., Zitzenbacher, G., Laengauer, M., Miethlinger, J., Steinbichler, G.: Evaluation of the melting behavior and the melt penetration into the solid bed using a model experiment. In: Proceedings of the 26th Annu. Reg. Meet. of the Polym. Proc. Soc. PPS-2016, Graz, Austria, vol. 1779(1) (2016)

Interoperability Technology: Blockchain, IoT and Ontologies for Data Exchange

Integrating Processes, People and Data Management to Create a Comprehensive Roadmap Toward SMEs Digitalization: an Italian Case Study

Marco Spaltini$^{(\boxtimes)}$ ⓘ, Federica Acerbi ⓘ, Anna De Carolis ⓘ, and Marco Taisch ⓘ

Department of Management, Economics and Industrial Engineering, Politecnico di Milano, via Lambruschini 4/b, 20156 Milan, Italy
marco.spaltini@polimi.it

Abstract. Digitalization is increasingly gaining interest among manufacturers. Focusing on manufacturing, this interest already turned into a necessity that cannot be further postponed. This explains why digitalization or Industry 4.0 transition characterizes most of the strategies for manufacturers irrespectively of their size or maturity level. Nevertheless, size and maturity do matter in such kind of transition. Literature and practice advocate that Small and Medium Enterprises (SMEs)s face huge barriers to keep high the competitive advantage of their products, thus they first need to act on their processes and internal resources. Among them, lack of competencies and limited data exploitation are threatening the competitiveness for the medium-long term the most. Thus, the present contribution aims to present the results obtained from the integrated application of three maturity models focused on the skills gap, data management, and operations management within the scope of digital maturity. These maturity models were developed as stand-alone tools focusing the attention on a specific need (i.e. operations, people, or data). In this contribution, they were integrated to provide an overarching view to develop a unique roadmap towards social and economic sustainable industrial environments. The integrated model was applied to an Italian SME.

Keywords: Manufacturing · Maturity Models · Resource Management

1 Introduction

The wave of digitalization has spread in every branch of today's society as well as Industry. This led to the development and distribution of objects and services that are more and more often associated with the term "smart". Looking at the manufacturing sector, the digital transition has encompassed operations processes, monitoring and control, organizational structures, communication channels, HR management, and many other fields [1]. However, such dramatic change does not come without investments of practitioners in terms of financial resources for assets (both hardware and software) and time. In this

© IFIP International Federation for Information Processing 2024
Published by Springer Nature Switzerland AG 2024
C. Danjou et al. (Eds.): PLM 2023, IFIP AICT 702, pp. 75–84, 2024.
https://doi.org/10.1007/978-3-031-62582-4_7

sense, the lack of sufficient financial resources is one of the main barriers that damper the diffusion of digitalization, often referred to as Industry 4.0 (I4.0) [2]. However, other major causes regard skill and cultural aspects that characterize companies [3]. Indeed, [4] argued that a digital advancement of processes must be supported by a proportionate increase in the capability of the organization to use, manage and understand both technology and the data around them. This bifocal perspective appears even more critical when analyzing Small-Medium Enterprises (SMEs) as they traditionally struggle in both the dimensions aforementioned [5]. Hence, literature has proposed models and methodologies to support this transition starting from an understanding of the current situation in which the company lies. These tools, referred to as Maturity Models (MMs), are indeed defined as *"the state of being complete, perfect or ready"* [6]. From the definition, it can be argued that the field of applicability is wide and may encompass all three struggling points to the I4.0 transition, namely: Process-related [1], People-related [7], and Data-related [8]. However, although these 3 dimensions are often analyzed individually, a plan toward higher levels of Digital Maturity shall not tackle them as independent silos rather it must consider and integrate them holistically. Indeed, literature has provided a rich multitude of MMs for digital transformations focusing on specific aspects of it. However, few contributions have been lavished towards the integration of the different perspectives in one single method. For this reason, this paper proposes an integrated methodology to assess the digital maturity of manufacturing SMEs and, consequently, delineate a roadmap by considering holistically Process, People, and Data management dimensions. The study was conducted by focusing on 3 MMs already developed by the authors but not integrated into one single tool. The methodology presented has been validated in an Industrial case with an Italian SME (2 plants) and the results have been reported in this contribution.

The rest of the paper is structured as follows: Sect. 2 depicts the theoretical background of the MMs for each macro-dimensions; Sect. 3 describes the roadmapping methodology, named P.P.D. Digital Roadmapping Model proposed in this contribution. Then, Sect. 4 reports and discusses the results from the Industrial case, and finally in Sect. 5 conclusions and limitations of the study are disclosed.

2 Theoretical Background

The present section delineates the theoretical background of the three methodologies employed to create the overall assessment model covering three fundamental areas for the digital transformation of manufacturing companies: (i) processes (see sub-Sect. 2.1), (ii) people (see sub-Sect. 2.2), and (iii) data management (see sub-Sect. 2.3).

2.1 Process Digital Assessment

The digital transition was pushed especially to improve the efficiency of manufacturing processes [9]. Among all, I4.0 enabling technologies are gaining momentum [10] for instance the diffusion of the Internet of Things used to improve eco-efficiency in manufacturing [11] but also additive manufacturing, blockchain, advanced robotics, and artificial intelligence [12]. The technological-related opportunities for manufacturing companies' operations are many although these are also backed by some challenges such as resistance from the workforce and huge financial investments [12]. In addition, to pursue a structured and successful digital transformation, it is required to rely on a solid digital backbone through the integration of existing information and communication technologies.

In this context, to support manufacturing companies in evaluating their current digital readiness, [1] developed a maturity model. In particular, the model aims to perform first a descriptive assessment to highlight the key strengths and weaknesses of all the functions covering the operations, to then add prescriptive suggestions based on the current state assessed. This maturity model is based on a 5-level maturity scale (i.e. initial, managed, defined, integrated and interoperable, digitally oriented) covering the following manufacturing areas: Design and Engineering, Production, Quality management, Maintenance management, and Logistics management (inbound, internal, and outbound).

2.2 People Digital Assessment

Digital transformation inevitably requires the involvement of the entire workforce, both operators [13] and managers [14] to ease the introduction of new technologies in production plants. Indeed, new skills and specific job profiles are emerging since the introduction in manufacturing companies of I 4.0 enabling technologies [15]. These new job profiles must cover soft and hard skills but also need to have the minimum knowledge in terms of Information and Communication Technologies (ICT) to be competitive [16]. In this context, the capability to assess to current digital maturity level of people becomes essential for companies to understand which profiles they need to invest in to align people and technologies within a unique path toward digitalization.

In this regard, [17], subsequently extended in [18], proposed a maturity model aiming at assessing the digital maturity of people, both at operative and managerial levels, operating in manufacturing companies. More in detail, the assessment is performed for every single profile (i.e., managers and operators of the following functions: design & engineering, production, quality, maintenance, logistics, and the data science and ICT managers) across 5 levels of maturity (i.e., basic, aware, practiced, competent, proficient) assessing the maturity for soft skills, hard skills, and ICT literacy skills. The authors opted for a prescriptive approach of the MM (in contrast with a comparative or descriptive one) as it enables the company to identify structured plans to undertake a digital path based on the emerged company's needs.

2.3 Data Assessment

The spreading of I4.0 enabling technologies in the industrial domain is pushing companies in exploiting data gathered from the field [19]. In this context, hence, data-driven decision-making processes can be established for asset management [20] and also for product life cycle management [21] to support both managers and operators in their daily activities. Indeed, proper data management by manufacturing companies can expedite the establishment of sustainable and circular strategies, for example, facilitating the adequate usage of natural resources in industrial processes [22].

For these reasons, researchers are trying to evaluate how to measure data productivity in data-driven decision-making processes [8]. Moreover, still in this view, data management assessment models, more in detail maturity models, have been developed to support manufacturing companies in accelerating their digital transformation from initial levels to optimized ones [23, 24]. The assessment is performed through a 5-levels maturity scale by keeping into account several dimensions, like processes, people, and operations. This broad perspective aims to create the right awareness in manufacturing companies on the evaluation of how good they are at exploiting the data gathered inside their plants to facilitate their decision-making process.

3 P.P.D. Digital Roadmapping Model Development

The above-mentioned maturity models are specifically focused on the current state assessment of a certain aspect of manufacturing companies while keeping a silos perspective. The proposed integrated version of these models, the P.P.D. Digital Roadmapping Model, aims to provide a comprehensive view of the three elements (i.e., process, people, and data management) which are considered fundamental to be kept into account to create a structured roadmap facilitating the digital transformation of manufacturing companies. The P.P.D. Digital Roadmapping Model, hence, aims to investigate separately the three aspects to assess the maturity level of the company in a detailed way. Indeed, separated interviews are firstly performed with the area managers and their teams of the functions covering all the operations (i.e., design & engineering, production, quality, maintenance, logistics, and supply chain) to perform the process and data management assessments. More specifically, regarding process management the model proposed by [1] is used, while for data management those proposed by [23, 24] have been used and revised to be aligned with the maturity scale and the dimensions covered by [1]. More specifically, also for the data management maturity model, the areas covered are design & engineering, production, quality, maintenance, logistics, and supply chain [1]. This is evaluated through the maturity scale of 5 levels looking at (i) data collection and cleaning, (ii) data storage, (iii) data sharing, (iv) data analysis, and (v) data exploitation for decision-making processes. Then, all the managers and operators operating in the above-mentioned functions are asked to answer a self-assessment questionnaire to perform the skills maturity assessment (the one proposed by [17]). These three separate assessments generate separate radar charts which cover all the operations of the company. Then, based on these separate assessments, it is possible to perform a deep analysis of the overall maturity of the company and its concrete needs by keeping into account the current state from several perspectives. Indeed, the 5-level maturity scales of the three models are the same, even though they employ a different perspective. This alignment ensures to have a consistent and systematic assessment which sets the basis to create a unique roadmap. To report an example, the lack of sensorized machines on the shop floor is easily reflected in a lack of data available to perform deep analyses and also a lack of knowledge of the operators in exploiting data from the field. However, some of the operators might have the right knowledge about data exploitation, but they cannot exploit it due to the lack of specific technologies in the company. Thus, through the assessment emerges the potentialities these people have in becoming mentors for other colleagues in case that specific technology would be considered useful for the business of the company after the assessment on process and data sides. The following paragraph describes in detail the key characteristics of the models adopted.

3.1 The Methodology

As argued by [25], the first step of digitalization consists of an understanding of the status quo (AS-IS) of the firm.

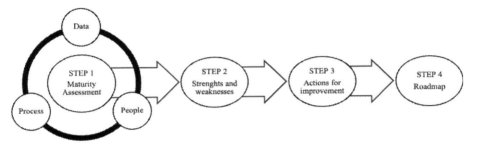

Fig. 1. Steps of the P.P.D. Digital Roadmapping methodology

Hence, as depicted in Fig. 1, the researcher will exploit the information collected to identify both strengths and weaknesses that characterize the processes analyzed and rank them according to a priority order. This phase turns out to be crucial for the effectiveness of the roadmap since it summarizes all the main elements that might be tackled and, in this sense, it must be strongly aligned with the company's business objectives and sector's needs (identified during the assessment phase). The prioritization of weaknesses is based on an initial sorting of them into root causes of the criticality and effects or rather those weaknesses that prove to be a consequence of other pain points. Then causes and effects are mapped and linked in graphical form and, starting from the map, each relationship is translated into a numerical matrix counting how many times a cause impact on an effect [26]. This mapping process, given its crucial role in the whole process, must be conducted with or validated by the company's experts [27]. The third step consists of the identification of the key actions for improvement to mitigate effects or directly eliminate the root causes detected. Step 4 aims to spread such actions along a given time horizon. This last phase proves to be crucial as well since it allows the beneficiary of the roadmap (the firm) to balance the effort needed and understand which are the logical sequence that linked actions might have [1].

3.2 Areas Covered

The overall methodology, as previously mentioned, aims to cover 3 main dimensions (Process, People, and Data) however, this aim is too broad and generic when dealing with manufacturing contexts. For this reason, the 6 areas proposed by [1] (i.e. Design and Engineering, Production, Quality management, Maintenance management, Logistics management (inbound, internal and outbound), and Supply Chain management) are investigated to better contextualize the analysis performed. Each area is then further detailed into sub-areas (e.g. Production Planning) to maximize the capillarity of the assessment. It is worth highlighting that, since the areas are designed with a modular approach, they might be faced independently and some of them might be omitted as well. This last case applies to those realities that do not manage or are not interested in tackling all the departments of their organizations but rather want to rationalize resources only on specific key areas. Last, a seventh area that bonds all the previous 6 is given by Digital Backbone or rather an area that aims to assess and eventually improve IT-related departments which are usually cross-departmental or staff to the whole organizational structure. Regarding the People side of the methodology, focused on the evaluation and reinforcement of people's skills, the authors have introduced a 3 layers classification common for all the areas abovementioned. They are ICT literacy, Hard Skills (specific for each department), and Soft Skills.

3.3 The Scale

To provide a comprehensive result, all the 3 dimensions of the assessment have been evaluated according to a 5-scale Likert scale. The selection of such a scale represents a standard for MM [28–30] as it allows to give a quantitative evaluation of multi-dimensional factors thus providing a comparable and homogeneous value. In particular, the authors opted to rely on scales already available in the literature and validated in previous studies as reported in the theoretical background (i.e. [1] for Process, [17, 18] for People, and [23, 24] for Data).

4 P.P.D. Digital Roadmapping Model Application

The multi-perspective MM model presented was applied to an Italian manufacturing SME operating in the prefabricated building sector. The study was conducted lasted 3 months and involved the authors (as facilitators and evaluators), the heads of the areas abovementioned, and a sample of operators. The results are shown in Fig. 2.

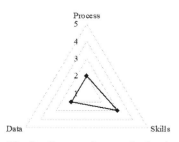

Fig. 2. Company's maturity level

The firm's processes and data management, distributed in 2 plants, presented an overall Maturity level of 2 as the processes resulted in partially controlled and mainly managed by the high experience technicians and managers that do not properly store and use data acquired on the field. The systems in use were not advanced, not integrated among the different functions, not suitable for the functionalities that were required to fulfill, and misaligned with the needs of the process built over time. Moreover, these shortcomings resulted in a proliferation of local documents and files, especially Excel, for the management of processes and, given their unreliability regarding the quality of the data recorded in them, a progressive distrust of current information systems. Regarding the skills analysis, the overall state of propensity to digitization showed a high level of knowledge of the internal systems (level 3) Soft skills were adequately high (around 4), assuming a future openness to the demand for change. On the other hand, as far as hard skills are concerned, regarding I 4.0, there was a medium-low level of competence (2) requiring ad hoc training courses based on the technological investments selected. Indeed, the level was consistent with the technologies currently present in the company, and it does not facilitate the identification of a mentor or a digital leader for the colleagues. To better identify the main pain points, the authors mapped the main criticalities and effects through the PCIM (Prioritized Criticalities based on Impact Matrix) framework [26]. This allowed us to visually and quantitively identify the main areas of improvement of the company.

More specifically, the most impacting criticalities referred to: limited knowledge of all the managers about the potentialities derived from I4.0 technologies, limited data exploitation for the decision-making process, lack of quality and maintenance control plants, paper-based data sharing across functions, lack of integration among available information systems, lack of specific information systems (e.g. warehouse management systems). Given this AS-IS scenario, the methodology allowed to depict a roadmap to Digital Transformation (Fig. 3) which consisted of the following aspects: (i) **Process**, review of business (green circles) processes and IT systems (purple circles); (ii) **Skills** (red circle), identification of training and education modules for each category of worker assessed; and (iii) **Data** (dark blue circles), identification of collectible data and design of a balanced scorecard for each process assessed. Such projects were plotted based on the expected effort (y-axis), defined with the interviewees, and time (x-axis).

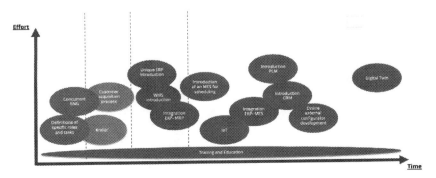

Fig. 3. Roadmap to digital transformation

5 Conclusions

This research aimed at presenting a methodology to support manufacturing SMEs in their Digital Transformation process. The methodology was designed to tackle in a holistic and integrated way 3 main viewpoints that must be taken carefully into account to ensure the success of the transition to I4.0, namely: Process, Skills, and Data management. To do it, the authors conducted a descriptive review of current MMs for Digital Transformation. It emerged that they were focused only on one out of the three dimensions abovementioned. Then, they focused their attention on three MMs. Hence, the new methodology was developed based on the extant scientific literature and validated in an Industrial case that involved Italian SMEs operating in the prefabricated construction sector. The methodology was designed to cover all the main steps of roadmapping: Assessment, Critical analysis, Solutions proposal, and Planning. The results of the research are deemed relevant from both a theoretical and managerial perspective. Indeed, it proposes an integrated and multi-dimensional MM and roadmapping methodology for manufacturing contexts; secondly, it provides tangible support to managers and C-levels in the definition of balanced and time-distributed investments and actions for improvement throughout their whole digital transition strategy deployment. The research presents some limitations as well given mainly by the reliance on a single-case study which will be overcome through future reiteration of the methodology and possible refinement based on feedback from companies. Additionally, the People-oriented MM adopted does not assess a valuable dimension such as the resistance to the chance of the organization.

References

1. de Carolis, A., Macchi, M., Negri, E., Terzi, S.: A maturity model for assessing the digital readiness of manufacturing companies. IFIP Adv. Inf. Commun. Technol. **513**, 13–20 (2017). https://doi.org/10.1007/978-3-319-66923-6_2
2. Castelo-Branco, I., Cruz-Jesus, F., Oliveira, T.: Assessing Industry 4.0 readiness in manufacturing: evidence for the European Union. Comput. Ind. **107**, 22–32 (2019). https://doi.org/10.1016/j.compind.2019.01.007
3. Orzes, G., Rauch, E., Bednar, S., Poklemba, R.: Industry 4.0 implementation barriers in small and medium sized enterprises: a focus group study. In: IEEE International Conference on Industrial Engineering and Engineering Management, pp. 1348–1352, December 2019. https://doi.org/10.1109/IEEM.2018.8607477
4. Gökalp, E., Martinez, V.: Digital transformation maturity assessment: development of the digital transformation capability maturity model. Int. J. Prod. Res. (2021). https://doi.org/10.1080/00207543.2021.1991020
5. Mittal, S., Khan, M.A., Romero, D., Wuest, T.: A critical review of smart manufacturing & Industry 4.0 maturity models: implications for small and medium-sized enterprises (SMEs). J. Manuf. Syst. **49**(October), 194–214 (2018). https://doi.org/10.1016/j.jmsy.2018.10.005
6. Mettler, T., Rohner, P.: Situational maturity models as instrumental artifacts for organizational design. In: Proceedings of the 4th International Conference on Design Science Research in Information Systems and Technology - DESRIST 2009, p. 1 (2009). https://doi.org/10.1145/1555619.1555649
7. Spaltini, M., Acerbi, F., Pinzone, M., Gusmeroli, S., Taisch, M.: Defining the Roadmap towards Industry 4.0 : the 6Ps maturity model for manufacturing SMEs. In: 29th CIRP Life Cycle Engineering Conference (2022)

8. Miragliotta, G., Sianesi, A., Convertini, E., Distante, R.: Data driven management in Industry 4.0: a method to measure data productivity. IFAC-PapersOnLine **51**(11), 19–24 (2018). https://doi.org/10.1016/j.ifacol.2018.08.228

9. Khajavi, S.H., Holmström, J.: Manufacturing digitalization and its effects on production planning and control practices. In: Umeda, S., Nakano, M., Mizuyama, H., Hibino, N., Kiritsis, D., von Cieminski, G. (eds.) Advances in Production Management Systems: Innovative Production Management Towards Sustainable Growth. APMS 2015. IFIP Advances in Information and Communication Technology, vol. 459. Springer, Cham (2015). https://doi.org/10.1007/978-3-319-22756-6_22

10. Chen, B., Wan, J., Shu, L., Li, P., Mukherjee, M., Yin, B.: Smart factory of industry 4.0: key technologies, application case, and challenges. IEEE Access **6** (2017). https://doi.org/10.1109/ACCESS.2017.2783682

11. Miragliotta, G., Shrouf, F.: Using internet of things to improve eco-efficiency in manufacturing: a review on available knowledge and a framework for IoT adoption. In: Emmanouilidis, C., Taisch, M., Kiritsis, D. (eds.) Advances in Production Management Systems. Competitive Manufacturing for Innovative Products and Services. APMS 2012. IFIP Advances in Information and Communication Technology, vol. 397. Springer, Berlin, Heidelberg (2013). https://doi.org/10.1007/978-3-642-40352-1_13

12. Olsen, T.L., Tomlin, B.: Industry 4.0: opportunities and challenges for operations management. Manuf. Serv. Oper. Manag. **22**(1), 113–122 (2020). https://doi.org/10.1287/msom.2019.0796

13. Fantini, P., Pinzone, M., Taisch, M.: Placing the operator at the centre of Industry 4.0 design: modelling and assessing human activities within cyber-physical systems. Comput. Ind. Eng. **139**, January 2020. https://doi.org/10.1016/j.cie.2018.01.025

14. Gfrerer, A., Hutter, K., Füller, J., Ströhle, T.: Ready or not: managers' and employees' different perceptions of digital readiness. Calif. Manage. Rev.**63**(2), 23–48, February 2021. https://doi.org/10.1177/0008125620977487

15. Pinzone, M., Fantini, P., Perini, S., Garavaglia, S., Taisch, M., Miragliotta, G.: Jobs and skills in Industry 4.0: an exploratory research. In: Advances in Production Management Systems. The Path to Intelligent, Collaborative and Sustainable Manufacturing . APMS 2017, pp. 282–288 (2017). https://doi.org/10.1007/978-3-319-66923-6_33

16. Acerbi, F., Assiani, S., Taisch, M.: A research on hard and soft skills required to operate in a manufacturing company embracing the industry 4.0 paradigm. In: Proceedings of the Summer School Francesco Turco, pp. 1–12 (2019)

17. Ameri, F., Stecke, K.E., von Cieminski, G., Kiritsis, D. (eds.): APMS 2019. IAICT, vol. 567. Springer, Cham (2019). https://doi.org/10.1007/978-3-030-29996-5

18. Acerbi, F., Rossi, M., Terzi, S.: Identifying and assessing the required I4.0 skills for manufacturing companies' workforce. Front. Manufact. Technol. **2**, July (2022). https://doi.org/10.3389/fmtec.2022.921445

19. Tao, F., Qi, Q., Liu, A., Kusiak, A.: Data-driven smart manufacturing. J. Manuf. Syst. **48**, 157–169 (2018). https://doi.org/10.1016/j.jmsy.2018.01.006

20. Polenghi, A., Roda, I., Macchi, M., Pozzetti, A.: Conceptual framework for a data model to support asset management decision-making process. In: Ameri, F., Stecke, K., von Cieminski, G., Kiritsis, D. (eds) Advances in Production Management Systems. Production Management for the Factory of the Future. APMS 2019. IFIP Advances in Information and Communication Technology, vol. 566. Springer, Cham (2019). https://doi.org/10.1007/978-3-030-30000-5_36

21. Zhang, Y., Ren, S., Liu, Y., Sakao, T., Huisingh, D.: A framework for Big Data driven product lifecycle management. J. Clean Prod. **159** (2017). https://doi.org/10.1016/j.jclepro.2017.04.172

22. Acerbi, F., Taisch, M.: Towards a data classification model for circular product life cycle management. In: Nyffenegger, F., Ríos, J., Rivest, L., Bouras, A. (eds.) Product Lifecycle Management Enabling Smart X. PLM 2020. IFIP Advances in Information and Communication Technology, vol. 594. Springer, Cham (2020). https://doi.org/10.1007/978-3-030-62807-9_38

23. Zitoun, C., Belghith, O., Ferjaoui, S., Gabouje, S.S.D.: DMMM: data management maturity model. In: 2021 International Conference on Advanced Enterprise Information System (AEIS), pp. 33–39, June 2021. https://doi.org/10.1109/AEIS53850.2021.00013

24. Pörtner, L., Möske, R., Riel, A.: Data management strategy assessment for leveraging the digital transformation. In: EuroSPI 2022: Systems, Software and Services Process Improvement, pp. 553–567 (2022). https://doi.org/10.1007/978-3-031-15559-8_40

25. Hansen, C., Daim, T., Ernst, H., Herstatt, C.: The future of rail automation: a scenario-based technology roadmap for the rail automation market. Technol. Forecast Soc. Change **110**, 196–212 (2016). https://doi.org/10.1016/j.techfore.2015.12.017

26. Acerbi, F., Spaltini, M., De Carolis, A., Taisch, M.: Developing a roadmap towards the digital transformation of small & medium companies: a case study analysis in the aerospace & defence sector. In: Noël, F., Nyffenegger, F., Rivest, L., Bouras, A. (eds.) Product Lifecycle Management. PLM in Transition Times: The Place of Humans and Transformative Technologies. PLM 2022. IFIP Advances in Information and Communication Technology, vol. 667. Springer, Cham (2023). https://doi.org/10.1007/978-3-031-25182-5_28

27. Kerr, C., Phaal, R., Probert, D.: Cogitate, articulate, communicate: the psychosocial reality of technology roadmapping and roadmaps. R and D Manag. **42**(1), 1–13 (2012). https://doi.org/10.1111/j.1467-9310.2011.00658.x

28. Marcucci, G., Antomarioni, S., Ciarapica, F.E., Bevilacqua, M.: The impact of operations and IT-related Industry 4.0 key technologies on organizational resilience. Prod. Plan. Control, 1–15 (2021). https://doi.org/10.1080/09537287.2021.1874702.

29. Pirola, F., Cimini, C., Pinto, R.: Digital readiness assessment of Italian SMEs: a case-study research. J. Manuf. Technol. Manag. **31**(5), 1045–1083 (2020). https://doi.org/10.1108/JMTM-09-2018-0305

30. Rossi, M., Terzi, S.: CLIMB: maturity assessment model for design and engineering processes. Int. J. Prod. Lifecycle Manag. **10**(1), 20–43 (2017). https://doi.org/10.1504/IJPLM.2017.082998

Development of IoT Solutions According to the PLM Approach

Francesco Serio[1] , Ahmed Awouda[1] , Mansur Asranov[1,2] ,
and Paolo Chiabert[1]([✉])

[1] Politecnico di Torino, Corso Duca degli Abruzzi, 24, 10129 Torino, (TO), Italy
paolo.chiabert@polito.it
[2] Turin Polytechnic University in Tashkent, Kichik Halqa Yuli 17, 100095 Tashkent, Uzbekistan

Abstract. The Industrial Internet of Things (IIoT) is one of the nine enabling technologies of Industry 4.0, which in recent years has seen an exponential increase in its applications. New production devices that are naturally equipped with this technology and the retrofitting solutions for industrial devices already installed in our industries, promote the demand of IIoT solutions. The Internet of Thing is often associated with Product Lifecycle Management (PLM) due to its ability to provide data which, when appropriately analyzed, feed the PLM system allowing for the tracking of the product along its life cycle.

In this paper, the point of view is reversed: the IIoT solution, which is designed, implemented and maintained in an industrial system, is the product that must be managed with a PLM approach.

IIoT solutions have characteristics that require the use of a PLM approach: they must meet complex requirements, they must adhere to standards and be compatible with the company's existing IT infrastructure, they are complex systems that interface many other systems and have a long lifecycle during which they are subject to innumerable modifications and extensions.

It is therefore justified, from a research point of view, to investigate the characteristics that a PLM approach must have to support the development of an IIoT solution.

This paper, based on the theory and evidences from industries and academies, traces a reference framework for the development of an IIoT solution supported by the PLM approach.

To test the validity of the proposed guidelines, the paper illustrates their application in the development of a simple IoT solution dedicated to teaching and training.

Keywords: Product Lifecycle Management Guidelines · Industrial Internet of Things solutions · Knowledge Management

1   Introduction

Nowadays, insiders are well aware of both the nine technologies of Industry 4.0 (I4.0) and their relevance in renewing manufacturing strategies [1]. Among these nine, Internet of Things (IoT) has been considered a key technology due to its potential to be a

© IFIP International Federation for Information Processing 2024
Published by Springer Nature Switzerland AG 2024
C. Danjou et al. (Eds.): PLM 2023, IFIP AICT 702, pp. 85–95, 2024.
https://doi.org/10.1007/978-3-031-62582-4_8

game changer [2]. In addition, the most recent production devices that are naturally equipped with this technology and the retrofitting solutions for industrial machines already installed in our industries, promote the diffusion of Industrial Internet of Things (IIoT) solutions.

On the other hand, Product Lifecycle Management (PLM) is "a business strategy for creating and sustaining such a product-centric knowledge environment. It is rooted not only in design tools and data warehouse systems, but also on product maintenance, repair and dismissal support systems. A PLM environment enables collaboration between – and informed decision making by – various stakeholders of a product over its lifecycle" [3].

The IoT is often associated with PLM for its ability to provide data which, when properly analyzed, feed the PLM system allowing product traceability along its life cycle. In this paper, we change this prospective considering the IIoT solution as the final product to manage with a PLM approach because the scientific and technical literature does not show solid evidence of structured approach for developing IIoT solutions. For this purpose, we investigate the characteristics that a PLM approach must have to support the development of an IIoT solution taking into account its requirements, the standards to adhere with, the necessity to interface with the existing IT infrastructure and the need to update the IIoT solution during its lifecycle. An IIoT solution is not a single electronic product but a framework that includes hardware, software and infrastructural components. Differently from other electronic solutions, IIoT is inherently an open framework that dramatically changes along the time, where components are often upgraded with relevant new characteristics, in order to fulfil dynamic customer's demand or exploit technological innovation.

After this section, that introduces and contextualize the research work, you will find the literature review about IoT, IIoT, PLM and their interaction. Subsequently, we propose a reference framework for the development of an IIoT solution supported by the PLM approach and we test it on a simple IoT solution dedicated to teaching and training. Finally, the last section states the conclusions and the possible future works related to these topics.

2   Research Context

2.1   IoT and IIoT

For some years now, the Internet of Things has begun to influence and change our lives and habits. IoT is composed of devices that have sensors to gather and communicate data through network protocols. The availability of huge amount of data, that can be crossed and analyzed, provides valuable information for customers and producers. In [4, 5], the concepts are described in detail.

The declination of the IoT in the manufacturing industry is called the Industrial Internet of Things [6], where production machinery, conveyors, products, semi-finished products, raw materials and lately wearables worn by operators constantly send information about the production line or the entire plant in order to identify and correct errors promptly.

Furthermore, again thanks to IoT technologies, manufacturers can check the products leaving the factory to understand how and where they were distributed, consequently derive the demand rate in a given area and potentially eliminate out of stock and overproduction scenarios. Finally, thanks to some IoT technologies, such as RFID and barcodes, distributors can trace the arrival of products from the various manufacturers and manage warehouses, accounting and many other business areas.

2.2 PLM Approach

Taking into account the maturity of this topic, PLM has been defined from many experts until now. Among the most important contributions are worth mentioning those of Michael Grieves [3], with its definition of PLM model and of Mirrored Spaces Model, the one by Saaksvuori and Immonen [7] and the one by John Stark [8], who introduced an approach to correctly implement PLM in a firm following a series of ten steps. According to Terzi et al. [9], PLM generally refers to three distinct periods of the life of any product (see Fig. 1): Beginning of Life (BOL), Middle of Life (MOL), and End of Life (EOL). Given the complexity of each of these phases, many papers in the literature focuses only on one of them. In addition, the BOL is certainly the most complicated one because it has its origins in the product development (PD) process consisting of several phases that are constantly increasing their complexity. However, although the product development process is different for each company and above all for each industry, it is usual to define some typical common phases. Examples are given by the models proposed by Pahl & Beitz' [10], Ulrich & Eppinger (U&E) [11] and by the ASME\ANSI standards [12]. This paper considers the PD process defined by U&E as reference model for its intuitiveness and because it effectively supports the didactic purpose. The U&E model consists of six phases (see Fig. 2) that starting from many product concepts reduces the number of alternatives to the best one, which must pass all the different tests and must meet technical specifications and customer needs. The decision to go more into detail on the PD process is dictated by its significant impact on production and product monitoring and, consequently, on the cost of the final product.

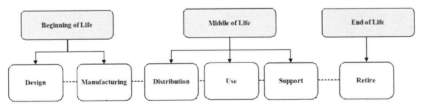

Fig. 1. Product Lifecycle periods [3]

2.3 IoT for PLM

In order to know the state of the art regarding the interaction between IoT and PLM it has been used as starting point the Systematic Literature Review (SLR) presented by

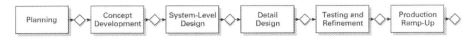

Fig. 2. The six phases of the generic development process [11]

Barrios et al. [13] and referred to November 2021. The research was then updated on both Scopus and Web of Science (WOS) searching for the same key words. The results were similar to those reported in [13], because none of the latest paper considers PLM and IoT at the same time, but generally, they refer more at one of the two. In addition, it clearly emerges in all of these cases as the prospective goes always from IoT to PLM, which translates in the focus of creating the best IoT solution to feed the PLM information system and not in using PLM to support the development of an IoT solution. As a logical implication, the absence of this perspective in theory reflects the situation in practice, where IIoT solutions are not developed with a structured approach because they rely on industrial best practices that are rarely shared to preserve the intellectual properties. The only noteworthy article is "Edge-Computing and Machine-Learning-Based Framework for Software Sensor Development" [14], where the authors propose a PLM approach to manage software sensors and in particular Machine Learning (ML) Algorithm. However, ML algorithms should be considered within another technology in I4.0 umbrella, which is Data Analytics rather than IoT solutions.

For these reasons, the focus of this manuscript is inverted compared to all the other papers because it proposes the creation of an IIoT solution with a PLM approach, where the IIoT solution is seen as a product itself.

2.4   Why PLM for IoT Solution

The main reason why to use PLM is the complexity management declined in many areas such as product design, time to market (TTM), standards and rules, supply chain extension, geographical distribution of design and manufacturing centers, different and complex software with many hardware and software architecture available.

On the other hand, the product complexity is constantly increasing due to both the technological evolution of manufacturing processes and the number of components. The latter is especially true for some assembled products such as cars or electronics. For all these reasons is now common to refer to "smart products" as defined by Kristis [15]. Further parameter that increases the complexity are the time to store and retrieve information and, as said, the TTM reduction, i.e. the need to develop a product solution as fast as possible in order to anticipate all the competitors. For what concern the standards and rules to handle during the product lifecycle we can just cite some of them: Vision2000 [16], End of Life Vehicles Directive [17] and TREAD Act for the automotive industry [18], WEEE (EU Waste Electrical and Electronic Equipment) for the electronics industry [19] and the RoHS (Restriction of Hazardous Substances) in Electrical and Electronic Equipment Directive [20].

Given that IIoT and IoT solutions in general share these characteristics and problems, it was natural for us to think of these solutions as real products whose entire life cycle can be managed. Other reasons that motivate the usage of PLM in developing IIoT solution is the significant amount of failures in IoT projects, mainly related to obscured business aims, overlooked technological problems, unforeseen company organizational issues, and finally, customer and vendors misalignment [21]. In general, all these topics are properly managed when the project is driven by a PLM approach.

3 The Proposed Theoretical Framework

The theoretical framework proposed in this chapter is based on PLM phases, defined by Terzi et al. [9], and depicted in Fig. 1.

The BOL corresponds to the application of the U&E model to IIoT solutions development, which consists of the six phases illustrated in Fig. 2 and categorized into three main enterprise functions: marketing, design, and manufacturing. Table 1, structured starting from the U&E model, shows a general overview of the IIoT development framework for the BOL phase, the various functions involved in the different phases and how these functions are characterized in terms of core functions (Marketing, Design, Manufacturing, Other).

The planning phase starts with the identification of opportunities, which is led by business strategy and includes an evaluation of market goals and technological advancements. The project mission statement, which details the product's target market, business objectives, major presumptions, and restrictions, is the result of the planning phase.

The target market's demands are determined during the concept development phase, and one or more concepts are chosen for further development and testing after being developed and evaluated. The form, purpose, and features of a product are described in a concept, which is typically supported by a set of requirements, a comparison to similar items, and an economic justification for the project.

In the system-level design phase, the product architecture is defined, the product is broken down into subsystems and components, essential components are provisionally designed, and responsibility for the detail designs is distributed among internal and external resources. During this stage, initial designs for the manufacturing system and final assembly are often developed as well. The output of this phase includes functional design specifications for each subsystem and should include architectures and geometric layout. The comprehensive definition of the structure, components, and properties of all the distinctive sections of the product as well as the identification of standard parts provided by suppliers, completes the detail design process. Each item has a process plan and control documentation, consisting of several files, which completely define the item, its interfaces with the system and its manufacturing or acquisition process.

The testing and refinement phase involves the construction and assessment of several preproduction versions of the product. The final phase is the production ramp-up phase where the product is made using the intended production system. Since these final phases have been partially applied in the case study the authors will not dive into details.

The MOL is based on three steps to be followed by the IIot solution users, comprehending the parts depicted in Fig. 3, which shows the activities dependency. According to the Fig. 3, the Theoretical part is focused on developing and assessing fundamental knowledge and skills, the Practical part supports hands-on experience and finally the survey part collects necessary feedback from both the previous parts to improve the product.

Table 1. Product Development Phases and Functions for developing an IoT product (BOL)

	Planning	Concept Development	System-Level Design	Detail Design	Testing and Refinement
Marketing	Articulate market Opportunity *Industry 4.0 opportunity, Education & Professional Training Market Opportunity*	Collect customer needs. *Needs in terms of Component functionality, scalability, network coverage, energy, open source* Finance: Facilitate Economic analysis. *Compare various vendors, SW & HW options, Manufacturing, and assembly options*	Develop plan for product options and extended product family *In case of different product variants develop taxonomy for product variation*		Facilitate Field testing *Prepare environment for testing HW, SW, Network, interface, cyber-security*
Design	Consider product platform and Architecture *Which IoT platforms & architectures, Software Options*	Investigate feasibility of product concepts. *Software & Hardware compatibility, Structural feasibility*	Develop product architecture. *Create SW & HW design, System architecture, structural design*	Define items and interfaces *Complete structural design and CAD/CAE models*	Test overall performance, reliability, and durability. *Test HW, SW, Network, interface, cyber-security*
	Research & Find available technologies *Technologies related to: protocols, hardware and software, manufacturing methods*		Define major subsystems and interfaces & Preliminary component engineering. *Define HW, SW and structural subsystems and how they interact*	Define system and subsystem design. *Hardware design, Software and application design*	Implement design changes
Manufacturing	Identify production & assembly Constraints. *Physical constraints, software & hardware constraints*	Estimate Product & Manufacturing cost *BOM, cost breakdown for product components*	Identify suppliers for key components *Identify Platform, HW and SW vendors*	Define production processes. Define quality assurance processes	

Notably, since the IIoT solution is one-of-a-kind product, in the authors approach the distribution phase is not considered.

The EOL, finally, takes into account all the actions needed to efficiently reuse all the IIoT solution components, in this case there will be two possible scenarios: (1) all the components are obsolete for their re-usage; (2) just some of the components are reusable. In the first scenario of the EOL, the solution will be donated as it is to a school for minor didactical activities. In the second scenario the reusable components will be employed to build the future prototype of the updated solution, instead the obsolete parts will be employed to manufacture a simpler solution to donate at the same target of the first scenario.

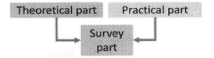

Fig. 3. PLM approach for a IoT solution (MOL)

4   Development of an IoT Solution with a PLM Approach

The need to mitigate common problems in implementing IoT solutions [21] stimulated authors to investigate the adoption of the proposed PLM approach in developing an IoT solution and testing its applicability with a didactic experiment. The latter highlighted the teaching and training activities aimed for reskilling IoT developers/designers. Main activities related to the development were organized according to the six phases of the U&E model included in the Beginning of Life period, as depicted in Figs. 1 and 2.

4.1   Product Design

The U&E model phases were customized for developing an IoT solution and performed according to the PLM approach. Each phase has been supported by the corresponding module available in PLM framework.

Planning. The aim of this phase is to conceive the concept mission statement through the opportunities identification. The objective of IoT solution is the development of a sandbox system, which could give the opportunity for data generation and acquisition, process monitoring and control. The solution should meet the following requirements: (1) Components should be commercial and widely available, (2) Allow for low-cost crafts-men manufacturing, (3) based on open-source solutions, (4) Fast response to change, (5) Minimal maintenance.

Concept Development. The planning of the IoT solution starts considering the main objective of the didactic activities: the interaction with sensors/actuators represents the customer needs, and professor guidelines for teaching activities defines the technical requirements. The concept of the system should meet the following categories of requirements: (1) Device level requirements: installation of the sensors/actuators for data collection and device control should be simple for future replication. Moreover, variables from the devices should be simple and understandable, and reading/controlling should be easily performed; (2) Openness: hardware and software should be independent from vendor with wide integration opportunities; (3) Simplicity: software components should be simple to configure, deploy and verify, according to the low-code/no-code Approach; (4) Scalability: the IoT solution should be scalable and portable.

System-Level Design. The concept requirements impose specific challenges in terms of system-level design. The Fig. 4 depicts system-level design of IoT solution, which consists of the following sub-systems: (1) Structure/body – structural enclosure, which supports all other sub-systems; (2) Thermal and environmental subsystem – the components that generate thermal and environmental scenarios; (3) Sensors and actuators – various sensors for measuring variables, describing and monitoring thermal and environmental scenarios and actuators for controlling thermal and environmental sub-system components; (4) Logical subsystem – the components for data processing, actuators management, and users' interaction.

Leftstream arrows in Fig. 4 represent datasets of the physical and dynamical variables, while rightstream arrows represent control information, which enforces changes in the thermal and environmental subsystem.

Fig. 4. System-Level Design of IoT solution

Detail Design. Based on system level design, the detailed design process was broken down to the following engineering activities, performed by different design teams and operated according to concurrent engineering paradigm supported by PLM software:

1. Structural design – The objective of this project activity is to design the physical structure of the IoT solution, consisting of a modular polystyrene box, through the utilization of a CAD software. Deliverables of this activity are product data of the structure/body design, such as 2D drawings, 3D models, materials description, rendering, assembling instruction and simulation, cost analysis, etc.. Several modifications occurred with respect to the initial design requirements due to some criticalities related to the assembly phase and to the usability of the devices. The IoT sandbox consists of two environments separated by interchangeable solid or holed walls to allow different degrees of interaction. The covers of these two environments can be independently opened for maintenance and upgrading activities. Moreover, considerations on the integration of IoT components and constraints imposed by other teams, according to the flowchart in Fig. 5, influenced the final design.

2. Thermal and environmental design – This activity designed the components for thermal and environmental scenarios, along with the required power supply. Two Peltier cells were used to cool and heat the air inside the two chambers of the IoT sandbox for temperature and humidity variation. Three sets of lights were included to create lightning variation. Additionally, a thermodynamic model, also used for enabling a Cyber-Physical System, was developed to describe the heat exchange inside the chambers of the IoT sandbox. The outputs of this activity consist of Peltier cells and lights connection, power supply schemes and thermodynamic model description.

3. Sensors subsystem design – This activity developed sensors network for measuring the variation of humidity, lightning, current, voltage, and temperature inside the chambers, as well as outside of the IoT sandbox. Correspondingly, the sensors network design includes the sensors for temperature, humidity, light intensity, current and voltage. Additionally, several relays were added to the system for controlling Peltier cells and lights, thus, generating a complex interaction between the teams for

thermal and sensors subsystems design. Output of this activity are the sensors and relays connection schemes.

4. Software and Control subsystem - The activity developed the IT system for collecting and managing the data obtained from the sensors network and for defining the control of thermal and light actuators. Raspberry Pi microcomputers were included in the design to handle computing workloads related to data processing, visualization, and control. Software part of the subsystem consists of MQTT broker service, which supports data transmission, and NodeRed responsible for data processing, visualization, and control logic. This bundle of software was containerized using Docker system to ensure portability and scalability. This activity generated important product data for the IoT solution, such as Raspberry Pi, sensors network and relays connection scheme, as well as NodeRed flows and MQTT broker configuration information. According to the PLM approach the product data generated in this activity must be stored, shared and upgraded according to the evolution of the IoT solution.

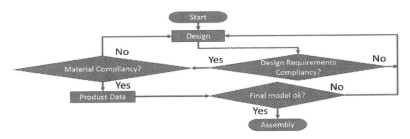

Fig. 5. Structural design methodology flowchart.

Testing and Refinement. To ensure usability, reliability and durability of the IoT solution, two set of testing procedures of subsystems and functions were performed. The first was an inter subsystem test (integration between different subsystems) where signal and information exchange between subsystems, logical functionality of the software and cybersecurity resilience were tested. The latter was an intra subsystem test where each subsystem was tested individually against designed features and functionalities.

4.2 The IoT Solution Created

The product developed, is the Smart IoT box, a device with sensors and actuators that allows for data acquisition, processes monitoring and control. The device gives the opportunity to improve learning process and develop fundamental skills of building and managing IoT/IIoT solutions, one of the enablers for Industry 4.0 Paradigm.

Since the IoT solution is one-of-a-kind product, no comprehensive production planning is required. Furthermore, during the design process, several physical experiments (see Fig. 6) have been carried out to verify critical aspects IoT solutions and provide the final proposed solution.

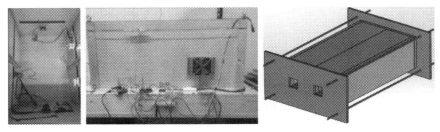

Fig. 6. Preliminary prototype of IoT sand box (left) e final proposal (right)

Taking into account the MOL, the Smart IoT box is intended to be used during practical activities in several courses offered for bachelor, master and PhD levels in field of Industry 4.0. Instead, during the EOL stage of the product lifecycle, the device is planned to be used as showcase for explaining disposal and recycling processes for HW and SW components of IoT products.

5   Conclusions

The paper analyzes the complexity of IIoT solutions and proposes a theoretical framework to manage their lifecycle with a PLM approach. Moreover, it reports a partial practical application of the proposed framework involving the Beginning Of Life based on the Ulrich & Eppinger model. The focus on the BOL period is justified by the relevance of the preliminary design decisions on overall costs, upgradability and scalability of the IIoT solution. However, the Middle Of Life and the End Of Life are not reported in the practical application because the MOL of our solution will start in the next didactic period.

The development of an IoT solution for didactic activities, to be performed in a didactic environment, demonstrates the applicability of the proposed approach and produces an effective One-of-Kind product. Nevertheless, open questions arise from the Middle of Life period where maintenance, upgrading and scalability are strictly required. Finally, open issues affect the End of Life period because it is not clear the role that HW and SW components will play into the Circular Economy model. Further investigations will address the open questions and issues that will be firstly verified in practice and secondly reported in future works.

References

1. Yin, Y., Stecke, K.E., Li, D.: The evolution of production systems from Industry 2.0 through Industry 4.0. Int. J. Prod. Res. **56**(1–2), 848–861 (2018)
2. Garrido-Hidalgo, C., Olivares, T., Ramirez, F.J., Roda-Sanchez, L.: An end-to-end internet of things solution for reverse supply chain management in industry 4.0. Comput. Ind. **112**, 103127 (2019)
3. Grieves, M.: Product Lifecycle Management: Driving the Next Generation of Lean Thinking: Driving the Next Generation of Lean Thinking. McGraw Hill Professional (2005)

4. Atzori, L., Iera, A., Morabito, G.: The Internet of things: a survey. Comput. Netw. **54**(15), 2787–2805 (2010). ISSN 1389-1286. https://doi.org/10.1016/j.comnet.2010.05.010
5. McEwen, A., Cassimally, H.: Designing the Internet of Things. John Wiley & Sons Inc., UK (2013)
6. Singh, I., Centea, D., Elbestawi, M.: IoT, IIoT and cyber-physical systems integration in the SEPT learning factory. Proc. Manufact. **31**(116–122) (2019). ISSN 2351-9789. https://doi.org/10.1016/j.promfg.2019.03.019
7. Saaksvuori, A., Immonen, A.: Product Lifecycle Management, 3rd edn. Springer, Heidelberg (2008). https://doi.org/10.1007/978-3-540-78172-1
8. Stark, J.: Product Lifecycle Management, 3rd edn. Springer (2015)
9. Terzi, S., Bouras, A., Dutta, D., Garetti, M., Kiritsis, D.: Product lifecycle management—from its history to its new role. Int. J. Prod. Lifecycle Manag. **4**(4), 360 (2010). https://doi.org/10.1504/IJPLM.2010.036489
10. Pahl, G., Beitz, W., Feldhusen, J., Grote, K.: Engineering Design: A Systematic Approach, 3rd edn. Springer, London (2006). https://doi.org/10.1007/978-1-84628-319-2
11. Ulrich K., Eppinger S.: Product Design and Development. 6th edn. McGraw-Hill Education (2016)
12. https://www.asme.org/codes-standards/y14-standards
13. Barrios, P., Danjou, C., Eynard, B.: Literature review and methodological framework for integration of IoT and PLM in manufacturing industry. Comput. Indust. **140** (2022). https://doi.org/10.1016/j.compind.2022.103688
14. Hanzelik, P.P., Kummer, A., Abonyi, J.: Edge-computing and machine-learning-based framework for software sensor development. Sensors **22**, 4268 (2022). https://doi.org/10.3390/s22114268
15. Kiritsis, D.: Closed-loop PLM for intelligent products in the era of the internet of things. Comput.-Aided Des. **43**(5), 479–501 (2011). https://doi.org/10.1016/j.cad.2010.03.002
16. https://www.iso.org
17. https://eur-lex.europa.eu/legal-content/EN/ALL/?uri=CELEX:32000L0053
18. https://www.govinfo.gov/content/pkg/PLAW-106publ414/html/PLAW-106publ414.htm
19. https://environment.ec.europa.eu/topics/waste-and-recycling/waste-electrical-and-electronic-equipment-weee_en
20. https://environment.ec.europa.eu/topics/waste-and-recycling/rohs-directive_en
21. https://www.softwareag.com/content/dam/softwareag/global/marketing-material/en/analyst-reports/cumulocity/ar-why-iot-projects-fail-summary-beecham-research-en.pdf.sagdownload.inline.1629789733035.pdf

Development of a Multi-plant Cross-Function Roadmapping Tool: An Industrial Case in Food & Beverage Sector

Elena Beducci$^{(\boxtimes)}$ ⬥, Federica Acerbi ⬥, Marco Spaltini ⬥, Anna De Carolis ⬥, and Marco Taisch ⬥

Department of Management, Economics and Industrial Engineering, Politecnico di Milano, via Lambruschini 4/b, 20156 Milan, Italy

{elena.beducci,federica.acerbi,marco.spaltini,anna.decarolis, marco.taisch}@polimi.it

Abstract. The sector of Food & Beverage (F&B) has experienced a remarkable increase in the diversification of products, leading to a consequential increase in variability, which, if not properly managed, would result in tampering profitability. Assuming that this trend cannot be reversed, manufacturers must foster their operations' efficiency, thus sustaining their competitiveness. Industry 4.0 (I4.0) paradigm is recognized worldwide as one of the preferred paths to address this issue. By following the digitalization path, companies may encounter different barriers and will require support to overcome them. Specifically, MNCs require to exploit existing synergies among different plants and build common infrastructures. This paper presents a methodology aimed at creating a unified roadmap (TRM) for multi-plant manufacturing companies. The methodology is then applied through an industrial case study to enable the desired digital transformation of a whole MNC's country branch, starting from the MNC's objectives and the digital maturity assessment of the single plants. The industrial case, the Italian branch of a Swiss MNC (3 plants), allowed to support C-Levels in identifying the priority areas at subsidiary level and in allocating resources accordingly. Regarding the theoretical implications, the research allowed to shift from a single-plant view, adopted by several TRM studies, to a cross-plant cross-function integrated approach.

Keywords: Food & Beverage · Manufacturing · Digital Transition Roadmapping · Industry 4.0 · Assessment

1 Introduction

The digital transformation is spreading worldwide, influencing the entire society and, in the industrial field, the influences are reflected in the operations of several manufacturing sectors [1]. Among all, Industry 4.0 (I4.0) technologies are considered great drivers for improving manufacturing companies' efficiency and this is true especially

© IFIP International Federation for Information Processing 2024
Published by Springer Nature Switzerland AG 2024
C. Danjou et al. (Eds.): PLM 2023, IFIP AICT 702, pp. 96–106, 2024.
https://doi.org/10.1007/978-3-031-62582-4_9

for the food and beverage (F&B) sector [2]. The F&B sector is being asked to improve its sustainable oriented performances to ensure a balanced and proper resource management willing to exploit the value of the waste generated along the entire value chain [3]. To accomplish these goals in an efficient manner, the F&B sector is required to rely on the technological advancements by selecting the right technologies according to the specific needs [4]. The choice of the right technology is fundamental for both small and medium companies, to prudently use financial resources for technological investments but also for Multinational Companies (MNC)s [5] which are required to balance the investments in their dispersed plants, sometimes acquired from other companies along the years, by tailoring the technological introduction on the needs of each plant. In this context, it is fundamental to design a unified roadmap tailored on the strategic goals of the company as a whole while keeping into account the specific needs and digital maturity levels of its dispersed plants. Nevertheless, at the best of the authors knowledge, few contributions proposed multi-plant perspective. In this regard, the present contribution aims at developing a roadmapping tool to support digital transition of multi-plant manufacturing companies. The proposed methodology has then been validated through an industrial case study, involving a multi-plant company operating in the F&B sector. To address the research objective, the following research question has been formulated "How to support multi-plant manufacturing companies in the transition towards industry 4.0?". The developed roadmapping tool has been designed through the integration of a literature review on maturity models and roadmapping tools with an action research approach. Both the development methodology for the roadmapping tool and the results obtained out of the application of the tool are showed in this contribution highlighting the key practical and theoretical implications. The remainder of the paper is structured as follows. Section 2 describes the theoretical background used as a basement to develop the tool. Section 3 elucidates material and methods applied to the analyzed case study. Section 4 shows application of the model. Section 5 discusses the results obtained, while Sect. 6 concludes the contribution paving the way to future research opportunities.

2 Theoretical Background

Digital transformation and Industry 4.0 are relevant trends impacting manufacturers' operations. Looking at the barriers faced by manufacturers in digitalization effort, lack of competences and direction stand amongst the major ones [6] thus highlighting that a concrete support is essential to enable the change [7]. In this direction, Technology Roadmaps (TRMs) prove to be a valuable and sounded solution [8]. Although the concept seems quite established in literature, as it was first introduced in the 1960s [9], it has been evolving over time in parallel with the industry's technological advancement [10]. As a matter of fact, a clear and shared definition is difficult to detect even though, considering the scope of the research proposed, the authors adopted the one coined by [11] which define TRM as "*a process that mobilizes structured systems thinking, visual methods and participative approaches to address organizational challenges and opportunities, supporting communication and alignment for strategic planning and innovation management within and between organizations at firm and sector levels*". From this definition it can be noticed that TRM lays its value in the process of creation rather than the tool itself. For this reason the methodology should present certain characteristics such as: being shared by the company involved [12], being multi-functional [13], being aligned with corporate strategy [14] and being easy to learn [15]. In particular, [16] highlighted that the success of a transformation roadmap in manufacturing context might be maximized whenever a multi-dimensional approach is adopted including, among the others, operations processes and people and skills. The roadmapping process is constituted by a series of consecutive steps that lead to the development of the TRM. Again, literature has proposed plenty of methodologies which differ in relation to the steps to reach the outcome and, to some extent, also in terms of considered dimensions. In this direction, a subset of the most rated methodologies that differs by steps and dimensions adopted is provided in Table 1.

Table 1. Main TRM methodologies' dimensions considered and phases

Reference	Dimensions	Phases
[17]	Technology, product, and market	5
[18]	Not provided	8
[19]	Not provided	4
[20]	Partners, resources, supplementary technologies, core technologies, product, and market	3
[14]	Not provided	6
[15]	Technology, product and market	4
[21]	Customer, benefit, value-creation, partners, and finance	5
[22]	Markets, product, technology, regulations/standards, organization, and value goals	-
[23]	Product conformance, product performance, and objectives & target	5

Although roadmapping is the application of multiple phases, literature has focused its attention mostly on only one preliminary phase: the maturity assessment analysis. The theory behind Maturity Models (MMs), sometimes referred as Readiness Assessments, is hence rich and covers multiple fields of research, including digital transformation in manufacturing. MMs are defined as *"evaluation tools to analyze and determine the level of preparedness of the conditions, attitudes, and resources, at all levels of a system, needed for achieving its goal(s)"* [24].

Among the MMs developed so far, [25] proposed a methodology, the DREAMY4.0, designed specifically to assess the most critical manufacturing process, namely R&D, Production, Quality management, Maintenance, Logistics and Supply Chain, to be kept under control to facilitate the digital transformation of the company. In particular, the methodology, based on a 5-level ranking, aims at providing a score of the above-mentioned functions and evaluate them under 4 different dimensions. The four dimensions assessed are: technology, in relation to the advancement of software and hardware used, Execution, intended as the correctness in carrying out the core activities,

Monitoring and Control, to assess the capability to supervise efficiently and effectively the processes and Organization, to assess the alignment between the organizational structure and the processes governed.

3 Methodology

The authors took inspiration from existing literature to address the research objective of developing a roadmapping tool to support digital transition of multi-plant manufacturing companies. From the analyzed literature the authors have extrapolated 6 main roadmapping phases which are:

1. The understanding of the needs: which aims at collecting the necessary information to understand to strategic direction that the company is willing to undertake
2. The maturity assessment, or AS-IS analysis: which aims at understanding the current level of digital and sustainability maturity level within the subject of analysis and define the goals
3. The identification of problems and criticalities: which aims at critically analyzing processes to detect weaknesses that impede the company to achieve the desired goals and threatens performances
4. The definition of actions for improvement: which aims at identifying feasible solutions to overcome the weaknesses and support firms in achieving the desired scenario
5. The roadmapping, or planning: which aims at defining the necessary transformation journey to support organizations to achieve the stated goals
6. The control: which aims at monitoring the outcomes of the projects proposed and validate the roadmap or, on the contrary, adjust it

The roadmapping tool applied to the case study follows these phases. As suggested by [26], the roadmapping process was conducted through workshops with managers and operators of the main areas analyzed (i.e., Production Planning, Production, Energy Monitoring and Maintenance, Logistics).

Figure 1 shows the steps representing the roadmap generation.

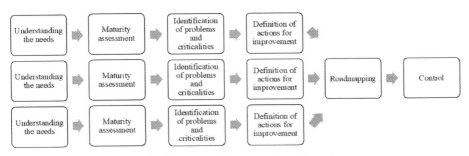

Fig. 1. Steps of the research conducted

Stage 1 was conducted through a an extensive interview with the top management teams, when the main strategic objectives of the firms were delineated. The AS-IS analysis (stage 2) was conducted by exploiting the DREAMY4.0 assessment model [25].

The identification of criticalities and their prioritization (stage 3) employed a simplified PCIM (Priority Criticality Index Mapping) approach proposed by [7] which encompassed a score assignment to the problems identified through face-to-face interviews with managers and process owners. In particular, the authors adopted a 2-dimensional compounded Likert scale based on 3 level, from 1 (low) to 3 (high). The 2 dimensions identified for the prioritization of criticalities adopted were (i) the frequency of occurrence of the problem arose and (ii) the impact registered on operations processes if or when the problem would manifest (see Fig. 2).

F/I	1	2	3
1	1	2	3
2	2	4	6
3	3	6	9

Fig. 2. Ranking system for criticality prioritization (F: frequency; I: impact)

Furthermore, to better guide the solution development process, each criticality comprised a classification procedure aimed at identifying 3 main types, namely: (i) Technology-based criticalities ("Tec" in Table 3), (ii) Process-related criticalities ("Pro" in Table 3) and (iii) Criticalities due to organizational incoherencies with processes or technologies adopted ("Org" in Table 3). Regarding stage 4, the definition of actions for improvements were validated by manufacturers' experts and managers. In stage 5 the actions proposed are valuated according to objective criteria decided with the company, then, they have been plotted in a roadmap and distributed over the given time horizon.

4   Roadmapping Tool Application

The methodology, illustrated in the previous section, has been applied to a multi-plant company operating in the F&B sector. The company owns three plants located in Northern Italy, which were used as industrial cases study.

According to stage 1 of the proposed methodology, a preliminary interview was conducted with the top management to clarify the strategic objectives of the group. The identified objectives are the following: (i) increase efficiency, (ii) reduce costs, (iii) increase data reliability to improve short- and long-term decision making and (iv) achieve the homogeneity of processes and architecture while preserving the product specialization of each plant. During this meeting, the IT manager of the group's Italian branches was also involved to clarify the peculiarities of the IT infrastructure. The three plants are characterized by different digitalization levels and have different and not integrated IT architectures. The digital infrastructure of the plants appears currently inadequate to address the global needs of the company: certain required systems are missing, and others do not meet the requirements of companies' operations. The inadequacy of the IT architecture could prevent the company to achieve the strategic goals set.

The authors then dedicated a full 2-days session for each plant (see Table 2) to conduct the interviews aimed at addressing stage 2 and stage 3 of the roadmapping methodology to assess the current AS-IS.

Table 2. Structure of the interview sessions

Step	Activity	Participants	Duration
1	Corporate strategy: objectives / main challenges / ongoing and future projects	Top management and IT manager	3 h
1	Plant operational strategies: objectives / main challenges / current and future projects	Plant managers	2,5 h
2	Factory tour + Interviews to key process managers	Key process managers	3 days (1 each plant)
3	Workshop to identify and map main current criticalities	Key process managers	3 days (1 each plant)
4	Identification and Prioritization of improvement projects	Company referees	1 day
5	Digital transformation plan discussion	Top management	4 h

Through the separated interviews, it was possible to highlight the presence of common criticalities that are reported in Table 3. Indeed, the criticalities sometimes resulted to be specifically linked to the inefficiencies of a single plant, but, in other cases, these were shared among the plants due to some transversal similarities. Based on the PCIM, using the prioritization scale reported in Fig. 2, the common criticalities were considered the most impactful and thus were used to set the basis for the unified roadmap. The list of the most impacting criticalities and the area of responsibility are reported in Table 3.

Table 3. Criticalities and their prioritization

Area	Criticality	PCIM Prioritization			Type
		Plant 1	Plant 2	Plant 3	
Planning	Production orders entered by hand on management system	Not present	9	6	Tec
Planning	Ineffective scheduling carried out on Excel	Not present	9	9	Tec
Planning	Information on the availability of non-system personnel	Not present	6	6	Tec
Planning	Paper-based shared production order	6	Not present	3	Org/Tec

(continued)

Table 3. (*continued*)

Area	Criticality	PCIM Prioritization			Type
		Plant 1	Plant 2	Plant 3	
Production	Manual progress of production monitoring	6	Not present	3	Tec
Logistics	No storage policies are defined	9	6	9	Tec/Pro
Maintenance	Shortage of KPIs for process monitoring	9	6	6	Org/Tec
Maintenance	Unstructured interventions History	9	Not present	6	Proc/Tec
Maintenance	No on-call assignment policy	6	3	Not present	Proc
Maintenance	Limited data usage	6	Not present	6	Tec

As shown in Table 3 and reported in the methodology section, the criticalities can be reconducted to three main types: process, organizational or technological. Numerous criticalities are connected to processes implemented over time to overcome the inadequacies of the IT systems. Other criticalities are related to the activities that, although might be automated, are currently carried out manually because of lack of proper technologies and/or lack of competencies. This issue increases inefficiency and forces employees to perform non-value-added activities. The lack of adequate digital supports creates problems such as that relevant information is communicated verbally or via paper-based documents, fostering possible mistakes due to information loss. Moreover, all the analyzed plants appeared to not be able to punctually collect and analyze all relevant data related to certain company's activities, such as production, maintenance, and logistics. These inefficiencies are identifiable from the most common criticalities shared between plants (collected in Table 3).

The production planning areas operate on IT systems inadequate to plants' needs. To overcome IT limits, the production orders are inputted manually on the ERP and are then shared on paper-based documents. Moreover, in those plants where the planning software is missing (i.e., plant 2 and plant 3), the production plan is redacted on an Excel file.

Due to IT infrastructure's limitations, the production area is not able to monitor in real time the production's progress, which has to be updated manually.

Similar criticalities can be identified in the maintenance area: few KPIs are collected and monitored, and the record of maintenance activities is not structured, preventing the managers to conduct punctual analysis.

In the logistic area, storage criteria are not defined and those which are defined are not shared coherently among all people operating in that area.

Despite the differences between single plants, it was possible to observe that distinct criticalities could be addressed by shared solutions. Indeed, specific solutions could be reconnected to transversal projects, shared by all plants. Improvement projects have been identified and prioritized together with the continuous improvement team and, thanks to this effort, the strategic unified digital roadmap has been developed and shared with the top management to implement the prioritized projects in a conscious and structured manner.

In Fig. 3 is shown the unified roadmap covering the criticalities emerged during the assessment. The projects focus both on processes' and IT architecture's improvements. The roadmap is structured along a timeline, and the first projects aim at solving processes' inefficiencies to facilitate the adoption and implementation of the adequate IT systems. Taking into consideration the relevance of human resources in the digital transition, certain projects focus on fostering, by mapping and improving the skills' baseline of people, the appropriate digital culture required to operate in the new system. In accordance with the review of the internal processes, it is suggested to identify the requirements of the new IT systems needed to answer the needs of processes and plants. The projects related to IT infrastructure aim at reinforcing the digital backbone of the firm, proposing adequate tools shared between plants, and at fostering the adoption of industry 4.0 standards and technologies at group level, creating an infrastructure for data collection, analysis and usage.

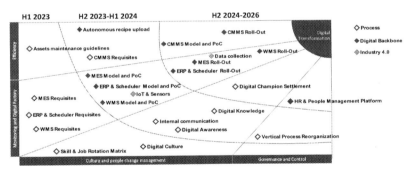

Fig. 3. Proposed Roadmapping output

5   Discussions

This contribution highlights the possibility to propose a unified roadmap based on the the assessment of multiple plants across several functions. The methodology enables to overcome relevant criticalities, either plant-specific or transversal between all the plants. Therefore, this research has twofold implications: both theoretical and practical. From a theoretical perspective, the presented roadmapping methodology shows the possibility to move from the single-plant view, on which most of the TRM studies are focused, to achieve an approach aiming at integrating both functions and plants within a single roadmap. The research, hence, highlights how a cross-plant and cross-function approach can answer the requirements of MNC companies, which desire to achieve a higher digitalization level across multiple plants in the same timeframe and in a structured manner.

From a practical perspective, the methodology supports C-levels in the appropriate allocation of resources dedicated to the processes' digitalization. By analyzing at the same time different plants, with different characteristics in terms of culture, processes, and IT infrastructure; the proposed methodology allows the identification of shared criticalities, impacting of the overall inefficiencies of the group, and the development of a common roadmap. The proposal of a unified roadmap guides the company in the definition and selection of transversal projects, optimizing efforts and invested resources (both economic and human), reaching in multiple plants similar digitalization levels. Moreover, it facilitates the implementation of homogeneous processes and IT architecture, and the transfer of best practices among the different plants. Nonetheless, the methodology applied has limitations: it has been applied to a single company case study and future research should validate the procedure in other manufacturing industries. Moreover, step 6, "control", has not being implemented. The boundaries of the case study were temporarily limited, and the outcomes' monitoring activity was not included. Future research should apply to a broader timeframe.

6 Conclusions

The objective of this research was to propose a roadmapping methodology to guide multi-plant companies in the creation of a unified roadmap, shared between multiple plants with distinct characteristics. Indeed, existing literature mostly focuses on TRM with a single plant-view, while the authors observed the need of overcoming this silos approach to support MNCs in creating a unique roadmap shared between multiple plants. The authors proposed a cross-plant cross-function methodology and applied it to an industrial case in the F&B sector. The contribution illustrates that the methodology can answer multi-plant companies' requirements: assessing distinct criticalities, identifying improvement projects, and proposing an unified roadmap, shared between multiple plants and functions. The research presents limitations such as the application to a single case study and the lack of a control phase. Future research could investigate whether the proposed methodology is effective on companies in different sectors, presenting a higher number of plants or with higher diversity between the plants themselves.

References

1. Olsen, T.L., Tomlin, B.: Industry 4.0: opportunities and challenges for operations management. Manuf. Serv. Oper. Manag. **22**(1), 113–122 (2020). https://doi.org/10.1287/msom.2019.0796
2. Demir, Y., Istanbullu Dincer, F.: The effects of Industry 4.0 on the food and beverage industry. J. Tourismol. 133–145, August 2020. https://doi.org/10.26650/jot.2020.6.1.0006
3. Mirabella, N., Castellani, V., Sala, S.: Current options for the valorization of food manufacturing waste: a review. J. Clean. Prod. **65**, 28–41 (2014). https://doi.org/10.1016/j.jclepro.2013.10.051
4. Bai, C., Dallasega, P., Orzes, G., Sarkis, J.: Industry 4.0 technologies assessment: a sustainability perspective. Int. J. Prod. Econ. **229**, November 2020. https://doi.org/10.1016/j.ijpe.2020.107776

5. Szász, L., Demeter, K., Rácz, B.G., Losonci, D.: Industry 4.0: a review and analysis of contingency and performance effects. J. Manuf. Technol. Manag. **32**(3), 667–694 (2021). https://doi.org/10.1108/JMTM-10-2019-0371
6. Schmitt, P., Schmitt, J., Engelmann, B.: Evaluation of proceedings for SMEs to conduct I4.0 projects. Proc. CIRP **86**, 257–263 (2020). https://doi.org/10.1016/j.procir.2020.01.007
7. Acerbi, F., Spaltini, M., De Carolis, A., Taisch, M.: Developing a roadmap towards the digital transformation of small & medium companies: a case study analysis in the aerospace & defence sector. In: Noël, F., Nyffenegger, F., Rivest, L., Bouras, A. (eds.) Product Lifecycle Management. PLM in Transition Times: The Place of Humans and Transformative Technologies. PLM 2022. IFIP Advances in Information and Communication Technology, vol. 667. Springer, Cham (2023). https://doi.org/10.1007/978-3-031-25182-5_28
8. Vinayavekhin, S., Phaal, R., Thanamaitreejit, T., Asatani, K.: Emerging trends in roadmapping research: a bibliometric literature review. Technol. Anal. Strateg. Manag, September 2021. https://doi.org/10.1080/09537325.2021.1979210
9. Willyard, C.H., McClees, C.W.: Motorola's technology roadmap process. Res. Manage. **30**(5), 13–19 (1987). https://doi.org/10.1080/00345334.1987.11757057
10. Carvalho, M.M., Fleury, A., Lopes, A.P.: An overview of the literature on technology roadmapping (TRM): contributions and trends. Technol. Forecast Soc. Change **80**(7), 1418–1437 (2013). https://doi.org/10.1016/j.techfore.2012.11.008
11. Park, H., Phaal, R., Ho, J.Y., O'Sullivan, E.: Twenty years of technology and strategic roadmapping research: a school of thought perspective. Technol. Forecast Soc. Change **154**, 119965, February 2020. https://doi.org/10.1016/j.techfore.2020.119965
12. Kerr, C., Phaal, R., Probert, D.: Cogitate, articulate, communicate: the psychosocial reality of technology roadmapping and roadmaps. R and D Manag. **42**(1), 1–13 (2012). https://doi.org/10.1111/j.1467-9310.2011.00658.x
13. Bellantuono, N., Nuzzi, A., Pontrandolfo, P., Scozzi, B.: Digital transformation models for the i4.0 transition: lessons from the change management literature. Sustainability (Switzerland), **13**(23) 2021. https://doi.org/10.3390/su132312941
14. Cotrino, A., Sebasti, M.A., González-Gaya, C.: Industry 4.0 roadmap : implementation for small and medium-sized enterprises. Appl. Sci. (Switzerland) (2020)
15. Hansen, C., Daim, T., Ernst, H., Herstatt, C.: The future of rail automation: a scenario-based technology roadmap for the rail automation market. Technol. Forecast Soc. Change **110**, 196–212 (2016). https://doi.org/10.1016/j.techfore.2015.12.017
16. Spaltini, M., Acerbi, F., Pinzone, M., Gusmeroli, S., Taisch, M.: Defining the roadmap towards industry 4.0 : the 6Ps maturity model for manufacturing SMEs. In: 29th CIRP Life Cycle Engineering Conference (2022)
17. Holmes, C., Ferrill, M.: The application of operation and technology roadmapping to aid Singaporean SMEs identify and select emerging technologies. Technol. Forecast Soc. Change **72**(3) SPEC. ISS., 349–357 (2005). https://doi.org/10.1016/j.techfore.2004.08.010
18. Bildosola, I., Río-Bélver, R.M., Garechana, G., Cilleruelo, E.: TeknoRoadmap, an approach for depicting emerging technologies. Technol. Forecast Soc. Change **117**, 25–37 (2017). https://doi.org/10.1016/j.techfore.2017.01.015
19. International Energy Agency, Technology Roadmap: A guide to development and implementation (2014)
20. Caetano, M., Amaral, D.C.: Roadmapping for technology push and partnership: a contribution for open innovation environments. Technovation **31**(7), 320–335 (2011). https://doi.org/10.1016/j.technovation.2011.01.005
21. Schallmo, D., Williams, C.A., Boardman, L.: Digital transformation of business models-best practice, enablers, and roadmap. Int. J. Innov. Manag. **21**(8), 1–17 (2017). https://doi.org/10.1142/S136391961740014X

22. Harmon, R.R., Demirkan, H., Raffo, D.: Roadmapping the next wave of sustainable IT. Foresight **14**(2), 121–138 (2012). https://doi.org/10.1108/14636681211222401
23. Donnelly, K., Beckett-Furnell, Z., Traeger, S., Okrasinski, T., Holman, S.: Eco-design implemented through a product-based environmental management system. J. Clean. Prod. **14**(15–16), 1357–1367 (2006). https://doi.org/10.1016/j.jclepro.2005.11.029
24. Benedict, N., et al.: Blended simulation progress testing for assessment of practice readiness. Am. J. Pharm. Educ. **81**(1), 14 (2017). https://doi.org/10.5688/ajpe81114
25. De Carolis, A., Macchi, M., Negri, E., Terzi, S.: A maturity model for assessing the digital readiness of manufacturing companies. IFIP Adv. Inf. Commun. Technol. **513**, 13–20 (2017). https://doi.org/10.1007/978-3-319-66923-6_2
26. de Alcantara, D.P., Martens, M.L.: Technology Roadmapping (TRM): a systematic review of the literature focusing on models. Technol. Forecast Soc. Change **138**, 127–138 (2019), July 2018. https://doi.org/10.1016/j.techfore.2018.08.014

A Preliminary Reflection Framework of Sustainability, Smart Cities, and Digital Transformation with Effects on Urban Planning: A Review and Bibliometric Analysis

Andreia de Castro e Silva[✉] ⓘ, Elpidio Oscar Benitez Nara ⓘ,
Marcelo Carneiro Gonçalves ⓘ, Izamara Cristina Palheta Dias ⓘ,
Camila Vitoria Piovesan ⓘ, and Gabrielly dos Santos Domingos ⓘ

Industrial and Systems Engineering Post-Graduation Program, Pontifical Catholic University of Parana (PUCPR), Curitiba 80215-901, Brazil
castro.andreia@pucpr.edu.br

Abstract. This study aims to identify high-impact articles and citations involving urban planning based on a systemic approach to sustainable practices, digital transformation, and smart cities to propose a preliminary construct to support the sustainable performance of urban planning focusing on the Brazilian reality. In this context, a theoretical essay was carried out based on the systematic review of current articles on the database Scopus. To achieve the objective, firstly, a systematic literature review (RSL) was carried out using the Biblioshiny tool developed in the R software. Then, an analysis of descriptive statistics of the main bibliographic metrics was presented. A Sankey Diagram was generated relating different bibliographic factors (countries, journals, and keywords). Furthermore, a word cloud and a Co-citation network were proposed through bibliographic coupling analysis using simple centrality algorithms. Finally, a study was carried out on a sample of 86 scientific articles to identify the relationship between the topics of interest. Therefore, it requires a version of a systemic approach of parameters and criteria commonly accepted in its dimensions, seeking to broaden and deepen the themes once the analysis made it possible to identify a relative scarcity of scientific research on the subject and its dissemination in Brazil. As a result, in addition to RSL, it was possible to propose a preliminary construct to support the sustainable performance of urban planning considering a systemic approach.

Keywords: Sustainable practices · Smart cities · Digital Transformation · Urban Planning

1 Introduction

The impacts caused by rapid urbanization demand that sustainable practices be incorporated into urban planning. Lessons from past and future challenges can help to identify practices for each situation, also considering the Triple Bottom Line, formed by the economic, social, and environmental dimensions. Planning and make sustainable strategic

© IFIP International Federation for Information Processing 2024
Published by Springer Nature Switzerland AG 2024
C. Danjou et al. (Eds.): PLM 2023, IFIP AICT 702, pp. 107–118, 2024.
https://doi.org/10.1007/978-3-031-62582-4_10

decisions to motivate the transformation of cities into more friendly or intelligent environments. The smart city as a concept provides a potential approach to finding solutions to the various sustainability issues arising from rapid urbanization [1].

Smart City is a theme that involves many areas of research converges and is considered a social and economic phenomenon driven by environmental and human well-being issues [2]. The smart city paradigm is strictly connected to sustainability aspects, and local context influences the implementation of smart cities [3, 4]. As depicted in the Sustainable Development Goal 11 of the United Nations is important that of building safer, more resilient, sustainable, and inclusive cities [5]. This will necessitate a step change in performance goals and tangible solutions [6]. The smart city industry plays a key role not only in the sustainable city, but also in the growth of the national economy, the reduction of environmental impact of urban activities, the optimized management of energy resources, and the design of innovative services and solutions for citizens [4, 7]. Is a concept that has multiple definitions that vary based on the elements that a city needs to be seen as smart, the resources it needs, its characteristics, objectives, purpose, and scope [8].

A large volume of relevant work has been published in different directions, proposing solutions, services, frameworks, and applications based on these technologies. The bibliometric analysis indicates that Smart Cities are emerging as a fast-growing topic of scientific inquiry, and much of the knowledge generated about them is uniquely technological [2]. But in the analysis of the available systematic reviews of the literature, the gap was identified that the studies do not pay attention to the links between the concepts of sustainability, smart cities, digital transformation and urban planning.

In this context, the aim of the study is a theoretical essay was carried out, based on the systematic review of current articles on the database Scopus. The study found in a preliminary way that the proper functioning and the effects and impacts between the variables are much more complex. Therefore, it requires a vision of a systemic approach of parameters and criteria that are commonly accepted in its dimensions, seeking to broaden and deepen the themes, since public policies actions, and initiatives are still very incipient. The analysis made it possible identify a relative scarcity of scientific research on the subject and its dissemination in Brazil.

2 Background

It is not today that there is a need for an urgent break, facing the challenges of sustainability, because we need, according to [9], to understand that the breach is in the relationships, in the way of thinking, and not only in the technology. That without natural resources, business will not survive, and that we need to get out of our comfort zone and understand that ethical behavior brings economic gains. Aware that this cannot be done just for financial reasons, but for the awareness of man's existence and survival. Research must seek solutions or innovations for the problems and needs of the current society [10].

Understanding that there is still much to do realize that the environment is not something that serves only to exploit and generate wealth. Therefore, organizations implement various strategies, according to the interests of their stakeholders and best practices to make their processes environmentally efficient and socially and economically viable [11].

The aim of sustainable urban development is to increase the quality of life of citizens through sustainability-oriented innovations (SOIs) [1]. All in all, a connection between the smart city concept, sustainability and urban development exists on several levels, and there is a need to further increase our understanding of the linkages between the various components of sustainability-oriented innovation in this context to enable systemic urban development in cities using digital transformation. It is needed to interact directly with the community where it operates, through products or services of any nature [12].

The socio-economic impacts on cities caused by rapid urbanization demand that sustainable practices be incorporated into urban planning [5] and develop management knowledge to start dealing with internal and external uncertainties [13]. Transformative changes are required for a 21st century sustainable urban planning transition involving multiple interconnected domains of energy, water, transport, waste, and housing and capability based on learning allows to adapt to changes in the economic and social context and technology [14, 15]. According to [16] Smart city technologies have recently become the subject of extensive research and development in the literature.

The concept of "smart city" has several definitions: knowledge city, sustainable city, and digital city. Until the 1990s, "digital cities" was the most used term, today, the most frequent is "smart cities" [17]. It is a recent concept, but it is already consolidated in the debates about sustainable development and technological solutions. According to the European union, Smart Cities are systems and people interacting and using energy, materials, services, and finance to catalyze economic development and improve quality of life. A characteristic situation of these process changes in organizations is quite visible [18]. [19] also talks about the systemic effect of processes on the sustainability of institutions.

These interaction flows are considered smart because they make strategic use of infrastructure and services and of information and communication with urban planning and management to respond to the social and economic needs of society. [20] says that a smart city has some characteristics that differentiate it from others, such as: (i) the use of network infrastructure to improve economic and political efficiency and enable social, cultural and urban development [21]; (ii) it has an underlying emphasis on business-led urban development [21]; (iii) a strong focus on the goal of achieving social inclusion of various urban residents in public services [22]; (iv) an emphasis on the crucial role of high-tech and creative industries in long-term urban growth [22]; (v) deep attention to the role of social and relational capital in urban development [22]; and (vi) social and environmental sustainability as an important strategic component [22].

With innovative technologies, companies must keep up with rapid technological evolution and offer innovative and increasingly sustainable products, ensuring a competitive advantage [23]. Digital transformation comprises more than the set of technologies usually addressed in the literature [23]. Can boost sustainable development strategies, providing opportunities to accelerate change [24]. In timeliness, the companies also face internal and external pressures on sustainable development practices [25].

The digital transformation process of smart cities has different aspects [26] and other publications highlight the digital components and architectures of smart cities [27, 28].

The digital transformation of cities is ecosystem specific [29, 30] and its pace varies according to the ecosystem: it evolves quickly in energy and transport and is slower in the housing and traditional industry sectors [31].

The high variability of options in the digital transformation process enforces a higher complexity level in configuring and setting up objectives and goals based on cities' needs; hence, a systematic approach is required to assist decision-makers in better and more sustainable transformation [32]. Additionally, enterprise architecture fosters digitalization towards achieving system alignment and data integration in cities to support the urban environment as they digitally transform services provided to citizens [31].

3   Method

A systematic review of the literature starts by defining appropriate keywords and the appropriate database for the research topic [33]. For this study, the Scopus database was used as a source, as it is considered a reliable database by many researchers [33–35]. The chosen keywords were used in the sequence shown in (Fig. 1), with the strategy of starting with words that generated the most comprehensive results to the least comprehensive ones.

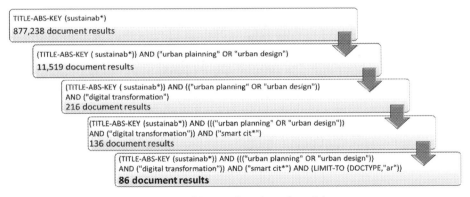

TITLE-ABS-KEY {sustainab*}
877,238 document results

{TITLE-ABS-KEY (sustainab*)} AND ("urban plainning" OR "urban design")
11,519 document results

{TITLE-ABS-KEY (sustainab*)} AND (("urban planning" OR "urban design")}
AND ("digital transformation")
216 document results

{TITLE-ABS-KEY {sustainab*}} AND ({("urban planning" OR "urban design"}}
AND ("digital transformation"}} AND ("smart cit*"}
136 document results

{TITLE-ABS-KEY {sustainab*}} AND ({("urban planning" OR "urban design"}}
AND ("digital transformation"}} AND ("smart cit*"} AND {LIMIT-TO {DOCTYPE,"ar"}}
86 document results

Fig. 1. Sequence of keywords and number of documents.

Bibliometric analysis techniques are based on a quantitative approach designed to identify, describe, and evaluate published research [36]. The use of transparent and reproducible search and review processes increases the reliability of the results and reduces the subjective bias of literature reviews [37, 38]. Bibliometrics can reliably connect publications, authors, or journals, identify research sub streams, and produce maps of published research [39]. For data visualization and analysis, the Bibliometrix package in R was used. Table 1 presents the main data of the sample of articles.

In Table 1, it is possible to identify that the database has 86 articles, distributed among 49 peer-reviewed journals. Ten articles were published by a single author. The sample had a total of 345 authors, which generates an index of authors per article of 4,05, obtained by dividing the number of authors (345) by the total number of articles in the sample (86).

Table 1. Main information about data.

Description	Results
Period	2016:2023
Sources (Journals, Books, etc.)	49
Documents	86
Annual Growth Rate %	16,99
Document Average Age	1,84
Average citations per doc	12,84
References	9026
DOCUMENT CONTENTS	
Keywords Plus (ID)	502
Author's Keywords (DE)	391
AUTHORS	
Authors	335
Authors of single-authored docs	10
AUTHORS COLLABORATION	
Single-authored docs	10
Co-Authors per Doc	4,05
International co-authorships %	33,72
DOCUMENT TYPES	
Article	86

4 Observations and Analysis

The following section presents the metadata analysis and insights, has done based on 86 of current articles on the database Scopus.

4.1 Metadata Analysis

This section presents the descriptive statistics based on the metadata of 86 papers. The metadata analysis contained publication of 86 papers by years, institutions, countries, journals, keywords, and co-citation network.

Publications by Year

Figure 2 shows that publications have grown since 2020. Moreover, the trend line also indicates an increasing pattern, which implies that the literature on Smart Cities and other keywords applied in this study is still growing.

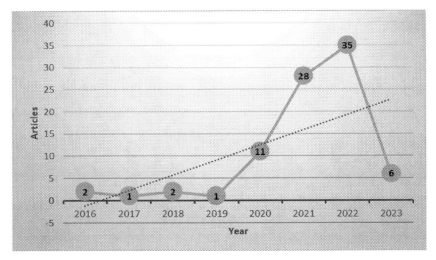

Fig. 2. Articles by year.

For example, in the year 2021, 28 papers were published, which is significantly the highest number of documents as compared to previous years. This concludes that there are increasing concerns and interests in the Smart Cities topic, parallel with sustainability issues, digital transformation, and urban planning.

Publications by Institutions

Figure 3 illustrates Smart Cities and other keywords publications by authors' affiliations, considering the top ten list of the institutes The figure shows that Aristotle University of Thessaloniki in Greece, published the highest number of papers in the Smart Cities literature.

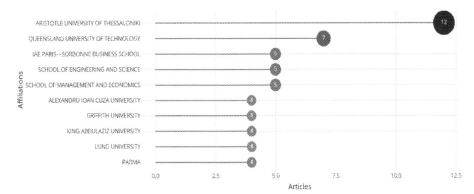

Fig. 3. Articles by affiliations.

Sankey Diagram – Countries x, Key Words, and Journal

The Sankey diagram (Fig. 4) is considered an advanced graphical way to display bibliometric data of the characteristics of related articles [40, 41], and presents the proportions of three analyzed topics [42].

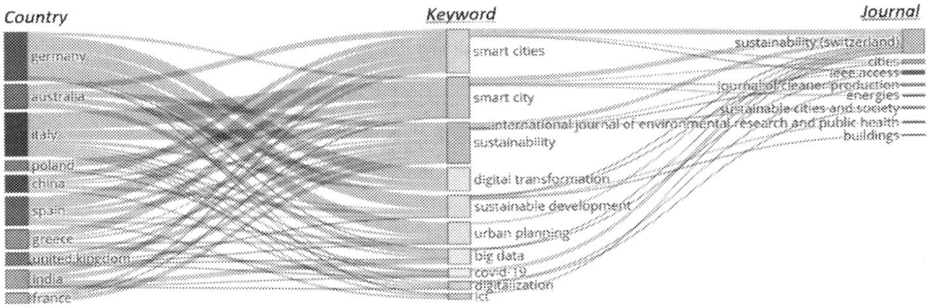

Fig. 4. Sankey Diagram.

From this diagram, a 3-field graph was generated among the 10 most important countries, 10 most important keywords, and 10 most important journals. The figure demonstrates that the Germany has the largest proportion of publications on the theme "digital transformation", with 16 works published in the sample. The "smart cities" and "smart city" themes are cited in all 10 countries and the journal that publishes the most on the themes is "Sustainability (Switzerland)".

Most Common Keywords Used

It is found that the most common word used in keywords is Smart City, showing 30 times, followed by 'sustainability,' 'urban development,' 'urban planning,' and so on. Figure 5 represents the word cloud derived from the software "*Biblioshine*", highlighting the most common words in bigger fonts while other relatively fewer common words appear in smaller fonts. This word cloud is straightforward for identifying common words in a complex environment [43]. Thus, it can identify the most common theme and keywords used in publications.

Contribution by Co-citation Network

Academic research depends on prospecting to retrieve the most relevant research studies and establish links with authors from key international research groups [44]. The existence of co-citation of 2 articles occurs when these 2 articles are cited in a third one [45].

Figure 6 shows the clusters identified by different colors, composed of four groups of articles and 33 nodes. The selected layout was Multi-Dimensional Scaling, on which the more significant the dimension of the circle, the greater the frequency of co-cited reports. Louvain's heuristic algorithm was used to make the grouping seeking to maximize the modularity in the network. This clustering method is agglomerative if the input is a weighted network with "N" vertices [46].

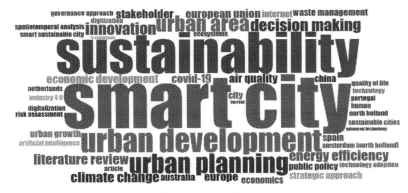

Fig. 5. Word Cloud.

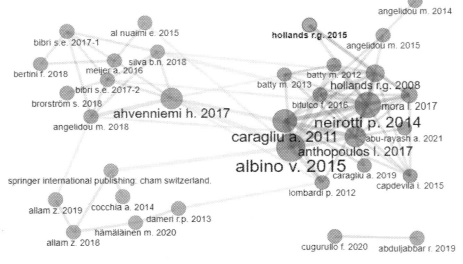

Fig. 6. Co-citation Network.

In the picture is easy viewings to notice the cluster with the highest frequency of co-cited articles is identified by the red color (upper right side), composed of 16 documents, followed by the purple collection (upper left side), with nine articles. The two smaller ones are written of six items in green color (bottom left) and two pieces in blue color (base right), respectively.

A Conceptual Framework for Smart Cities

After analyzing the existing literature on the possible relationships between the theme's sustainability, smart cities, digital transformation, and urban planning, it is possible to explore and deepen studies on the articles and the potential relationships between the variables.

In this sense, Fig. 7 presents the proposal of a conceptual framework for analyzing the variables using a systemic approach to study the possible relationships and interactions that may or may not happen between the variables used in the smart city's context. Finally, we identify some hypotheses (H1 and H2) and suggest that they be evaluated in future studies.

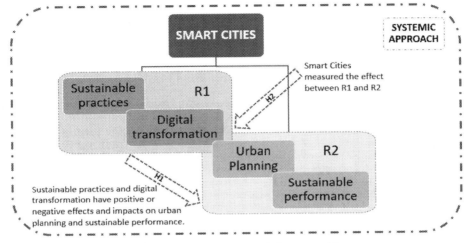

Fig. 7. Conceptual framework for smart cities.

H1: Sustainable practices and digital transformation have positive or negative effects and impacts on urban planning and sustainable performance.

H2: Smart Cities measured the effect between relation R1 (Sustainable practices and digital transformation) and relation R2 (urban planning and sustainable performance).

5   Conclusions and Future Research Potentials

The paper systematically reviews the extant literature and presents a descriptive analysis based on metadata analysis as well as offers insights based on content analysis. The data has been collected from remedies databases. Metadata analysis reveals influential authors, popular journals, publications by year, and top contributing countries and institutions.

This study tells that Aristotle University of Thessaloniki in Greece is the most published institution about Smart Cities. Moreover, Sustainability was the most popular journal regarding its: impact and the number of papers it publishes in this area. The study also reveals that Germany and Italia dominate this discipline in many publications. Additionally, the study finds that the business management discipline holds a large share of the literature on Smart Cities.

One limitation of this paper is that using the Scopus database only. Scopus is an extensive database of management and science journals [34]. However, the collection does not contain all peer-reviewed articles; as a result, a few important papers may have been missed.

In addition to the literature review, we proposed a conceptual framework based on the research themes and identified some hypotheses that can be tested in future studies, intending to investigate the possible relationships between the variables that were part of the literature analysis.

Acknowledgments. The authors would like to thank by research financing support the Industrial and Systems Engineering Graduate Program at Pontifical Catholic University of Parana (PPGEPS/PUCPR), The National Council for Scientific and Technological Development – Brazil (CNPq) and the Coordenação de Aperfeiçoamento de Pessoal de Nível Superior - Brazil (CAPES) - Finance Code 001.

References

1. Tura, N., Ojanen, V: Sustainability-oriented innovations in smart cities: a systematic review and emerging themes. Cities 103716 (2022)
2. Pérez, L.M., Oltra-Badenes, R., Oltra Gutiérrez, J.V., Gil-Gómez, H.: A bibliometric diagnosis and analysis about smart cities. Sustainability **12**(16), 6357 (2020)
3. Aina, Y.A.: Achieving smart sustainable cities with GeoICT support: the Saudi evolving smart cities. Cities **71**, 49–58 (2017)
4. Belli, L., et al.: IoT-enabled smart sustainable cities: challenges and approaches. Smart Cities **3**(3), 1039–1071 (2020)
5. Moreno, C., Allam, Z., Chabaud, D., Gall, C., Pratlong, F.: Introducing the "15-Minute City": Sustainability, resilience and place identity in future post-pandemic cities. Smart Cities **4**(1), 93–111 (2021)
6. Newton, P., Frantzeskaki, N.: Creating a national urban research and development platform for advancing urban experimentation. Sustainability **13**(2), 530 (2021)
7. Kim, K., Jung, J.K., Choi, J.Y.: Impact of the smart city industry on the Korean national economy: input-output analysis. Sustainability **8**(7), 649 (2016)
8. Toli, A.M., Murtagh, N.: The concept of sustainability in smart city definitions. Front. Built Environ. **6**, 77 (2020)
9. Baierle, I.C., Siluk, J.C.M., Gerhardt, V.J., Michelin, C.D.F., Junior, Á.L.N., Nara, E.O.B.: Worldwide innovation and technology environments: research and future trends involving open innovation. J. Open Innov. Technol. Market, and Complex. **7**(4), 229 (2021)
10. Aydin, R., Badurdeen, F.: Sustainable product line design considering a multi-lifecycle approach. Resour. Conserv. Recycl. **149**, 727–737 (2019)
11. Moraes, J.d., Schaefer, J.L., Schreiber, J.N.C., Thomas, J.D., Nara, E.O.B.: Algorithm applied: attracting MSEs to business associations. J. Bus. Indust. Market. **35**(1), 13–22 (2020). https://doi.org/ez433.periodicos.capes.gov.br/10.1108/JBIM-09-2018-0269
12. Uemura Reche, A.Y., Canciglieri, O., Estorilio, C.C.A., Rudek, M.: Green supply chain management and the contribution to product development process. Int. Bus. Trade Inst. Sustain. 781–793 (2020)
13. Ivascu, L.: Measuring the implications of sustainable manufacturing in the context of industry 4.0. Processes **8**(5) 585 (2020)
14. Ferreira, P.G.S., Lima, E.P., Costa, S.E.G., Monteiro, N.J., Castro e Silva, A.D.: Key performance measurement capabilities for managing distributed teams. Total Qual. Manag. Bus. Excellence, 1–25 (2022)

15. Megahed, N.A., Abdel-Kader, R.F.: Smart Cities after COVID-19: building a conceptual framework through a multidisciplinary perspective. Sci. Afr. **17**, e 01374 (2022)

16. João, B.N., Souza, C.L., Serralvo, F.A.: A systematic review of smart cities and the internet of things as a research topic. Cadernos EBAPE. BR **17**, 1115–1130 (2019)

17. Schaefer, J.L., Baierle, I.C., Sellitto, M.A., Siluk, J.C.M., Furtado, J.C., Nara, E.O.B.: Competitiveness scale as a basis for Brazilian small and medium-sized enterprises. Eng. Manag. J. **33**(4), 255–271 (2021)

18. Storch, L.A., Nara, E.O.B., Kipper, L.M.: The use of process management based on a systemic approach. Int. J. Product. Perform. Manag. **62**(7), 758–773 (2013)

19. Lazzaretti, K., Sehnem, S., Bencke, F.F., Machado, H.P.V.: Smart Cities: Insights and Contributions from Brazilian Research. Urbe. Revista Brasileira de Gestão Urbana, 11 (2019)

20. Hollands, R.G.: Will the real smart city please stand up? City: Anal. Urban Trends Cult. Theory Policy Action **12**(3), 303–320 (2008)

21. Caragliu, A., Del Bo, C., Nijkamp, P.: Smart cities in Europe. J. Urban Technol. **18**(2), 65–82 (2011)

22. Castro e Silva, A.D., et al.: Evaluation and choice criteria of sustainable suppliers in the construction industry: a comparative study in Brazilian companies. Sustainability **14**(23) 1–17 (2022)

23. Lerman, L.V., Benitez, G.B., Müller, J.M., de Sousa, P.R., Frank, A.G.: Smart green supply chain management: a configurational approach to enhance green performance through digital transformation. Supply Chain Manag. **27**(7), 147–176 (2022)

24. Reisdorfer-Leite, B., Marcos de Oliveira, M., Rudek, M., Szejka, A.L., Canciglieri Junior, O.: Startup definition proposal using product lifecycle management. In: Nyffenegger, F., Ríos, J., Rivest, L., Bouras, A. (eds) Product Lifecycle Management Enabling Smart X. PLM 2020. IFIP Advances in Information and Communication Technology, vol. 594. Springer, Cham (2020). https://doi.org/10.1007/978-3-030-62807-9_34

25. Muñoz, L.A., Bolívar, M.P.R.: Different levels of smart and sustainable cities construction using e-participation tools in European and Central Asian countries. Sustainability **13**(6), 3561 (2021)

26. Mora, L., Bolici, R., Deakin, M.: The first two decades of smart-city research: a bibliometric analysis. J. Urban Technol. **24**(1), 3–27 (2017)

27. Stratigea, A., Papadopoulou, C.A., Panagiotopoulou, M.: Tools and technologies for planning the development of smart cities. J. Urban Technol. **22**(2), 43–62 (2015)

28. Angelidou, M.: The role of smart city characteristics in the plans of fifteen cities. J. Urban Technol. **24**(4), 3–28 (2017)

29. Komninos, N., Tsarchopoulos, P.: Toward intelligent Thessaloniki: from an agglomeration of apps to smart districts. J. Knowl. Econ. **4**(2), 149–168 (2013)

30. Abella, A., Ortiz-de-Urbina-Criado, M., De-Pablos-Heredero, C.: Information reuse in smart cities' ecosystems. Profesional de la información **24**(6), 838–844 (2015)

31. Anthony Jnr, B., Abbas Petersen, S., Helfert, M., Guo, H.: Digital transformation with enterprise architecture for smarter cities: a qualitative research approach. Dig. Policy Regul. Govern. **23**(4), 355–376 (2021)

32. Inac, H., Oztemel, E.: An assessment framework for the transformation of mobility 4.0 in smart cities. Systems **10**(1), 1 (2021)

33. Malviya, R.K., Kant, R.: Green supply chain management (GSCM): a structured literature review and research implications. Benchmark Int. J. **22**(7), 1360–1394 (2015)

34. Tronnebati, I., El Yadari, M., Jawab, F.: A Review of green supplier evaluation and selection issues using MCDM MP and AI Models. Sustainability **14**, 16714 (2022)

35. Fahimnia, B., Sarkis, J., Davarzani, H.: Green supply chain management: a review and bibliometric analysis. Int. J. Prod. Econ. **162**, 101–114 (2015)

36. Bretas, V.P.G., Alon, I.: Franchising research on emerging markets: bibliometric and content analyses. J. Bus. Res. **133**, 51–65 (2021)

37. Maditati, D.R., Munim, Z.H., Schramm, H.J., Kummer, S.: A review of green supply chain management: from bibliometric analysis to a conceptual framework and future research directions. Resour. Conserv. Recycl. **139**, 150–162 (2018)

38. Apriliyanti, I.D., Alon, I.: Bibliometric analysis of absorptive capacity. Int. Bus. Rev. **26**(5), 896–907 (2017)

39. Zupic, I.. Čater, T.: Bibliometric methods in management and organization. Organ. Res. Methods **18**(3), 429–472 (2015)

40. Li, M.-J. Chien, T.-W., Liao, K.-W., Lai, F.-J.: Using the Sankey diagram to visualize article features on the topics of whole-exome sequencing (WES) and whole-genome sequencing (WGS) since 2012: bibliometric analysis. Medicine **101**(38), 30682 (2022)

41. Yang, T.-Y., Chien, T.-W., Lai, F.-J.: Citation analysis of the 100 top-cited articles on the topic of hidradenitis suppurativa since 2013 using Sankey diagrams: bibliometric analysis. Medicine **101**(44), 31144 (2022)

42. Riehmann, P., Hanfler, M., Froehlich, B.: Interactive sankey diagrams. In: Conference paper in IEEE Xplore, Germany (2005)

43. Birko, S., Dove, E.S., Ozdemir, V.: A delphi technology foresight study: mapping social construction of scientific evidence on metagenomics tests for water safety. PLoS ONE **10**(6), 0129706 (2015)

44. Nara, E.O.B., Schaefer, J.L., de Moraes, J., Tedesco, L.P.C., Furtado, J.C., Baierle, I.C.: Sourcing research papers on small - and medium-sized enterprises' competitiveness: an approach based on authors' networks. Revista Española de Documentación Científica **42**(2), e230 (2019)

45. Aria, M., Cuccurullo, C.: Bibliometrix: an R-tool for comprehensive science mapping analysis. J. Informet. **11**(4), 959–975 (2017)

46. Aires, V., Nakamura, F.: Aplicação de Medidas de Centralidade ao Método Louvain para Detecção de Comunidades em Redes Sociais. XLIX Simpósio Brasileiro de Pesquisa Operacional, pp. 1–8 (2017)

Development of a Human-Centric Knowledge Management Framework Through the Integration Between PLM and MES

Giovanni Marongiu$^{(\boxtimes)}$, Giulia Bruno , and Franco Lombardi

Politecnico di Torino, Corso Duca degli Abruzzi 24, 10129 Torino, Italy
giovanni.marongiu@polito.it

Abstract. Data management systems represent a powerful tool for collecting and managing real-time data in a company. One-of-a-Kind Production (OKP) firms, which leverage on their knowledge about past products and manufacturing processes to reduce lead time, are the ones which can potentially benefit the most from deploying systems like PLM and MES.

However, the benefits arising from data management systems are still limited due to the scarce integration among them. In particular, for OKP companies it is crucial to be able to collect and formalize the knowledge generated by operators during the manufacturing process, in order to constantly improve the design process. In this way, the iterations of trial and error can be reduced, saving time and costs.

This paper proposes a framework (i.e., a Knowledge Base System) for the integration of a PLM software used for product design and production cycle definition, and a MES software able to collect operators' feedbacks about manufacturing criticalities. The integration is realized by means of a centralized database, able to receive data from both systems and to relate them to generate useful information. Finally, a case study developed on a car prototyping company is presented to exemplify the advantages of systems integration.

Keywords: PLM · MES · Industry 4.0 · Knowledge Management · Human-centric production

1 Introduction

The many challenges faced by society at a global level, (i.e., the climate change, the Covid pandemic and Countries' political tensions) demonstrated the fragility of existing production models. As a consequence, the European Union deemed Industry 4.0's paradigm (I4.0) inadequate for the achievement of its strategic goals, including the 2030 Sustainable Development Goals [1]. In fact, despite being born to satisfy economic and ecological requirements, I4.0 evolved over time to become technology-focused and efficiency-driven, proving to be unsustainable in the long term from all critical points of view: social, economic, and environmental [2].

© IFIP International Federation for Information Processing 2024
Published by Springer Nature Switzerland AG 2024
C. Danjou et al. (Eds.): PLM 2023, IFIP AICT 702, pp. 119–129, 2024.
https://doi.org/10.1007/978-3-031-62582-4_11

For this reason, in 2021 the European Commission elaborated a new paradigm called Industry 5.0 (I5.0). I5.0 builds upon I4.0's technologies while stressing the role played by innovation and research in the achievement of a competitive sustainability [2]. The most important element introduced by this paradigm is the triple bottom line: resilient value creation, human well-being and sustainable society [3].

The European Commission also recognized the role of industry as pivotal to the concretization of each of these three dimensions [2]: industry needs to be sustainable in the long-term, resilient to external shocks and human-centered. This last definition means that, differently from I4.0, technology is now required to adapt to human needs and not the other way around. In fact, in the previous paradigm human and technological components were considered separately and almost in conflict with each other. Taking Zero Defect Manufacturing (ZDM) as an example, most approaches are currently based on the idea of removing and substituting human workers with technologies like machine learning and robots.

The underlying philosophy is that operators contribute to most of the mistakes made in manufacturing (up to 90% of the total) while robots are able to perform repetitive tasks with lower variability. However, recent studies [4] suggest that superior performances are achieved not when removing humans from the production lines, but rather when integrating and empowering them using technologies like cobots and Augmented Reality. This is a perfect example of the purpose of the new European policy: it is not aimed at proposing a new set of technologies (especially because I4.0 technologies' adoption was still below 30% in most industries in 2020 as stated by IOT Analytics), but rather at redirecting their usage toward more sustainable applications.

One of the technologies introduced in the smart revolution is the Information Systems (ISs) integration. It consists in the creation of data exchange channels and interaction modalities between ISs to increase efficiency, foster collaboration and improve productivity, but also to integrate data from multiple sources to extract previously unknown or implicit knowledge [5]. In this way, integration assumes a strategic role because it may contribute to the creation of a sustainable competitive advantage based on the company's know-how.

The three most common and used production ISs are the Enterprise Resource Planning (ERP), used to manage and integrate different business functions and for resource planning, the Manufacturing Execution System (MES), used to monitor and control production and resource usage, and the Product Lifecycle Management (PLM), used to collect, manage and distribute product data [6].

Regarding the existing integrating architectures among these ISs, the PLM-MES integration is the least established, especially if compared to the ERP-MES which can already benefit the ISA 95 – IEC 62264 standard for the data structure [7]. However, thanks to the information flow it creates between design and manufacturing phases, it represents an interesting opportunity for those company which base their competitiveness on the experience and knowledge acquired over time (e.g., One-of-a-Kind-Production or OKP companies, whose products are customized on a single customer's demand).

This article will present a framework for PLM-MES integration inside an OKP company. The two ISs interact through a centralized database called Knowledge Based System (KBS) by using inter-systems interaction flows. The purpose is to present an industrial solution and to demonstrate how an I4.0's technology can be applied in an I5.0's perspective.

The rest of the paper is organized as follows. Section 2 presents the analysis of the existing literature on integration solutions and its benefits. Section 3 describes the proposed PLM-MES knowledge management framework and the inter-system interaction flows. Section 4 discusses a case study on the implementation of the system in an OKP firm and it also reports the benefits of the proposed system in the context of human centricity. Finally, Sect. 5 draws conclusions and outlines future research directions.

2 Related Works

Despite the growing interest on the topic, the literature produced over the past four years regarding integration between PLM and MES is surprisingly scarce. This may be partially due to the limited PLM diffusion (the "*Before PLM era*" will not end until 2040 [6]) which may hinder research on the topic. PLM's low adoption is also the reason why the standard for ERP-MES integration does not include PLM applications and delegates the product data management to the other two systems [7].

The most important reference on PLM-MES architecture is represented by Ben Khedher et al. (2011) [7]. Starting from the analysis of the lifecycles of the data exchanged (i.e., product object, product instance, manufacturing system and purchase order) and the types of activities, it defines the areas of responsibility for each system. The same work also proposes an ontology-based integration.

When analyzing the PLM-ERP integration for OKP companies Bruno et al. (2020) [8] proposes an integration based on a centralized database (i.e., the KBS) able to store and integrate the data produced by both systems and to redistribute it to workers and designers, implementing in this way the feedback mechanism.

The same topic is also analyzed in a study presenting an application of the technology to an aeronautical OKP company [9]. This implementation complements the data provided by the MES by means of technologies like RFID for product tracking and a control system.

An integration based on international standards for PLM and MES has also been proposed [10]. This research proposes a neutral information model for product, process, and resource data exchange. It also allows to transfer and comment 3D files between systems.

Similar systems minimize the chances for errors allowing for smoothly transferring and sharing information from product development to manufacturing, as demonstrated by Stolt and Rad [11]. Although this study focuses on the PLM-ERP integration, similar considerations can be made for the PLM-MES interconnection. However, data integration is only part of the benefits this framework delivers. In fact, this architecture facilitates the creation and management of new knowledge. The relationship between PLM as product data repository and Knowledge Management (KM) has been demonstrated by Folkard et al. [12]. In this study, researchers identified the PLM as a key enabler of KM because of the role it plays in knowledge reuse through the provision of repositories to enable collecting, sharing and distributing information.

All studies mentioned above explain the expected operational and knowledge management benefits created by this technology. However, little information is provided on the topic of human centricity. In order to study this aspect, it is necessary to start from an analysis of the I5.0's implications. Ivanov (2022) [3] analyzes the effects of the new policy on Operations and Supply Chain Management. In this study the major technological principles of I5.0 possible ways to enhance human centricity at the plant level can be analyzed.

Finally, a literature review on Human Centric ZDM [4] explains how technology can enhance operators' and engineers' capabilities; as such, it represents a useful example to lead future research on different applications.

3 Knowledge Management Framework

The proposed framework aims to integrate PLM and MES systems through the KBS, as shown in Fig. 1. A preliminary implementation of the framework was proposed by Bruno et al. (2020) [8]. The PLM application is used for the management of product, equipment, and process data, and for enhancing collaboration and coordination during the design and the industrialization phase. Inside the PLM, products and production routes can be defined and made available inside the KBS. Instead, the MES is employed for planning, monitoring, and controlling production, but also for collecting the operators' feedback. The KBS is a centralized database able to store and integrate data from both systems. It acts as a mediator and can make available information both to PLM and MES, thus allowing the reuse of knowledge. The allocation of the activities between the MES and the PLM is compliant with the solution proposed by Ben Khedher et al. (2011) [7].

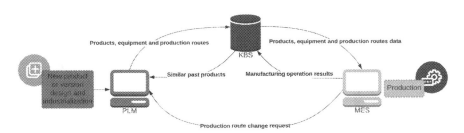

Fig. 1. Interaction flows between PLM, KBS and MES.

3.1 The KBS

The KBS data structure is shown in Fig. 2, according to the UML class diagram formalism [13]. It has been designed to meet the needs of an OKP company. In fact, for this type of organizations, it is necessary to correctly store information about all the product and production routes' versions, in order to be able to store a complete record of the design process history. In this diagram, blue classes belong to PLM and yellow ones to MES (the two green classes refer to ERP, where data related to the customer orders are inserted).

In the PLM, the process starts when the company receives a project, which is made of multiple customer orders. Each order corresponds to one product, one raw material and

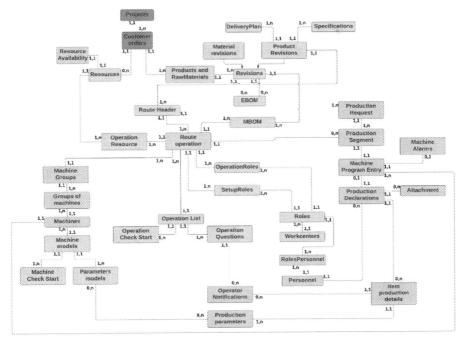

Fig. 2. UML class diagram of the KBS

multiple resources. Each record of product or raw material may have multiple versions, each with their own sets of specifications, BOM, delivery plan and versions of the production routes. Only one production route will be flagged as active for each revision of a product; however, the full revision history is stored in the database to be later used as a knowledge source. The production route is divided into route header, which stores the general information about a route, and route operation, which contains the sequence of operations required to produce a specific revision of a product. The route operation table is connected with revisions (in order to construct a MBOM), machine groups (an entity storing dynamic groups of machines), operation roles and setup roles (to identify the kind of operators required for the operation itself and for the setup), operation resource (where the resources required to perform a specific operation can be stored) and operation list. The operation list reports a list of generic operations; in this way, it is possible to define controls to be made before the operation (i.e., operation check start) and after (i.e., operation question).

In the MES, multiple production requests can be generated for each production route. The planning of each request is made using production segment, which is related to route operation. Machine program entry stores the specific machine used to perform the operation and is directly related to the production declarations, which contains information about the production progress. More detailed information can be found inside the item production details. This last table has associated the parameters used during the operation and the answers given by the operator to the standard questions provided inside operation questions, in this way a more detailed analysis of the problematic operation can be made.

The KBS, other than facilitating system intercommunication, has the role of repository of integrated data which can be later analyzed to extract knowledge. Therefore, it can be used for multiple purposes:

- to find the similarity between a new product and the existing ones, in order to provide information about the design process, the production route and the issues occurred in the past. This helps the company to anticipate design problems and solving them before they emerge, thus saving time and money.
- to run cluster analysis on products, defects and solutions used. In this way, the KBS can provide quantitative data to support the development of organizational best practices.
- to provide more accurate forecasts of time and cost thanks to the analysis of past experiences, which can benefit planning accuracy and customer relationships.
- to reduce the dependency on key knowledge-holders, a common problem for OKP companies. Since part of their knowledge can be stored and retrieved, the firm becomes able to contain the know-how loss in case one of their employees leaves.
- to identify potential areas of improvement both in design and manufacturing, enabling continuous improvement.

3.2 The Interaction Flows

One of the most crucial elements in PLM-MES integration is the definition of the interaction flows among the systems, especially the feedback flow from production to design. Two kinds of flows exist (see Fig. 1).

- Flows from PLM to MES. Product, equipment, and process data are defined in PLM and then exported to the KBS, where MES can read it.
- Flows from MES to PLM. The information about operation results is exported into the KBS and made available to the PLM, where designers can use them to improve products or equipment design and industrialization.

More precisely, the PLM-MES flow can be further divided into two types, depending on the timing. The first type transfers data concerning the type of operations, the machine models, the production parameters model, the checks to be made before and after the operation and a list of questions to be answered after each process. These can be considered static data, since they define the characteristics of the plant and the production, thus is not very frequent. The second type of flow is instead continuous, and it is designed to systematically insert product, equipment, and process data from PLM to MES. This data is highly dynamic and subject to endogenous and exogenous changes. In order not to limit the organization flexibility, information updates on a single production route must be communicated to the MES system as soon as possible. The many changes to be reported and the number of orders processed at once by companies allow us to consider this flow continuous.

The MES-PLM flow represents the most important advantage provided by the integration, because it makes possible the storage and reuse of the historical knowledge, by formalizing the trial-and-error process typical of an OKP company. The MES-PLM flow can be further classified in two types according to the path used for the interaction; both are necessary to deploy the feedback mechanism between manufacturing and design. The first type has the role of summarizing the data collected in production to produce information about the product revision; in particular, it moves data about production results (e.g., number of conforming, number of scraps, unsuccessful operations, issues found), from MES to KBS and, after a processing operation aimed at improving readability, ultimately to the PLM. The second one deploys a real-time mechanism to simultaneously stop the production from using the inappropriate cycle and to trigger a second industrialization phase in the PLM. Therefore, it is a more direct form of interaction between MES and PLM (but simultaneous with the previous one) and it is needed to deploy proper and timely change management processes. In an OKP company the production route may go under several changes during manufacturing; therefore, their revision and approval requires a formal process. This must be triggered by the MES (i.e., the production, where the route incompatibility is discovered); however, since the industrialization is part of PLM areas of responsibility, the change process must be managed inside the PLM. This problem is solved by using the concept of lifecycle, as described in the next section.

3.3 Entity Lifecycle

A lifecycle is a sequence of states an entity (i.e., a data object) may be in. The states, represented as nodes, are connected by arcs, which allow to move a data object from one state to another following one of the paths allowed. The changes of states are entirely managed by PLM as part of the product data. However, transition actions may be triggered by the MES system when needed.

The concept of the lifecycle has already been used for PLM-MES integration [7] focusing on the product macro phases. On the contrary, in this framework the focus is on

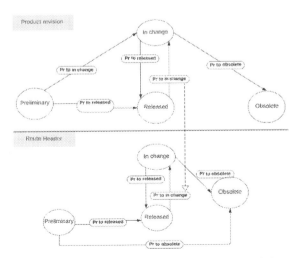

Fig. 3. Lifecycles of product revision, production route their interactions.

detailing the design and manufacturing phase. Two different but interacting lifecycles have been designed, one for the product (in this case, for the revision of the product) and one for the production route, as shown in Fig. 3. Given the business model of an OKP company, the Product revision (i.e., the customer's product specifications) dominates over the Route header; a change in the former's information creates the need for the latter's reevaluation.

Both the product revision and the production route are created in state Preliminary. A Route Header cannot be created until the Product revision's information is validated i.e., Released. Once the product is released, a Route Header can in turn be generated and released. When parts and routes are released, they cannot be modified without a formal change management process.

In case an issue emerges during production such that a change to the route is required, the MES immediately requests the promotion of the route to the In change state This action activates a change process, which in turn informs all the responsible entities and activate a formal workflow to coordinate the re-design of the route. Along with the promotion to the In change state, an indication about where to find in the KBS the specific production declaration reporting the problem can be included. Depending on the entity of changes required, the change process may lead to two alternative outcomes: if a minor change is needed (e.g., adding detailed work instructions to an operation) the Route Header can be simply released back; instead, in case of major changes (adding or removing resources, modifying the operations order) a new version must be generated.

When a product faces a revision (i.e., a new data object of type Product revision is generated), it must be first moved to the In change state to be evaluated; this causes the promotion of the latest Process Plan to In Change and, of course, the activation of a Change Workflow. Once accepted, a new version of Product Revision can be generated, while the old version is promoted to the Obsolete state. This promotion is triggered inside the PLM system and is essential to ensure instantaneous communication between the otherwise siloed departments of Design and Production.

4   Implementation and Discussion

4.1   Implementation of PLM-MES Integration Framework

The proposed framework has been implemented within an Italian OKP company which produces car body parts some of the most important global producers. The proposed solution employs open-source software for most of its components.

The PLM application is Aras Innovator, released by Aras Corporation [14]. Multiple data objects and relative forms were created, replicating the data structure shown in Fig. 2. Aras also allows to associate to each data object a lifecycle. Using this function, the lifecycles reported in Fig. 3 for Product revisions and Production route were implemented. The same lifecycles were connected by means of C# codes activated on the promotion event.

The KBS was implemented by using PostgreSql [15], according to the diagram shown in Fig. 2. The commercial MES system Jpiano was implemented by the AEC company [16].

The flow from PLM to KBS was deployed by means of Node-RED, a visual programming tool that allows online service communication [17]. For some entities this flow is activated periodically; in some cases, it can be activated when specific events (like the Product revision promotion on Aras) occur. The second activation mode is designed to improve the timeliness of communication among the systems; in fact, in this way the MES system can be readily informed in case a change occurs. The flow from MES to PLM exploits a HTTP request sent by Jpiano directly to Aras' server which activates the promotion of the item specified in the request body.

4.2   Contribution to Human-Centricity and Industry 5.0

The exposed integration contributes significantly to the achievement of human-centricity in production and to the transition toward I5.0. In fact, some of the most important I5.0's technological principles are collaboration, coordination, and communication [3]. Thanks to the data migration and the interaction flows, communication between two typically departmental steps of design can rise in frequency and quality, thus improving the coordination at the organizational level. A timely feedback and knowledge sharing between design and manufacturing is particularly critical for OKP companies since, thanks to this practice, they can significantly reduce the amount of design iterations, thus reducing the lead time and costs. Finally, the PLM itself represents an important tool for collaborative design.

The PLM-MES integration plays a peculiar role when analyzed under a human-centric perspective. It allows to formalize the manufacturing experience and knowledge created during operations (that would otherwise remain implicit in operators' heads) and to make the information about the problems occurred and the solutions adopted in the past directly available for the design of similar components. Therefore, this solution enhances designers' capabilities by enriching their personal experience with historical data about defects and choices made on similar products. To extend this concept, the knowledge reuse feature can complement the results obtained from traditional software tools such as simulations, by providing real manufacturing examples, thus overcoming software's inherent limitations, and providing better guidance for designers in their decision-making process.

Furthermore, the PLM-MES integration can be framed within the lean manufacturing approach, and in particular ZDM, which is now considered to be one of the key enablers of I5.0 [4]. By making immediately available to operators drawings, work instructions and stamping simulation results, it provides key information about possible failures and areas to focus on during quality control. In this way, a proactive approach can be taken by workers during manufacturing, who can implement the necessary actions to prevent defects before they occur, making a further step toward ZDM even in the OKP's unpredictable environment.. The lesson learning mechanism presented creates the conditions for the continuous improvement of both the worker and the process.

This integration also makes the whole organization more resilient to changes. As mentioned, the historical knowledge is formalized and stored in a database, so that it could be made available immediately to anyone. Traditionally, this form of information remains in workers' heads; as a result, when employees leave the organization, the latter loses an important part of its know-how which is a critical dimension of the contemporary competitive advantage. Thanks to this technology, the knowledge can not only be saved in similar situations but can also be distributed to new hires in order to reduce their learning curves. Overall, the whole system becomes more flexible and resilient.

5 Conclusions

In this paper, an improved framework for PLM-MES integration has been proposed. This framework presents multiple benefits for all kinds of production but especially for OKP companies, which in this way can formalize and store the knowledge produced during their activities and reuse it when new similar products must be designed and redistributed.

The framework has been implemented using a centralized database to convey data from both systems to integrate and reuse it. The designed interaction flows create immediate feedback between production and design, improving the effectiveness of this solution. The communication channel established between these two departments also demonstrates the conformity of the technology to I5.0's requirements.

Future research should focus on the introduction of the ERP system, whose functions are currently played by the PLM. The full integration benefits can only be generated when all information sources are integrated correctly in order to generate knowledge. Another area requiring further analysis is the formalization and representation of manufacturing knowledge; ontologies can be employed on this aspect for the definition of the concepts' taxonomy and relationships. Finally, it is necessary to understand how the data generated during production can be processed to present only the most meaningful and representative information to workers, to avoid information overload.

References

1. European Commission: Directorate General for Research and Innovation. Industry 5.0: towards a sustainable, human centric and resilient European industry. Publications Office, LU (2021)
2. European Commission: Directorate General for Research and Innovation. Industry 5.0: a transformative vision for Europe: governing systemic transformations towards a sustainable industry. Publications Office, LU (2021)
3. Ivanov, D.: The industry 5.0 framework: viability-based integration of the resilience, sustainability, and human-centricity perspectives. Int. J. Prod. Res. (2022)
4. Wan, P., Leirmo, T.: Human-centric zero-defect manufacturing: State-of-the-art review, perspectives, and challenges. Comput. Ind. **144** (2023)
5. Saihi, A., Awad, M., Ben-Daya, M.: Quality 4.0: leveraging Industry 4.0 technologies to improve quality management practices – a systematic review. Int. J. Qual. Reliab. Manag. **40**(2), 628–650 (2023)
6. Stark, J.: Product Lifecycle Management, 3rd edn. Springer, Cham (2015). https://doi.org/10.1007/978-3-319-17440-2_1
7. Ben Khedher, A., Henry, S., Bouras, A.: Integration between MES and product lifecycle management. In: ETFA2011, Toulouse, pp. 1–8. IEEE (2011)
8. Bruno, G., Faveto, A., Traini, E.: An open source framework for the storage and reuse of industrial knowledge through the integration of PLM and MES. MPER **11**, 62–73 (2020)

9. D'Antonio, G., Sauza Bedolla, J., Chiabert, P., Lombardi, F.: PLM-MES integration to support collaborative design. In: Weber, C., Husung, S., Cantamessa, M., Cascini, G., Marjanovic, D., Venkataraman, S. (eds.) International Conference on Engineering Design ICED15, Milan, vol. 10, pp. 081–090 (2015)

10. Jauhar, T.: A Neutral Information Model for Data Exchange between Plm and Mes Based on International Standards (2022). https://ssrn.com/abstract=4213319

11. Stolt, R., Rad, M.A.: Transiting between product development and production: how IT in product development and manufacturing integrate. In: Noël, F., Nyffenegger, F., Rivest, L., Bouras, A. (eds.) PLM 2022. IFIP AICT, vol. 667, pp. 136–143. Springer, Cham (2023). https://doi.org/10.1007/978-3-031-25182-5_14

12. Folkard, B., Keraron, Y., Mantoulan, D., Dubois, R.: The need for improved integration between PLM and KM: a PLM services provider point of view. In: Rivest, L., Bouras, A., Louhichi, B. (eds.) PLM 2012. IFIP AICT, vol. 388, pp. 85–98. Springer, Heidelberg (2012). https://doi.org/10.1007/978-3-642-35758-9_8

13. OMG, About the Unified Modeling Language Specification version 2.5.1. https://www.omg.org/spec/UML/. Accessed 30 Jan 2023

14. Aras. https://www.aras.com/. Accessed 30 Jan 2023

15. PostgreSQL. https://www.postgresql.org/. Accessed 31 Jan 2023

16. Jpiano. https://www.aecsoluzioni.it/wp/en/jpianomessoftwareen/. Accessed 31 Jan 2023

17. Node-RED. https://nodered.org/. Accessed 31 Jan 2023

Blockchain Applications in the Food Industry: A Pilot Project Implementation in the Ancient Grains Industry

Bianca Bindi[1], Gloria Padovan[1], Giacomo Trombi[1], Niccolò Bartoloni[1], Virginia Fani[1], Marco Moriondo[2], Camilla Dibari[1], and Romeo Bandinelli[1(✉)]

[1] Univerrsity of Florence, Florence, Italy
`romeo.bandinelli@unifi.it`
[2] CNR, Florence, Italy

Abstract. The purpose of this paper is to present the results of a pilot project implementation of PLM solution based on blockchain technology in the ancient grains Italian Industry. After a literature review on previous experience of blockchain technology in the food Industry, a case study analysis has been done to identify the actors along the supply chain of durum wheat for pasta production, as well as to study the main regulations for wheat production. Moreover, possible weaknesses and strengths in the application of the blockchain for the ancient wheat supply chain have been highlighted. Finally, the main evidences of the pilot project are reported and a set of information systems able to manage the product lifecycle have been identified. The paper provides valuable insights to companies that are trying to implement such solution in the food Industry.

Keywords: Blockchain · Ancient Grain · Food Industry · PLM

1 Introduction

Recently, blockchain technology has been introduced in the field of food safety to trace food information. Blockchain is a distributed computing paradigm characterized by decentralization, network-wide recording, low cost, high efficiency, security, and reliability. It can reduce administrative costs, trading risks, improve information credibility, increase regulatory transparency, and implement trusted processes. Using blockchain in food supply chain can potentially eliminate information asymmetry, achieve the synchronous updating of information across all nodes of the supply chain, eliminate product quality problems caused by stakeholders, and strengthen information credibility.

The study aims to build an ancient grain Product Lifecycle Information (PLM) management solution based on blockchain technology. In detail, starting from a literature review, a single case study has been carried you to identify the key nodes and the actors of the durum wheat for pasta production supply chain. Moreover, main regulations and standards for wheat production have been analyzed. Last, a set of information systems have been identified to manage the product information along the whole product lifecycle.

C. Danjou et al. (Eds.): PLM 2023, IFIP AICT 702, pp. 130–139, 2024.
https://doi.org/10.1007/978-3-031-62582-4_12

The main contribution of the paper is the definition of a framework and a set of guidelines to replicate the pilot application of the blockchain in the wheat supply chain, using a "from field to fork" approach in order to manage the whole product lifecycle.

The paper is structured as follows: Sect. 2 presents a literature review on agri-food supply chains, with a focus on the ancient grain value chain. Section 3 introduces the pilot project, while Sect. 3.1 highlights some of the results. Finally, Sect. 4 presents the conclusions and future steps.

2 Industrial Background

2.1 The Ancient Grain Supply Chain

The grains industry has an important role in the worldwide economy, due to the large quantity produced and its role in the its role in the food supply chains of all countries. Within this industry, ancient grains have grown steadily over the last 10–15 years. The total gross value has been estimated at 457.35 M$ in 2022 and is expected to reach 6.3 B$ in 2026. In Italy, the cultivation of durum wheat continues to be a key element of our agriculture, mostly characterised by high quality and exclusive foreign markets.

From a global perspective, countries outside Europe (e.g. Turkey) have significantly increased their production capacity, offering products at competitive prices but with high quality or environmental standards. Traceability, quality and environmental soundness are critical factors for Italian pasta producers, both for domestic and international markets.

The ancient grain do not use fertilisers, pesticides and intensive labour, with lower costs and consequently lower production rates. As reported by several authors, its consumption brings health benefits to the consumer. Moreover, the rediscovery of old varieties of ancient cereals allows a better response to today's important challenge of activating responsible practices and sustainable consumption.

As reported in the literature, ancient grain is a specific type of cereal that is not suitable for the application of modern agricultural practices.

Although the supply chain of ancient grain is short and simple, it involves a series of coordinated actions and efforts carried out by farmers, millers and wheat processors (pasta makers, bakers and chefs). In this SC, the central role is played by the mill, which is responsible for the quality of the raw material, the sharing of knowledge and all those relationships that allow all actors to adopt a win-win strategy in order to achieve financial and non-financial mutual benefits.

2.2 Blockchain Application in the Agri-Food Industry

Blockchain food traceability is currently one of the most interesting topics in the global agri-food segment. Several contributions can be found regarding barriers and benefits of its application to trace different types of products.

[1] report more than 90 papers published between 2016 and 2021 on this topic. Perceived benefits of blockchain application relate to traceability, minimising food fraud and supply chain monitoring [2–4], increasing transparency [2], integrating IoT devices

[5] and eliminating intermediaries, reducing transaction costs [6]. Despite this, only a limited number of applications and case studies described as pilot projects can be found in the literature, and a smaller number of applications can be found in the market. [7] and [8] describe an application in the coffee industry, more cases can be found in the fish industry [9, 10], in wine, fresh food and beef [11–14], milk, eggs and pasta [6–10] and seafoods [14]. As reported in the case studies found in the literature, the application of blockchain technology in the agri-food industry is strictly related to the product and, consequently, to the supply chain. Different products require different information to be stored, different IoT and ICT tools needed to collect this data, and different industrial backgrounds. Thus, each specific industry, like that of ancient grains, requires vertical and focused business and technology models capable of facilitating the implementation of such tools, overcoming the entry barriers of low informatization and digital awareness of workers.

3   Pilot Project Description

3.1   Pilot Project Introduction

As mentioned in the previous section, the aim of the pilot project was to analyse the ancient grain supply chain in order to identify the critical points and the information to be exchanged between the different actors.

In order to achieve this result, a case study analysis was carried out within three actors involved in the pasta ancient grain supply chain in the Tuscany region of Italy, namely a farmer, a miller and a pasta maker. The aim of the case study was to understand the production processes, the Critical Control Points (CCP) and the documents exchanged between the different actors, using the Business Process Modelling Notation (BPMN) standards (Fig. 1).

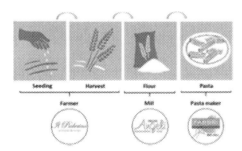

Fig. 1. Pilot Supply Chain representation.

3.2   Pilot Project Process and Information

In this section, the BPMN diagrams of the supply chain, the CCPs for every considered phase and the data and documents involved are reported. The first actor presented is the farmer, which is the most upstream of the supply chain analyzed within this project, producing ancient wheat. Then the second node of the supply chain is the mill, that deals in turning grain into flour and the last one is the pasta maker, which turn the flour into pasta representing the finished good downstream of the whole supply chain.

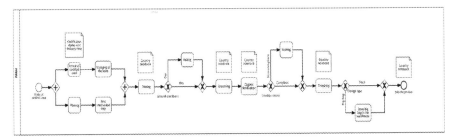

Fig. 2. BPMN diagram for farm process

The main activities of the farmer are shown in Fig. 2. It specialises in the cultivation of cereals, including ancient wheat. In order to obtain a high quality product, the farm only buys seeds certified by the *Agricultural Consortium of Siena*, founded in 1901 with the aim of supporting agriculture and increasing agricultural production in the area. The process begins with the purchase of the seeds, which require a certification stamp issued by the *Consortium of Siena*. In particular, the seed is certified by CREA, a national research organisation that certifies the authenticity of the variety purchased. This first step is followed by the sowing of the seeds. In the case of ancient wheat varieties, the amount of seed to be sown is around 160–180 kg per hectare [kg/ha], as opposed to normal cereals which are sown at 200/250 kg per hectare [kg/ha]. This is followed by rolling to compact the soil and improve germination. The sowing, tilling, organic fertilising and threshing phases are recorded in a special register called the *country notebook*: a document that allows farms to record all the operations carried out in the field, the protocols followed and the techniques used. In particular for each phase the main treatments to be reported in the *country notebook* are the name of the crop treated and its extension expressed in hectares, the date of the treatment, the name of the product and the relative quantity used, expressed in kg or litres, the type of adversity that has made the treatment necessary and the dates of the most important phenological stages of each crop: sowing or trans-planting, the beginning of flowering and the harvest.

In the threshing phase thanks to the high digitalization of the combine harvester and the presence of GPS sensors, it is possible to obtain the geolocation of the crop and to record it in the blockchain. Once the grain has been harvested, there are two possible storage ways: the first is the truck where the grain is stored in the trailer and shipped directly to the customer and Big Bag where the grain is stored in containers with a maximum capacity of 1300–1400 kg, and then placed in the warehouse. The Big Bag will then be shipped to the next actor in the supply chain to process the grain into flour.

In summary, for the farmer, the main documents that must be uploaded to the blockchain to ensure food safety, traceability and quality are:

- seed certification;
- country notebook

- field geolocation;
- a document which records some information (quantity and batch number) in order to trace the batch of cereals which leaves the farmer.

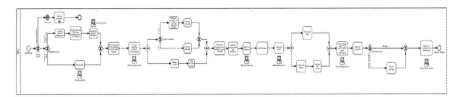

Fig. 3. BPMN diagram for mill process

Going forward the supply chain, the second actor is represented by the mil. The main activities of the mill are shown in Fig. 3. Specifically, the mill specialises in the production of stone-ground flour. This type of processing produces a flour with an irregular grain size, a higher proportion of bran and complete preservation of the germ. Stone-ground flours are much tastier, more nutritious and easier to digest. The slow and careful processing preserves the nutritional properties without overheating the flours, resulting in a high value final product. The mill buy raw materials by different reliable suppliers. The main raw materials are: soft wheat, cereals, pulses and ancient wheat.

When raw materials arrive, they are accompanied by a transport document containing information such as the document number and date, the quantity and type of raw material and the grain batch number. The incoming raw materials are first subjected to a visual inspection. This is one of the most critical stages in the process. In fact, it often happens that the farmer does not comply with the correct conditions for the storage of the grain, which leads to the detection of non-compliance. The next step is to take a number of samples and send them to a certified laboratory for analysis. If any defects are found, the supplier is immediately contacted and the goods are returned. After the first control point, the raw material can be stored in the silos at the mill warehouse or in an external storage facility.

At this point the production process begins. Almost every day, the production schedule is drawn up. The first phase is the cleaning of the grain. After the silos, the grain is put into big bags. These are then transported to the cleaning machine by forklift truck. After that stage, the product is put back into the big bag and transported to the grinding stone by means of a stacker. It is from this point that the grinding process can be started.

The final stage of the production process is the sieving stage. The product is transferred from the millstone to the sieve by means of an aspiration system. The sieve mesh is set according to the type of flour required. At the same time, the waste from this process is collected. On average, the rejects from this stage are around 15%. In summary, the average scrap of the whole production process, from the cleaning phase to the sieving phase, is about 18–20%. At the end of the transformation process, the flour is returned to the big bags.

Periodically, a test is carried out on the finished product using Chopin's Alveograph to evaluate some technical aspects of the flour, such as the strength, toughness and extensibility of a dough of flour, by plotting the behaviour of a dough of flour on a graph called an alveogram. Before being packaged, the flour is stored for approximately 7–10 days.

At this stage the flour is ready for packaging. Three types of packaging are used: the big bag for sale to large customers, the 25 kg bag and the 10 kg bag. They are stored in the finished product warehouse.

In summary, for the mill, the documents that can be stored into the blockchain are:

- the *mass balance* (that contains information regarding on the input and output of the raw material for that phase);
- the date of arrival of the ancient grain;
- the type and quantity of wheat;
- the number of the transfer document relating to the wheat purchased;
- the processing dates;
- the quantity of wheat obtained and the batch number;
- the scraps and the yield, that can be easily calculated from the ratio output quantity to the input quantity (the *bass balance*).

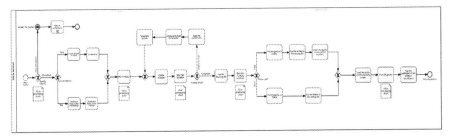

Fig. 4. BPMN diagram for pasta maker process.

Finally, the activities of the pasta factory are shown in Fig. 4. This actor of the SC is specialised in the production of pasta according to ancient processes and criteria. The production process must be carried out at low temperatures (below 38 °C), with the aim of preserving the organoleptic characteristics of the product in order to guarantee a high level of quality and a unique taste. In order to guarantee the high quality of the product, the factory relies exclusively on a portfolio of reliable suppliers. The suppliers are personally selected by the pasta maker, who visits them to assess the quality of the product and the production process.

Prior to ordering, the pasta plant asks its suppliers for raw material samples to perform gluten tests and evaluate the quality of the product. If the test is passed, the factory proceeds to order the raw material. The raw materials arrive with the transport document, which contains some information such as the document number and date, the quantity and type of raw material and the batch number. As soon as the raw materials arrive, they are subjected to documentary control and visual inspection. The initial control phase is critical to detect any non-conformities. If any non-conformities are found, the supplier is immediately contacted, some pictures are sent to them and the goods are returned.

In general, the number of non-conformities is low thanks to the strong relationship and trust between the pasta factory and the selected suppliers.

The raw material can arrive in tanks or in bags (25 kg). In the former case, it is sent directly to the silos, which have a capacity of 10 tonnes, while the bags are stored in the warehouse and then manually loaded into the hopper.

This is the point at which the production process is ready for start. The production schedule is set every 15 days and the production lead time is approximately 2 weeks.

The first activity is to mix the flours to obtain the desired type of pasta. The mixture is then sent to a long kneader (4 m) to start the kneading phase. This phase is the most critical of the whole process, as it is the one in which any non-conformity could be found. This is possible thanks to the expertise of the operator, who can detect anomalies simply by observing the dough. If a non-conformity is detected, it's necessary to stop the line, empty the mixer, discard the dough, pass organic semolina to clean the line to avoid contaminating the following steps, empty the line and restart production.

Then, depending on the type of pasta to be made, the bronze drawing die is set and the drawing process begins. This phase of the production takes about an hour and a half. There are two different ways to store pasta. In particular, long pasta is placed on trolleys, while short pasta is placed on frames. This is followed by the final phase of the production process, the drying process. The trolleys and frames are moved by hand to the drying cells, where they remain for 3–6 days. This is the most important phase of the process, as it requires very long drying times at low temperatures in order to obtain a quality finished product. After this drying period, the long pasta, unlike the short pasta, is cut before being packed.

During the production process, several controls are carried out, such as visual inspection, morphological, temperature, humidity, colour, cutting and drying controls.

Humidity is the most important parameter to control because drying is strongly influenced by humidity. Once the drying process is complete, the finished product passes through the vibrating sieve. After this phase, the pasta is ready to be packed and stored in the warehouse for a minimum of 15 days and a maximum of 6 months. The average waste in the whole process is about 10–20%.

In summary, for the pasta maker, the documents that can be stored into the blockchain are:

- the *mass balance* (that contains information regarding on the input and output of the raw material for that phase);
- the date of arrival of the flour;
- the type and quantity of flour;
- the number of the transfer document relating to the flour purchased;
- the processing dates;
- the quantity of pasta obtained and the batch number;
- the scraps, that can be easily calculated from the ratio output quantity to the input quantity (the *bass balance*);
- the production process parameters.

3.3   Pilot Project PLM Implementation and Data Management

Starting from the data defined in the previous section, a data model has been defined and a framework for data storage and exchange has been developed, using different commercial software in a product lifecycle management architecture.

As mentioned above, mockups representing the user interfaces where each of the three actors will upload data, information and documents that will subsequently be entered into the blockchain, have been developed.

This section provides some screenshots and a brief description of the mockups. Due to space limitations, only the interface of the first actor, the farm, will be described in detail. First, the user must log in using the ID and password used to register on the platform. Once the user has logged in, the main menu appears on the screen. It contains four items, which represent the documents that must be uploaded to the blockchain in order to guarantee food safety, traceability and quality. In particular, they are: Seed certification, Country notebook, Field geolocation and Transport document (Fig. 5).

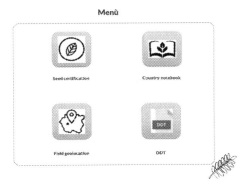

Fig. 5. Farmer menu interface.

The first document to be uploaded is the seed certificate. This is a document issued by the consortium to the farmer at the time of the sale of the seed, guaranteeing the authenticity of the characteristics of the seed itself. The user has to enter in the platform some information contained in the seed certification stamp (Fig. 6), such as

– Seed variety
– Place of production
– Date of packaging
– Germination
– Purity
– Gross weight

Once this part has been completed, the platform returns you to the home screen, in which the item you just completed appears with a green tick, while the remaining missing items appear with a red tick (Fig. 7). The user is free to continue with the loading of the missing data.

Fig. 6. Farmer seed certification interface.

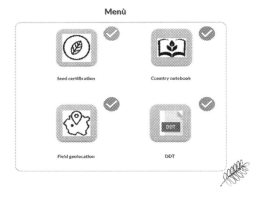

Fig. 7. Farmer seed certification interface completed.

Once the data on the seeds has been acquired, it is transferred to the blockchain linked to the software, without any action by the user and without the possibility of modifying the existing uploaded data. In the same way, the interfaces have been developed until the end of the supply chain.

4   Conclusion and Future Steps

The purpose of this paper was to present the results of a pilot project implementation of PLM solution based on blockchain technology in the ancient grains Italian Industry. After a review of the application of blockchain in the agri-food industry, a description of the SC has been presented. Due to its peculiarities, the CCP of the three main actors, the farm, the mill and the pasta maker, have been identified and the information data needed to be recorded on a blockchain using specific software have been listed. From a practical point of view, the paper is a useful tool for companies wishing to implement a traceability solution to guarantee food safety and quality. From a scientific point of view, the paper contributes to the definition of a framework for the application of blockchain in the agri-food industry, with a particular focus on the ancient grain sector.

References

1. Srivastava, A., Dashora, K.: Application of blockchain technology for agrifood supply chain management: a systematic literature review on benefits and challenges. Benchmarking **29**(10), 3426–3442 (2022). https://doi.org/10.1108/BIJ-08-2021-0495
2. Kayikci, Y., Subramanian, N., Dora, M., Bhatia, M.S.: Food supply chain in the era of industry 4.0: blockchain technology implementation opportunities and impediments from the perspective of people, process, performance, and technology. Prod. Plan. Control, 1–21 (2020). https://doi.org/10.1080/09537287.2020.1810757
3. Saberi, S., Kouhizadeh, M., Sarkis, J., Shen, L.: Blockchain technology and its relationships to sustainable supply chain management. Int. J. Prod. Res. **57**(7), 2117–2135 (2019)
4. Yadav, S., Singh, S.P.: Blockchain critical success factors for sustainable supply chain. Resour. Conserv. Recycl. **152**, 104505 (2020)
5. Feng, H., Wang, X., Duan, Y., Zhang, J., Zhang, X.: Applying blockchain technology to improve agri-food traceability: a review of development methods, benefits and challenges. J. Clean. Prod. **260**, 121031 (2020)
6. Bumblauskas, D., Mann, A., Dugan, B., Rittmer, J.: A blockchain use case in food distribution: do you know where your food has been? Int. J. Inf. Manag. **52**, 102008 (2020)
7. Miatton, F., Amado, L.: Fairness, transparency and traceability in the coffee value chain through blockchain innovation. In: 2020 International Conference on Technology and Entrepreneurship - Virtual (ICTE-V), pp. 1–6 (2020)
8. Gashema, C.: Blockchain and certification for more sustainable coffee production – how can blockchain complement the sustainability certifications (2021)
9. Antonucci, F., Figorilli, S., Costa, C., Pallottino, F., Raso, L., Menesatti, P.: A review on blockchain applications in the agri-food sector. J. Sci. Food Agric. (2019)
10. Westerlund, M., Nene, S., Leminen, S., Rajahonka, M.: An exploration of blockchain-based traceability in food supply chains: on the benefits of distributed digital records from farm to fork. Technol. Innov. Manag. Rev. **11**(6), 6–19 (2021). https://doi.org/10.22215/timreview/1446
11. Gashema, C.: Blockchain and certification for more sustainable coffee production: how can blockchain complement the sustainability certifications. Second cycle, A2E. SLU, Department of Molecular Sciences, Uppsala (2021)
12. Galvez, J.F., Mejuto, J.C., Simal-Gandara, J.: Future challenges on the use of blockchain for food traceability analysis. TrAC Trends Anal. Chem. (2018)
13. Patelli, N., Mandrioli, M.: Blockchain technology and traceability in the agrifood industry. J. Food Sci. **85**(11), 3670–3678 (2020)
14. Andrew, J., Kennedy, W.: Food microbiology and food safety practical approaches food traceability from binders to blockchain (2019)

Protecting Manufacturing Supply Chains Through PLM - Blockchain Integration and Data Model Encapsulation

Abdelhak Belhi[1]([✉]) and Abdelaziz Bouras[2]

[1] Joaan Bin Jassim Academy for Defence Studies, Al Khor, Qatar
`abdelhak.belhi@jbj.edu.qa`
[2] CSE, College of Engineering, Qatar University, Doha, Qatar
`abdelaziz.bouras@qu.edu.qa`

Abstract. Product design is often a process that involves multiple parties collaborating with each other to design a final product. The involvement of multiple parties induces several risks associated with cyber security and intellectual property theft. These risks are hard to address especially in the case of traditional centralized platforms which may be prone to misconfiguration and software vulnerabilities. As a potential solution, we aim at addressing the issue of design data exchange through decentralized platforms such as the blockchain. Our solution leverages data formats that can segment product data models and gives the ability to control access to data through a decentralized platform which can be fully integrated with PLM processes through APIs. A proof of concept of this solution using the open-source Odoo PLM platform as well as the Hyperledger Fabric blockchain development platform is demonstrated.

Keywords: Product Lifecycle Management · Blockchain · Supply Chain Management · Odoo · Hyperledger Fabric

1 Introduction

Given the nature of PLM processes which involve dealing with multiple parties in most of the product lifecycle, securing the information that is shared among the parties involved has different objectives. The first and the most important is related to the protection of the product's intellectual property (IP) and this is usually two folds. On the one hand, the IP needs to be secured so that critical information is not threatened by mostly focusing on the aspects of confidentiality and integrity to the outside world. The second aspect is that some of the parties collaborating in some area of the product development might have some conflict of interest related to another area and this has indeed been the case in several legal battles between companies where for example, in a case involving two companies A and B, Company A outsourced some parts from company B to develop Product 1 and the two companies shortly found themselves in a court case of IP theft over Product 2 from Company B (Company A accused B of IP theft) [1]. For these reasons

© IFIP International Federation for Information Processing 2024
Published by Springer Nature Switzerland AG 2024
C. Danjou et al. (Eds.): PLM 2023, IFIP AICT 702, pp. 140–150, 2024.
https://doi.org/10.1007/978-3-031-62582-4_13

a lot of researchers dedicated efforts to developing solutions to address the security of product lifecycle processes.

In an aim to control shared product data in PLM context, in the paper, we propose a blockchain-based access control solution to control access to design data models. Our solution leverages a blockchain-PLM integration platform we developed, and adds an access control component to enable a role-based product development environment fully customizable. This solution uses a prototype we built to integrate information systems to blockchain platforms. Figure 1 below explains how our blockchain-based interaction works where all transactions between involved parties are managed by blockchain smart contracts.

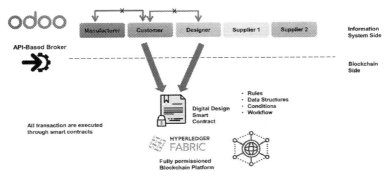

Fig. 1. Blockchain-Integration scenario with supply chain: the Odoo Hyperledger Fabric Integration

The rest of this paper is organized as follows. In Sect. 2, we present a review of works related to access control in product data models as well as the uses of blockchain in industrial applications. In Sect. 3, we present the methodology of our data model access control through blockchain integration, and in Sect. 4, we present our proof-of-concept blockchain-based access control approach and discuss its strengths and weaknesses. We draw our conclusion in Sect. 5 and give some future work perspectives.

2 Related Work

2.1 Encapsulation and Access Control on Data Models

From an access control point of view, multiple contributions had the goal to address concerns related to permissions. The authors of [2] proposed a role-based technique for the visualization of CAD models using security policies and innovative methods to reduce the fidelity of product component drawing to protect intellectual property. In [3], the authors use a similar approach but focus on providing users with more levels of access control rather than granting or denying access. This is achieved by reducing the fidelity of 3D models. A lot of approaches investigate the use of encryption to enforce access control in collaborative CAD environments. The authors of [4] propose two encryption approaches for CAD data focusing mainly on efficient 3D data compression and crypt-analysis resistance. In [5], a new parametric approach for CAD model encryption in cloud environments is proposed. The authors use an approach they call Enhanced Encryption Transformation Matrix to alter the structure of CAD models. Similar approaches relying on 3D model tampering were proposed in [6–8]. Although these model tampering approaches are innovative, traditional approaches for access control to CAD data are the most realistic and better suited for production environments. Most of these approaches [3, 9, 10] consider the fragment of the design data and applies access control policies on these fragments so that users with different roles can access assigned fragments.

2.2 Integration of Blockchain and Enterprise Information Systems

Blockchain integration was investigated in many contributions focusing on multiple aspects of collaboration in industrial applications. As an example, the authors of [11] proposed an empirical study that focuses on blockchain integration in the contexts of manufacturing and operation management. They depicted a detailed evaluation of the different disadvantages preventing effective blockchain integration. In [12], IoT and machine learning integration with blockchain were studied to improve smart manufacturing quality control. The authors of [13], proposed a new productivity improvement application based on blockchain in the context of auto manufacturing and spare parts manufacturing. In [14], the authors investigated the blockchain and supply chain integration impact on additive manufacturing. In a more practical fashion, in [15], the authors presented Block-SC, a new blockchain service for cloud manufacturing with the goal of establishing a service composition collaboration platform. In the context of manufacturing, the authors of [16] proposed an architecture based on blockchain for cyber-physical systems. The authors of [17] presented a shared-manufacturing blockchain-based framework. This framework enables resource sharing in a peer-to-peer fashion whilst maintaining trust using a consensus mechanism based on proof-of-participation and smart contracts [12].

Numerous applications in the supply chain management context were proposed. The authors of [18] proposed a blockchain and supply chain integration model. In [19], a blockchain-based supply chain is developed where the authors investigated the use of ontologies and design tools to translate traceability information to enforce domain constraints in smart contracts.

Blockchain technology was used by the authors of [20] to design an industrial platform with a product-centric information system used as a backbone.

In [21], the authors presented a blockchain integration study in the context of the aircraft parts supply chain. Beyond traceability, Blockchain was used to monitor performance and parts usage, thus impacting positively data integrity as well as safety.

In the era of Industry 4.0, emerging technologies such as IoT, cloud manufacturing, AI, and blockchain are making smart manufacturing a reality. To improve supply chain efficiency and attain business innovation, blockchain is adopted by big manufacturers such as Ford and BMW.

Blockchain is a secure and efficient platform that has multiple benefits and can be used for improving collaborative applications. However, its integration requires a lot of work in terms of development and integration. This technology received a lot of attention and interest in the context of manufacturing and supply chains as it enables a large variety of applications and introduces peer-to-peer trust [22].

3   Methodology

From a high-level standpoint, our solution consists of integrating blockchain technology with PLM systems and handling critical processes and confidential data through the blockchain platform. Both platforms will communicate with each other using an API broker. The data models' access control will be handled by blockchain smart contracts and for our proof of concept, we aim to mainly investigate the feasibility as well as the impact of blockchain-based access control in PLM data models.

3.1   Solution Idea

To address confidentiality and access control concerns to PLM systems' data, we propose to use a blockchain platform to serve as a secure and efficient alternative to traditional relational databases often used by PLM systems and enterprise information systems. In traditional and cloud-based PLM systems, data is often centralized in databases or cloud storage buckets. This approach has several disadvantages related to security, availability, tamper-proofness and access control. Our solution integrates permissioned blockchains with PLM systems to handle collaborative data in a decentralized paradigm. Blockchain will be used to store PLM transaction data as well as to handle these transactions using smart contracts. Additionally, the novelty in our approach comes in the form of enforcing access control policies to all the data present in the blockchain in multiple formats focusing mostly on formats directly related to product design such as CAD and mostly all portable documents. The access control module we integrate into our earlier prototype [23–25] serves as a blockchain-based reference that manages role- permission associations to PLM process assets meaning data visibility as well as actions permissions. Figure 3 highlights the overall architecture of our solution.

Due to the nature of the blockchain platform and given that the system data is present at all nodes, the need to enforce access control is tightly associated with data confidentiality. Thus, in our solution, encryption is used to enforce data visibility. Actors having the rights to a certain category of information (a role) will have keys to this information associated with their blockchain wallet.

3.2   Design and Implementation

Our solution is centered around the Odoo PLM as well as the Hyperledger Fabric Blockchain platform. The two platforms are mainly chosen due to their functionalities as well as being open source, thus allowing the maximum level of customization.

Odoo PLM focuses mostly on the lifecycle aspects of PLM but unfortunately does not allow the ability to preview advanced design models such as CAD models or portable documents. We thus had to integrate web-based plugins into Odoo in order to allow the visualization of these data models inside Odoo.

Although in the case of CAD, the interface only allows viewing the files, we aim in the future to integrate it with CAD editors such as CATIA, SolidWorks, and Revit.

Figure 2 below represents a laptop manufacturing scenario where several actors have to access different parts of the laptop design specifications. In this scenario, and to secure the data encapsulation, we leverage a blockchain-based solution through smart contracts to manage the access control to different design data depending on the role of the actor.

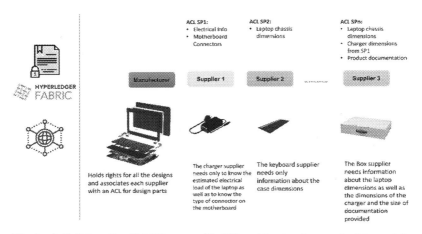

Fig. 2. A Collaborative PLM Process with different levels of access to data models

Blockchain – Odoo Integration

All communication between Odoo and HyperLedger fabric is managed through a broker component. This broker is implemented in Node.JS and its main task is to transmit transaction data back and forth between the platforms. It relies on the HyperLedger Fabric JavaScript SDK for the connection to the blockchain network. To allow reliable and secure communication between the two systems, a RESTful API is used for data and transaction transfer. Client interfaces such as our Odoo interface or a mobile logistics app (developed for real-time asset tracking for the Transporter actor) use this API to connect to the blockchain network through the broker.

Data Model Access Control

Realistically, we cannot focus on all data models, but we can prepare scenarios for the most used format in PLM. Our main idea behind this solution is that for a given product, the parties collaborating for its development focus on certain aspects of the product only. And thus, restricting access to the parts which are not needed is better from an Intellectual property standpoint to avoid issues such as the one highlighted in [1]. As proof of concept, we implement file-level access controls in our solution. In this scenario, we assume that for a given product, each major aspect of its design is represented by a design file (a CAD model). Each actor in the PLM network will then be assigned access rights to the parts he has privileges on. We achieve this by implementing an Odoo data model that is fully controlled by a blockchain smart contract. This model incorporates access control features for all the data files as well as the associated files. Securing access can be through visibility, meaning hiding the related parts from unauthorized users, or by encryption. When a user attempts to view a given part, the blockchain-based reference manages checks the user's privileges on that part and then either send him the data or the encryption keys. The users identity verification aspects are fully managed by the Hyperledger platform which provide excellent certificate-based identity management.

In the future, we aim at using formats that allow for inner encapsulation through encryption such as formats based on OBJ, etc., or by using approaches similar to the ones discussed in section two where the data is encrypted using different methods. By doing this, we ensure that a single model can be used to represent the product and can allow for different access levels securely.

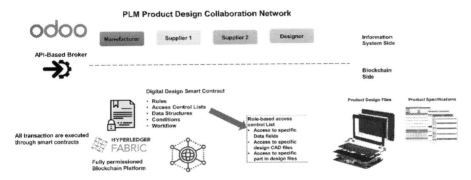

Fig. 3. Blockchain-based PLM data models access control

4   Results and Discussions

To evaluate and validate our proof-of-concept, we implemented a small collaborative design scenario involving one manufacturer and multiple suppliers. The scenario interface was implemented on Odoo using Python. All data handling and transaction management was shifted from Odoo to the blockchain through smart contracts. The scenario we investigate involves a laptop as a product and the manufacturer sets up a network with a different supplier for the parts. For each part, a 2D and a 3D data model is associated, and an access control list ACL specifies which other suppliers have the rights to see the data models. The access control list can be updated dynamically. We, unfortunately, could not cover partial access control due to the nature of the data models we used. We aim in the future to investigate using specific file and data formats to allow for more granularity in access control.

For our scenario, Fig. 4 represents a warning message for an unauthorized user trying to view a data model to which he has no rights. The data model will be hidden.

Figure 5 shows a screenshot of a user viewing a data model in 2D and 3D for a part he has the right to see.

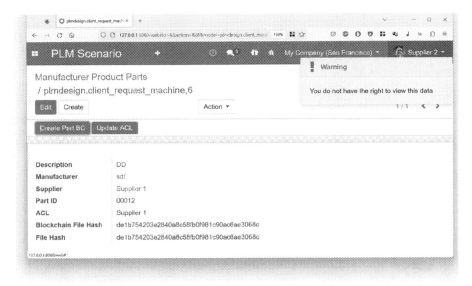

Fig. 4. Access Control enforcement scenario through blockchain integration

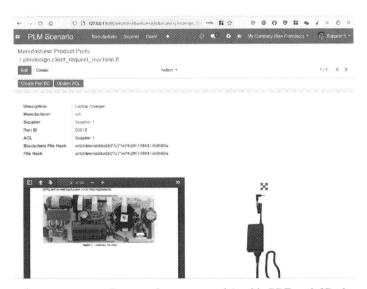

Fig. 5. A product component (Laptop charger example) with PDF and 3D data models + Blockchain-based access control

Figure 6 shows the interface of the manufacturer where he can add suppliers and parts to a product design while assigning dynamic access control rights (see Fig. 7).

Figure 8 and Fig. 9 show the blockchain log before and after an update to the access control lists ACLs.

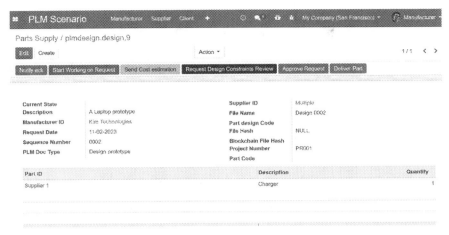

Fig. 6. Main product page and parts list (BOM)

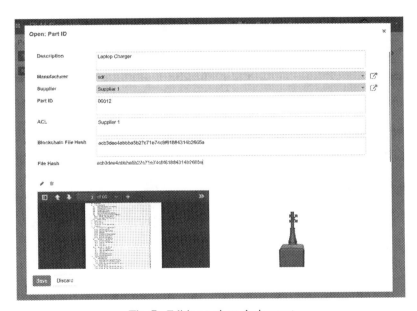

Fig. 7. Editing a given design part

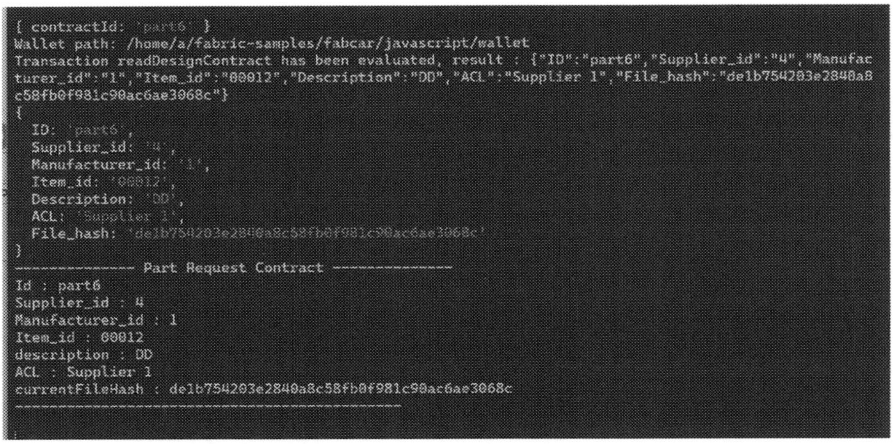

```
{ contractId: 'part6' }
Wallet path: /home/a/fabric-samples/fabcar/javascript/wallet
Transaction readDesignContract has been evaluated, result : {"ID":"part6","Supplier_id":"4","Manufac
turer_id":"1","Item_id":"00012","Description":"DD","ACL":"Supplier 1","File_hash":"de1b754203e2840a8
c58fb0f981c90ac6ae3068c"}
{
  ID: 'part6',
  Supplier_id: '4',
  Manufacturer_id: '1',
  Item_id: '00012',
  Description: 'DD',
  ACL: 'Supplier 1',
  File_hash: 'de1b754203e2840a8c58fb0f981c90ac6ae3068c'
}
-------------- Part Request Contract --------------
Id : part6
Supplier_id : 4
Manufacturer_id : 1
Item_id : 00012
description : DD
ACL : Supplier 1
currentFileHash : de1b754203e2840a8c58fb0f981c90ac6ae3068c
```

Fig. 8. Blockchain log for supplier part request smart contract

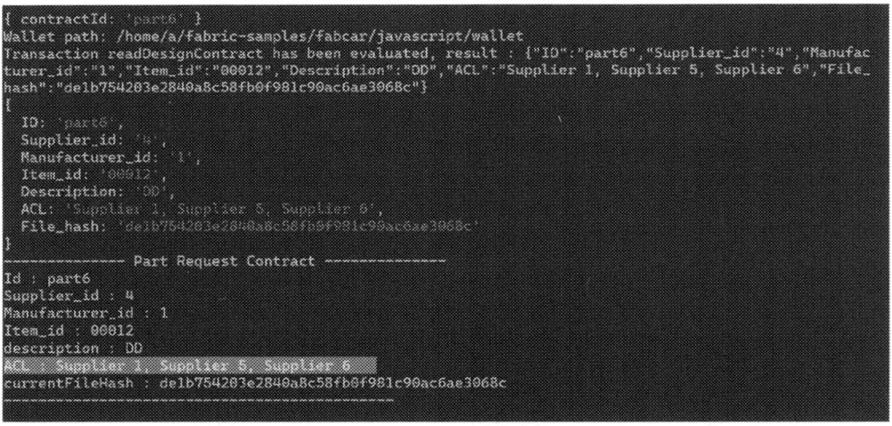

```
{ contractId: 'part6' }
Wallet path: /home/a/fabric-samples/fabcar/javascript/wallet
Transaction readDesignContract has been evaluated, result : {"ID":"part6","Supplier_id":"4","Manufac
turer_id":"1","Item_id":"00012","Description":"DD","ACL":"Supplier 1, Supplier 5, Supplier 6","File_
hash":"de1b754203e2840a8c58fb0f981c90ac6ae3068c"}
{
  ID: 'part6',
  Supplier_id: '4',
  Manufacturer_id: '1',
  Item_id: '00012',
  Description: 'DD',
  ACL: 'Supplier 1, Supplier 5, Supplier 6',
  File_hash: 'de1b754203e2840a8c58fb0f981c90ac6ae3068c'
}
-------------- Part Request Contract --------------
Id : part6
Supplier_id : 4
Manufacturer_id : 1
Item_id : 00012
description : DD
ACL : Supplier 1, Supplier 5, Supplier 6
currentFileHash : de1b754203e2840a8c58fb0f981c90ac6ae3068c
```

Fig. 9. ACL update blockchain log

5   Conclusion

Blockchain's impact on manufacturing and industrial applications was compared by some people to the wheel invention. This disruptive technology showed its potential in revolutionizing product design and manufacturing in many ways. In this paper, we investigated a new aspect of blockchain integration with industrial applications related to the access control of data models. We used Hyperledger Fabric as a permissioned blockchain platform to store the transaction data and manage the lifecycle of smart contracts. Hyperledger Fabric was mainly selected as it is an open-source platform used extensively in the product development environment. As for the information system, we used Odoo which is an open-source framework written in Python. Our access control approach assumes that for a given part of a product, an access control list is associated to allow actors in the network to view the part specification depending on their needs. Our approach relies fully on the blockchain to manage transactions between actors as well as to enforce access control for product parts data models. Although it would have been more beneficial to allow more granularity when setting access control policies for certain fields or areas in data models, in the future, we aim to address this aspect using advanced file format and data models.

Acknowledgment. This publication was made possible by NPRP grant NPRP11S-1227-170135 from the Qatar National Research Fund (a member of Qatar Foundation). The statements made herein are solely the responsibility of the authors (www.supplyledger.qa).

References

1. Saardchom, N.: Design patent war: Apple versus Samsung (2014). https://doi.org/10.1177/2277977914548341
2. Cera, C.D., Kim, T., Han, J.H., Regli, W.C.: Role-based viewing envelopes for information protection in collaborative modeling. Comput. Des. **36**, 873–886 (2004). https://doi.org/10.1016/J.CAD.2003.09.014
3. Qiu, Z.M., Kok, K.F., Wong, Y.S., Fuh, J.Y.H.: Role-based 3D visualisation for asynchronous PLM collaboration. Comput. Ind. **58**, 747–755 (2007). https://doi.org/10.1016/j.compind.2007.02.006
4. Elsheh, E., Ben Hamza, A.: Secret sharing approaches for 3D object encryption. Expert Syst. Appl. **38**, 13906–13911 (2011). https://doi.org/10.1016/J.ESWA.2011.04.197
5. Cai, X.T., Wang, S., Lu, X., Li, W.D., Liang, Y.W.: Parametric and adaptive encryption of feature-based computer-aided design models for cloud-based collaboration. Integr. Comput. Aided Eng. **24**, 129–142 (2017). https://doi.org/10.3233/ICA-160535
6. Cai, X.T., He, F.Z., Li, W.D., Li, X.X., Wu, Y.Q.: Encryption based partial sharing of CAD models. Integr. Comput. Aided Eng. **22**, 243–260 (2015). https://doi.org/10.3233/ICA-150487
7. Liang, Y., He, F., Li, H.: An asymmetric and optimized encryption method to protect the confidentiality of 3D mesh model. Adv. Eng. Inform. **42**, 100963 (2019). https://doi.org/10.1016/J.AEI.2019.100963
8. Cai, X., Li, W., He, F., Li, X.: Customized encryption of computer aided design models for collaboration in cloud manufacturing environment. J. Manuf. Sci. Eng. Trans. ASME **137** (2015). https://doi.org/10.1115/1.4030592/375274
9. Fabian, B., Kunz, S., Konnegen, M., Müller, S., Günther, O.: Access control for semantic data federations in industrial product-lifecycle management. Comput. Ind. **63**, 930–940 (2012). https://doi.org/10.1016/J.COMPIND.2012.08.015
10. Kim, T., Cera, C.D., Regli, W.C., Choo, H., Han, J.: Multi-level modeling and access control for data sharing in collaborative design. Adv. Eng. Inform. **20**, 47–57 (2006). https://doi.org/10.1016/J.AEI.2005.05.016
11. Lohmer, J., Lasch, R.: Blockchain in operations management and manufacturing: potential and barriers. Comput. Ind. Eng. **149**, 106789 (2020). https://doi.org/10.1016/j.cie.2020.106789
12. Shahbazi, Z., Byun, Y.-C.: Integration of blockchain, IoT and machine learning for multistage quality control and enhancing security in smart manufacturing. Sensors **21**, 1467 (2021). https://doi.org/10.3390/s21041467
13. Hasan, H.R., Salah, K., Jayaraman, R., Ahmad, R.W., Yaqoob, I., Omar, M.: Blockchain-based solution for the traceability of spare parts in manufacturing. IEEE Access **8**, 100308–100322 (2020). https://doi.org/10.1109/ACCESS.2020.2998159

14. Kurpjuweit, S., Schmidt, C.G., Klöckner, M., Wagner, S.M.: Blockchain in additive manufacturing and its impact on supply chains. J. Bus. Logist. **42**, 46–70 (2021). https://doi.org/10.1111/jbl.12231

15. Aghamohammadzadeh, E., Fatahi Valilai, O.: A novel cloud manufacturing service composition platform enabled by blockchain technology. Int. J. Prod. Res. **58**, 5280–5298 (2020). https://doi.org/10.1080/00207543.2020.1715507

16. Lee, J., Azamfar, M., Singh, J.: A blockchain enabled cyber-physical system architecture for industry 4.0 manufacturing systems. Manuf. Lett. **20**, 34–39 (2019). https://doi.org/10.1016/j.mfglet.2019.05.003

17. Yu, C., Jiang, X., Yu, S., Yang, C.: Blockchain-based shared manufacturing in support of cyber physical systems: concept, framework, and operation. Robot. Comput. Integr. Manuf. **64**, 101931 (2020). https://doi.org/10.1016/j.rcim.2019.101931

18. Francisco, K., Swanson, D.: The supply chain has no clothes: technology adoption of blockchain for supply chain transparency. Logistics **2**, 2 (2018)

19. Kim, H.M., Laskowski, M.: Toward an ontology-driven blockchain design for supply-chain provenance. Intell. Syst. Account. Finance Manag. **25**, 18–27 (2018)

20. Mattila, J., Seppälä, T., Holmström, J.: Product-centric information management: a case study of a shared platform with blockchain technology (2016)

21. Madhwal, Y., Panfilov, P.B.: Blockchain and supply chain management: aircrafts' parts' business case. In: Annals of DAAAM Proceedings, 28 (2017)

22. Belhi, A., Bouras, A., Patel, M.K., Aouni, B.: Blockchains: a conceptual assessment from a product lifecycle implementation perspective. In: Nyffenegger, F., Ríos, J., Rivest, L., Bouras, A. (eds.) PLM 2020. IFIP AICT, vol. 594, pp. 576–589. Springer, Cham (2020). https://doi.org/10.1007/978-3-030-62807-9_46

23. Belhi, A., Gasmi, H., Bouras, A., Aouni, B., Khalil, I.: Integration of business applications with the blockchain: Odoo and hyperledger fabric open source proof of concept. IFAC-PapersOnLine **54**, 817–824 (2021)

24. Belhi, A., Gasmi, H., Hammi, A., Bouras, A., Aouni, B., Khalil, I.: TMSLedger: a transactions management system through integrated Odoo hyperledger smart contracts. In: Bouras, A., Khalil, I., Aouni, B. (eds.) Blockchain Driven Supply Chains and Enterprise Information Systems, pp. 201–220. Springer, Cham (2023). https://doi.org/10.1007/978-3-030-96154-1_11

25. Belhi, A., Gasmi, H., Hammi, A., Bouras, A., Aouni, B., Khalil, I.: A broker-based manufacturing supply chain integration with blockchain: managing Odoo workflows using hyperledger fabric smart contracts. In: Canciglieri Junior, O., Noël, F., Rivest, L., Bouras, A. (eds.) PLM 2021. IFIP AICT, vol. 640, pp. 371–385. Springer, Cham (2022). https://doi.org/10.1007/978-3-030-94399-8_27

Smart Product-Service Systems: A Review and Preliminary Approach to Enable Flexible Development Based on Ontology-Driven Semantic Interoperability

Athon Francisco Staben de Moura Leite[1,2]([✉]) (iD), Matheus Beltrame Canciglieri[1] (iD), and Osiris Canciglieri Junior[1] (iD)

[1] Industrial and Systems Engineering Graduate Program, Pontifical Catholic University of Paraná, Curitiba, PR, Brazil
{athon.leite,matheus.beltrame,osiris.canciglieri}@pucpr.br
[2] Robert Bosch do Brasil, Curitiba, PR, Brazil

Abstract. This research conducts a systematic literature review on Smart Product-Service Systems (SPSS) to identify the current state of research and the gap related to flexibility issues by the views of semantics, requirements, and design methodologies. The review covers studies published between 2000 and 2022 in and finds a lack of systemic approaches on standardized formalization models, interoperability, context-aware systems, and self-adaptation. This gap in knowledge makes it difficult for companies to fully understand and implement a seamless and flexible SPSS development lifecycle. In response to this gap, the research proposes a preliminary approach for the implementation and management of SPSS, offering a possible solution for companies looking to understand and implement SPSS. The approach suggests that companies should technically focus on use of Artificial Intelligence (AI) technologies to support decision-making, and on formal standardized models to represent knowledge, helping to enable semantic interoperability across the development lifecycle. The proposed preliminary approach is a starting point for companies and for future research in the field.

Keywords: Smart Product-Service Systems · Ontology · Semantic Interoperability · Artificial Intelligence · Literature Review

1 Introduction

The concept of Product-Service Systems (PSS) originated as a way for manufacturers to offer a comprehensive after-sales service to their customers, ensuring that products continue to function optimally over their lifetime [1]. Over time, the concept of PSS has expanded to include a range of services that go beyond just after market applications, incorporating value-added services such as product upgrades, extended warranties, and even access to digital content, among other examples [2].

© IFIP International Federation for Information Processing 2024
Published by Springer Nature Switzerland AG 2024
C. Danjou et al. (Eds.): PLM 2023, IFIP AICT 702, pp. 151–162, 2024.
https://doi.org/10.1007/978-3-031-62582-4_14

Concurrently, with the advent of the Internet of Things (IoT) and advanced technologies, PSS has evolved into a smart and connected system that can collect and analyse data from products and services in use [3]. This data can be used to predict and prevent product failures, offer personalized services, and create new revenue streams for manufacturers [4].

Smart PSS represents a shift in the traditional PSS approach by increasing the degree of co-creation value proposition, where the focus is creating a superior customer experience and maximizing the value derived from the product whilst involving stakeholders across the entire development process [5, 6]. In this context, Smart Product-Service Systems (SPSS) have become a popular strategy for companies to improve their value offer and providing a more holistic deliver to attend customer needs. However, despite the growing interest in this field, there is still a lack of understanding when it comes to the implementation and management of SPSS when considering the dynamic nature of its requirements and design flexibility across its development lifecycle [7]. In this context, this research aims to identify the current state of research in the field of SPSS, with a specific focus on issues concerning semantics, requirements, and flexibility in design methodologies.

Following the study of this research gap, the paper proposes a preliminary framework for the implementation and management of SPSS considering the developmental issues in the areas approached by the literature review. This continuous and evolving framework can be a starting point for companies and academics looking to understand and implement flexible development of SPSS in their operations, and for future research in the field.

1.1 Methodology

The exploratory goals of this research are applied in nature and qualitative in approach. A systematic literature review and content analysis are the research methods [8]. The four steps of the methodical procedure are Problem Identification, Systematic Literature Review, Content Analysis, and Solution Proposal, as shown in Fig. 1.

Fig. 1. Methodological Procedure.

Problem identification (Detail 1 of Fig. 1) is a crucial step in the research process as it sets the foundation for the entire project. In the context of this research, the problem identification process involved a comprehensive analysis of the existing approaches to the development of Smart Product-Service Systems (SPSS), as approached in the introductory section of this research. To gather relevant information on SPSS and related fields, a systematic literature review is proposed (Detail 2 of Fig. 1). This review covers various databases, by a database aggregator engine, to collect and analyze the most relevant studies in the field. The literature review will help to identify the current state of the art in the theme and gaps in the current approaches.

The results of the literature review will be then analyzed using a qualitative content analysis technique to identify main contributions and limitations used in research (Detail 3 of Fig. 1). The content analysis will help identify some of the key factors that that have been addressed by research and general points for improvement. Based on the results of the content analysis, a preliminary framework for the development of more flexible SPSS will be proposed (Detail 4 of Fig. 1). The framework aims to provide a flexible and adaptable approach to the development of SPSS that can be adapted to different scenarios and dynamic requirements. The framework provides a systematic approach to the development of SPSS that can be used by engineers, designers, and other stakeholders in the development process.

2   Systematic Literature Review

2.1   Review Methodology

A systematic literature review will be conducted to gather relevant information on Smart Product-Service Systems (SPSS) and related fields by the perspective of issues related to design, requirements, and semantics. The review followed the PRISMA (Preferred Reporting Items for Systematic Reviews and Meta-Analyses) guidelines to ensure the quality and rigor of the process [9]. The methodology provides a standardized framework for reporting on the quality and rigor of the review process, being widely recognized and used in the scientific community to ensure the transparency and quality of systematic literature reviews (Fig. 2).

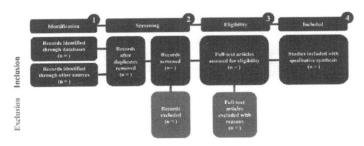

Fig. 2. PRISMA.

The review covers various databases, by using LENS.ORG, a scientific database aggregator comprising a vast number of bases such as Springer, Science Direct, Scopus, Emerald Insight, IEEEXplore, Taylor & Francis Online and many others. The search will be conducted using clusters of keywords related to each of the main issues studied in the review through a previous study [10] such as "Product-Service System", "Hybrid Product", "Smart Product-Service Systems" (for PSS design issues); "Requirements Elicitation", "Requirements Specification", "Requirements Validation" (for requirements issues); and "Semantic Interoperability", "Semantic Interop"(for semantic issues). The inclusion and exclusion criteria will be established across review phases to ensure that only relevant studies are included in the review.

The results of the literature review will be then analyzed to identify the current state of the art in the development of SPSS and to identify the gap in the current approaches. The final analysis will be conducted using content analysis techniques to identify the key themes, tools, contributions, and limitations in the research.

2.2 PRISMA

PRISMA (Preferred Reporting Items for Systematic Reviews and Meta-Analyses) is a guideline for conducting systematic literature reviews and meta-analyses, which focuses on delivering a standardized approach for reviews in many fields of research [11]. In this study the review will be split into the four main phases of PRISMA. In terms of objectives, this literature review should answer the following questions:

1. **What are the most relevant works, from the last 20 years, that approach identifying/reducing/solving issues related to semantics, design and requirements' structure in the development of Product-Service Systems?**
2. **What are the main contributions and limitations of these papers?**
3. **Which tools, methods and/or technologies from these works were applied to reduce/solve such issues?**

Identification

In this research, the identification phase started out by defining the criteria for surveying the literature, establishing period range, language, type of publications, and area of concentration/domain within area of concentration. A summary of the identification criteria is depicted in Table 1.

Table 1. Identification Criteria.

Search Criteria
From 2000 – 2022
English written papers
Related to Manufacturing Engineering AND/OR Systems Engineering and its adjacent areas
Journal Papers AND/OR Conference Papers AND/OR Book Chapters

Then, the querying method was established, based on the combinations of keyword clusters triple-wise (one cluster of each theme) and pairwise (a cluster from one theme combined with another). Additionally, 136 additional papers were added to the analysis based on papers discovered during the exploratory investigations. The results of this phase were the identification of n = 9112 papers for further screening, bringing an overview of more than two decades in the field of study and adjacent areas.

Screening

Using the aid of a spreadsheet software, the screening procedure began by eliminating any duplicates from the total number of papers discovered during the previous phase.

The unique registers of the total number of entries led to a total of n = 5901 publications being left for screening based on title and DOI.

Table 2. Screening Criteria.

Screening Criteria
Peer-Reviewed
Title or abstract approaches: (PSS Development) **AND/OR** (Flexible Design) **AND/OR** (Interoperability) **AND/OR** (Requirements) **AND/OR** (Semantics)
Paper in the field of interest and adjacent areas
Approach still in use (if applicable)

The screening process then, as shown in Table 2, examined the titles and abstracts of each of the distinct registries, using as a basis for exclusion papers that did not have peer review and/or did not have a defined strategy for resolving issues of interest and/or did not have outdated strategies and/or did not have familiarity with the field of interest and its surrounding areas. As a result of this analysis, n = 274 papers were chosen as eligible from the n = 5901 publications examined.

Eligibility

The Eligibility step of this literature review started by analyzing general aspects from selected papers from the screening process. From the 274 papers left after the screening process, 201 were eligible to be assessed, based on paper availability using the institutional access in research databases. Classification criteria for publications were established based on each of the major topics covered in the review. Based on the literature that was discovered and the primary research problem interests, an affinity score ranging from 0 to 3 was used to determine the affinity of the following criteria:

- Flexibility in Design: The degree to which the paper addresses how design choices can impact the flexibility of a system development.
- Flexible Requirements Management: The degree to which the paper presents methods to elicit or manage flexible requirements.
- Standardised and/or Well-defined Representation: Analyse the extent of how papers address the importance to standardised and/or well-defined knowledge representation (addressing semantics or not).

Included

Ultimately, the rules for paper inclusion and exclusion were determined, in order to determine the most relevant papers in the field concerning the main issues addressed by this literature review, being the Rule for acceptance a grade ≥ 2 in more than one criterion; and for rejection a grade $= 0$ (in any criteria) OR Grade < 2 in more than one criterion.

As seen in Table 3, this selection of papers answers the first question of the literature review:

1. **What are the most relevant works, from the last 20 years, that approach identifying/reducing/solving issues related to semantics, design and requirements' structure in the development of Product-Service Systems?**

Table 3. Selected Papers.

Id	Selected Paper	Id	Selected Paper	Id	Selected Paper
1	Khedr, M., & Karmouch, A. (2004) [12]	13	Maleki, E., Belkadi, F., Zhang, Y., Bernard, A. (2016) [35]	25	Liu, Z., Ming, X., Zhang, X. (2019) [4]
2	Durugbo, C., Hutabarat, W., Tiwari, A., Alcock, J. R. (2010) [14]	14	Estrada, A., Romero, D. (2016) [37]	26	Wang, Z., Chen, C. H., Zheng, P., Li, X., Khoo, L. P. (2019) [24]
3	Akmal, S., Batres, R. (2011) [16]	15	Lazoi, M., Pezzotta, G., Pirola, F., Margarito, A. (2016) [39]	27	Liu, Z., Ming, X., Qiu, S., Qu, Y., Zhang, X. (2020) [26]
4	Berkovich, M., Leimeister, J. M., Hoffmann, A., Krcmar, H. (2011) [18]	16	Trevisan, L., Brissaud, D. (2016) [41]	28	Chen, Z., Ming, X., Wang, R., Bao, Y. (2020) [28]
5	Berkovich, M., Leimeister, J. M., Krcmar, H. (2011) [20]	17	Wiesner, S., Lampathaki, F., Biliri, E., Thoben, K. D. (2016) [43]	29	Farsi, M., Erkoyuncu, J. A. (2020) [30]
6	Dong, M., Yang, D.; Su, L. (2011) [22]	18	Scholze, S., Correia, A. T., & Stokic, D. (2016)	30	Watanabe, K., Okuma, T., Takenaka, T. (2020) [32]
7	Berkovich, M., Leimeister, J. M., Hoffmann, A., Krcmar, H. (2012) [23]	19	Neves-Silva, R., Pina, P., Spindler, P., Pezzotta, et al. (2016) [45]	31	Zhang, X., Ming, X., Yin, D. (2020) [34]
8	Akasaka, F., Nemoto, Y., Chiba, R., & Shimomura, Y. (2012) [25]	20	Zhang, J., Ahmad, B., Vera, D., Harrison, R. (2016) [13]	32	Zuoxu, W., Xinyu, L., Pai, Z., Chun-hsien, C., Pheng, K. L., Pss, A. (2020) [36]
9	Akmal, S., Batres, R., & Shih, L. H (2013) [27]	21	Correia, A., Stokic, D., Siafaka, R., Scholze, S. (2017) [15]	33	Li, X., Chen, C. H., Zheng, P., Wang, Z., Jiang, Z., Jiang, Z. (2020) [38]

(*continued*)

Table 3. (*continued*)

Id	Selected Paper	Id	Selected Paper	Id	Selected Paper
10	Schmidt, D. M., Malaschewski, O., Fluhr, D., Mörtl, M. (2015) [29]	22	Wu, Y., Lee, J. H., Kim, Y. S., Lee, S. W., Kim, S. J., Yuan, X. (2017) [17]	34	Guillon, D., Ayachi, R., Vareilles, É., Aldanondo, M., Villeneuve, É., Merlo, C. (2021) [40]
11	Peruzzini, M., Marilungo, E., Germani, M. (2015) [31]	23	Wiesner, S., Westphal, I., Thoben, K. D. (2017) [19]	35	Rosa, M., Wang, W. M., Stark, R., Rozenfeld, H. (2021) [42]
12	Zhu, H., Gao, J., Cai, Q. (2015) [33]	24	Savarino, P., Abramovici, M., Göbel, J. C., Gebus, P. (2018) [21]	36	Yang, X., Wang, R., Tang, C., Luo, L., Mo, X. (2021) [5]
				37	Wu, C., Chen, T., Li, Z., Liu, W. (2021) [44]

Based on the rigorous analysis of literature from the past twenty years, it can be said that the papers selected in this literature review brought a holistic view on the studied issues.

Content Analysis

The topics discussed in these papers relate to the design, development, and evaluation of Product-Service Systems (PSS). These papers cover various aspects of PSS design, such as requirements analysis, ontology development, service selection, customer acceptance, and resource allocation. The authors proposed different approaches and methodologies for PSS design, ranging from QFD-based approaches to ontology-based approaches, and from structured methodologies to Multicriteria Decision-making methods. The Content Analysis thoroughly analyzed each of the 37 papers and found their contributions and limitations, and the main tools, methods and methodologies used in literature, answering the remaining questions of the literature review:

2. **What are the main contributions and limitations of these papers?**

These papers discuss many facets of design, requirements analysis, and semantics from the viewpoint of ontology development, as well as service selection, client acceptance, and resource allocation as additional areas that were discovered. It was also clear that cooperation and communication amongst the many actors involved in PSS development are important factors to take into account during development. Numerous tools and modular concepts were developed to manage the interfaces between different players in order to address this issue, however one of the major drawbacks noted in numerous articles was the absence of contemporary computational and standardized methodologies and implementations towards Industry 4.0 concepts, as well as clearly defined development methodologies. This impacts not only the scalability of such solutions but also hinders the validation of them under multiple scenarios.

3. **Which tools, methods and/or technologies from these works were applied to reduce/solve such issues?**

An evaluation of the tools and techniques discovered in the literature review has demonstrated that not all solutions explored are fully integrated or interoperable, needing additional exploration to overcome the limits of the research. By suggesting the integration of various technologies that could facilitate the development process and deal with the reconfiguration of Smart Product-Service Systems in response to the primary studied issues, the suggested solution for the research problem will complement the solutions offered in the found papers with new elements.

3   Preliminary Approach for Flexible Smart PSS Development

Based on the limitations and technologies found on current literature, a preliminary approach for flexible design in Smart Product-Service Systems, adapting/combining steps from products/services/systems' well explored development methodologies such as Integrated Product Development Process (IPDP), DevOps and Model-Based Systems Engineering (MBSE), is proposed, including tools found on selected literature based on Ontologies, Recommender Systems and Multicriteria Decision-Making (MCDM). A representation of the approach is depicted in Fig. 3.

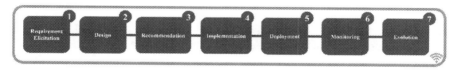

Fig. 3. Preliminary Approach.

The steps and tools of the proposed approach are:

1. Requirement Elicitation: Identifying the specific needs of the customers and the corresponding requirements for the Smart Product-Service System. This step defines the weights for development features (decision criteria), extracts information from customer and system feedback and identify patterns and trends through Natural Language Processing (NLP), Multicriteria Decision Making (MCDM) and chatbots.
2. Design: Approaches the conceptual design of the system (new or existent), including the product and service components, their interactions, and the overall system architecture in form of Ontology models that represent the domain knowledge, entities, and concepts of the system.
3. Recommendation: Analyze the impacts of new and existing requirements, providing a suggestion of best resource allocation based on resource limitations, recommends features based on gathered requirements, and suggests design alternatives for new features based on MCDM and Recommender Systems with aid of Ontologies.

4. Implementation: Building and testing the system, including the development of the software and hardware components, and the integration of the product and service components by the aspect of new features.
5. Deployment: Deploying the system in the field, including the installation and commissioning of the product and service components, and the training of users and maintenance personnel.
6. Monitoring: Continuously monitoring the system and its components, and performing maintenance and updates as needed to ensure optimal performance.
7. Evolution: Continuously adapting and evolving the system to meet changing customer needs and market trends.

This is still a preliminary version of the solution, containing some of the found tools in the systematic literature review and adding some new tools to the process (such as Natural Language Processing and Chatbots). The main focus of this approach is to provide a deeper analysis of the requirements and focusing on scalability and implementation of techniques in the scope of Industry 4.0 (the main identified gap between selected papers). This approach will need high synergy and connection with development methodologies to be fully embraced but has the potential to manage requirements in an easier, more complete and holistic manner.

4 Conclusion and Future Works

The study aimed to address the challenges faced in the design of Smart Product-Service systems by proposing a preliminary solution that addresses the issues of requirements, design methods, and semantics. It conducted an extensive literature review and analysed various works on the field of study. The results of this analysis provided a basis for the development of the proposed solution.

The review provided a deeper analysis of the current state of research on the field of flexibility in design for Smart Product-Service Systems while exploring the main contributions, limitations and tools/methodologies used by authors to overcome the challenges. Additionally, the preliminary approach promotes the integration of new technologies and innovations, allowing organizations to enhance their solutions and improve the customer experience. However, more studies are necessary, and a deeper analysis of the proposed solution is necessary to provide further insights.

For future works, based on the current limitations of the research, the application and further development of the approach are recommended, while updating the literature review to more recent works and methods. Another possible outcome is working on the potential complexity of the framework, particularly in the areas of technology integration and implementation.

Acknowledgment. The authors especially thank the financial support of Pontifical Catholic University of Parana (PUCPR) - Polytechnic School – Industrial and Systems Engineering Graduate Program (PPGEPS), Robert Bosch do Brasil, the Brazilian National Council for Scientific and Technological Development (CNPq) and the Coordination for the Improvement of Higher Education Personnel in Brazil (CAPES).

References

1. Valencia, A., Mugge, R., Schoormans, J.P.L., Schifferstein, H.N.J.: The design of smart product-service systems (PSSs): an exploration of design characteristics. Int. J. Des. **9**, 13–28 (2015)
2. Chowdhury, S., Haftor, D., Pashkevich, N.: Smart product-service systems (smart PSS) in industrial firms: a literature review. Procedia CIRP **73**, 26–31 (2018). https://doi.org/10.1016/j.procir.2018.03.333
3. Zheng, P., Wang, Z., Chen, C.H.: Smart product-service systems: a novel transdisciplinary sociotechnical paradigm. Adv. Transdiscipl. Eng. **10**, 234–241 (2019). https://doi.org/10.3233/ATDE190128
4. Liu, Z., Ming, X., Zhang, X.: A perspective on methodological framework integrating revised rough-DEMATEL to co-generate and analyze requirements for smart product-service system. In: ACM International Conference Proceeding Series, pp. 240–247 (2019). https://doi.org/10.1145/3312662.3312667
5. Yang, X., Wang, R., Tang, C., Luo, L., Mo, X.: Emotional design for smart product-service system: a case study on smart beds. J. Clean. Prod. **298**, 126823 (2021). https://doi.org/10.1016/j.jclepro.2021.126823
6. Abdel-Basst, M., Mohamed, R., Elhoseny, M.: A novel framework to evaluate innovation value proposition for smart product–service systems. Environ. Technol. Innov. **20**, 101036 (2020). https://doi.org/10.1016/j.eti.2020.101036
7. Zheng, P., Wang, Z., Chen, C.H., Pheng Khoo, L.: A survey of smart product-service systems: key aspects, challenges and future perspectives. Adv. Eng. Inform. **42**, 100973 (2019). https://doi.org/10.1016/j.aei.2019.100973
8. Gray, D.E.: Doing Research in the Real World. Sage (2013)
9. Liao, Y., Deschamps, F., de Freitas Rocha Loures, E., Ramos, L.F.P.: Past, present and future of industry 4.0 - a systematic literature review and research agenda proposal. Int. J. Prod. Res. **55**, 3609–3629 (2017). https://doi.org/10.1080/00207543.2017.1308576
10. Leite, A.F.C.S.M., et al.: Current issues in the flexibilization of smart product-service systems and their impacts in industry 4.0. Procedia Manuf. **51**, 1153–1157 (2020). https://doi.org/10.1016/j.promfg.2020.10.162
11. Canciglieri, M.B., et al.: A systematic literature mapping on the process reconfiguration of smart manufacturing systems with the integration of multi-criteria decision models and ontology based interoperability. In: Kim, K.Y., Monplaisir, L., Rickli, J. (eds.) FAIM 2022. LNME, pp. 647–654. Springer, Cham (2023). https://doi.org/10.1007/978-3-031-17629-6_68
12. Khedr, M., Karmouch, A.: Negotiating context information in context-aware systems. IEEE Intell. Syst. **19**, 21–29 (2004)
13. Zhang, J., Ahmad, B., Vera, D., Harrison, R.: Ontology based semantic-predictive model for reconfigurable automation systems. In: IEEE International Conference on Industrial Informatics (INDIN), pp. 1094–1099 (2016). https://doi.org/10.1109/INDIN.2016.7819328
14. Durugbo, C.: Integrated product-service analysis using SysML requirement diagrams. Syst. Eng. **14**, 111–123 (2013)
15. Correia, A., Stokic, D., Siafaka, R., Scholze, S.: Ontology for colaborative development of product service systems based on basic formal ontology. In: 2017 International Conference on Engineering, Technology and Innovation: Engineering, Technology and Innovation Management Beyond 2020: New Challenges, New Approaches, ICE/ITMC 2017 – Proceedings, January 2018, pp. 1173–1180 (2017). https://doi.org/10.1109/ICE.2017.8280014
16. Akmal, S., Batres, R.: Ontology-based semantic similarity measures for product-service system design. In: 2011 IEEE/SICE International Symposium on System Integration, SII 2011, pp. 932–937 (2011). https://doi.org/10.1109/SII.2011.6147574

17. Wu, Y., Lee, J.H., Kim, Y.S., Lee, S.W., Kim, S.J., Yuan, X.: A similarity measurement framework of product-service system design cases based on context-based activity model. Comput. Ind. Eng. **104**, 68–79 (2017). https://doi.org/10.1016/j.cie.2016.12.015
18. Berkovich, M., Leimeister, J.M., Krcmar, H.: Requirements engineering for product service systems; a state of the art analysis. Bus. Inf. Syst. Eng. **3**, 369 (2011)
19. Wiesner, S., Westphal, I., Thoben, K.D.: Through-life engineering in product-service systems - tussles for design and implementation. Procedia CIRP **59**, 227–232 (2017). https://doi.org/10.1016/j.procir.2016.09.006
20. Berkovich, M., Hoffmann, A., Leimeister, J.M.: It-enabled product service systems, 50–58 (2011)
21. Savarino, P., Abramovici, M., Göbel, J.C., Gebus, P.: Design for reconfiguration as fundamental aspect of smart products. Procedia CIRP **70**, 374–379 (2018). https://doi.org/10.1016/j.procir.2018.01.007
22. Dong, M., Yang, D., Su, L.: Ontology-based service product configuration system modeling and development. Expert Syst. Appl. **38**, 11770–11786 (2011). https://doi.org/10.1016/j.eswa.2011.03.064
23. Berkovich, M., Leimeister, J.M., Hoffmann, A., Krcmar, H.: A requirements data model for product service systems. Requir. Eng. **19**, 161–186 (2012). https://doi.org/10.1007/s00766-012-0164-1
24. Wang, Z., Chen, C.H., Zheng, P., Li, X., Khoo, L.P.: A graph-based context-aware requirement elicitation approach in smart product-service systems. Int. J. Prod. Res. **59**, 635–651 (2019). https://doi.org/10.1080/00207543.2019.1702227
25. Akasaka, F., Nemoto, Y., Chiba, R., Shimomura, Y.: Development of PSS design support system: knowledge-based design support and qualitative evaluation. Procedia CIRP **3**, 239–244 (2012). https://doi.org/10.1016/j.procir.2012.07.042
26. Liu, Z., Ming, X., Qiu, S., Qu, Y., Zhang, X.: A framework with hybrid approach to analyse system requirements of smart PSS toward customer needs and co-creative value propositions. Comput. Ind. Eng. **139**, 105776 (2020). https://doi.org/10.1016/j.cie.2019.03.040
27. Akmal, S., Batres, R., Shih, L.H.: An ontology-based approach for product-service system design. In: CIRP IPS2 Conference 2012 (2013)
28. Chen, Z., Ming, X., Wang, R., Bao, Y.: Selection of design alternatives for smart product service system: a rough-fuzzy data envelopment analysis approach. J. Clean. Prod. **273**, 122931 (2020). https://doi.org/10.1016/j.jclepro.2020.122931
29. Schmidt, D.M., Mörtl, M.: Product-service systems for increasing customer acceptance concerning perceived complexity. In: A-DEWS 2015 - Design Engineering in the Context of Asia - Asian Design Engineering Workshop, Proceedings, pp. 77–82 (2016)
30. Farsi, M., Erkoyuncu, J.A.: An agent-based model for flexible customization in product-service systems. Procedia CIRP **96**, 39–44 (2020). https://doi.org/10.1016/j.procir.2021.01.049
31. Peruzzini, M., Marilungo, E., Germani, M.: Structured requirements elicitation for product-service system. Int. J. Agile Syst. Manag. **8**, 189–218 (2015). https://doi.org/10.1504/IJASM.2015.073516
32. Watanabe, K., Okuma, T., Takenaka, T.: Evolutionary design framework for smart PSS: service engineering approach. Adv. Eng. Inform. **45**, 101119 (2020). https://doi.org/10.1016/j.aei.2020.101119
33. Zhu, G.N., Hu, J.: A rough-Z-number-based DEMATEL to evaluate the co-creative sustainable value propositions for smart product-service systems. Int. J. Intell. Syst. **36**, 3645–3679 (2021). https://doi.org/10.1002/int.22431
34. Zhang, X., Ming, X., Yin, D.: Application of industrial big data for smart manufacturing in product service system based on system engineering using fuzzy DEMATEL. J. Clean. Prod. **265**, 121863 (2020). https://doi.org/10.1016/j.jclepro.2020.121863

35. Maleki, E., Belkadi, F., Zhang, Y., Bernard, A.: Towards a new collaborative framework supporting the design process of industrial product service systems. In: Eynard, B., Nigrelli, V., Oliveri, S., Peris-Fajarnes, G., Rizzuti, S. (eds.) Advances on Mechanics, Design Engineering and Manufacturing. LNME, pp. 139–146. Springer, Cham (2017). https://doi.org/10.1007/978-3-319-45781-9_15
36. Zuoxu, W., Xinyu, L., Pai, Z., Chun-hsien, C., Pheng, K.L., Pss, A.: Smart product-service system configuration: a novel hypergraph model- based approach (2020)
37. Estrada, A., Romero, D.: A system quality attributes ontology for product-service systems functional measurement based on a holistic approach. Procedia CIRP **47**, 78–83 (2016). https://doi.org/10.1016/j.procir.2016.03.215
38. Li, X., Chen, C.H., Zheng, P., Wang, Z., Jiang, Z., Jiang, Z.: A knowledge graph-aided concept–knowledge approach for evolutionary smart product–Service system development. J. Mech. Des. Trans. ASME **142** (2020). https://doi.org/10.1115/1.4046807
39. Lazoi, M., Pezzotta, G., Pirola, F., Margarito, A.: Toward a PSS lifecycle management systems: considerations and architectural impacts. In: IESA (2016)
40. Guillon, D., Ayachi, R., Vareilles, É., Aldanondo, M., Villeneuve, É., Merlo, C.: Productϒservice system configuration: a generic knowledge-based model for commercial offers. Int. J. Prod. Res. **59**, 1021–1040 (2021). https://doi.org/10.1080/00207543.2020.1714090
41. Trevisan, L., Brissaud, D.: Engineering models to support product–service system integrated design. CIRP J. Manuf. Sci. Technol. **15**, 3–18 (2016). https://doi.org/10.1016/j.cirpj.2016.02.004
42. Rosa, M., Wang, W.M., Stark, R., Rozenfeld, H.: A concept map to support the planning and evaluation of artifacts in the initial phases of PSS design (2021). https://doi.org/10.1007/s00163-021-00358-9
43. Wiesner, S., Lampathaki, F., Biliri, E., Thoben, K.D.: Requirements for cross-domain knowledge sharing in collaborative product-service system design. Procedia CIRP **47**, 108–113 (2016). https://doi.org/10.1016/j.procir.2016.03.118
44. Wu, C., Chen, T., Li, Z., Liu, W.: A function-oriented optimising approach for smart product service systems at the conceptual design stage: a perspective from the digital twin framework. J. Clean. Prod. **297**, 126597 (2021). https://doi.org/10.1016/j.jclepro.2021.126597
45. Neves-Silva, R., et al.: Supporting context sensitive lean product service engineering. Procedia CIRP **47**, 138–143 (2016). https://doi.org/10.1016/j.procir.2016.03.103

A Preliminary Discussion of Semantic Web Technologies and 3D Feature Recognition to Support the Complex Parts Manufacturing Quotation: An Aerospace Industry Case

Murillo Skrzek[1], Anderson Luis Szejka[1(✉)] [iD], and Fernando Mas[2,3] [iD]

[1] Industrial and Systems Engineering Graduate Program (PPGEPS),
Pontifical Catholic University of Paraná (PUCPR), Curitiba, Brazil
{murillo.skrzek,anderson.szejka}@pucpr.br
[2] Mechanical and Manufacturing Engineering Department, University of Seville (US),
Seville, Spain
fmas@us.es
[3] M&M Group, Cadiz, Spain

Abstract. The complexity of airplane parts is constantly increasing. For example, the airplane fuselage comprises more than 700.000 parts with different dimensions, geometries, tolerances complexity, and multiple materials. This complexity scenario requires an advanced manufacturing park with conventional and non-conventional machining, rapid manufacturing machining, 3D measuring machines, and others. In parallel, reducing these parts' cost and production time challenges the aerospace industry suppliers, ensuring high manufacturing quality. Therefore, this paper explores a preliminary discussion of the Semantic Web Technologies and 3D feature recognition application to assist the complex parts manufacturing pricing and manufacturing planning since this process is made manually by a manufacturing engineer. During all quotation processes (3D part analysis, manufacturing planning, and manufacturing price definition), the engineer spends more than one week to quote a part depending on its complexity. This paper contributes to a Computer Integrated Manufacturing (CIM) domain since it discusses the application of Semantic web technologies and 3D feature recognition technologies to automate extracting the information from the part modelled in 3D and plan the manufacturing process to help to identify the price to manufacture it. Finally, the main research result concerns the identification of the contributions and limitations of the related works in this domain and the research opportunity to cover this research gap.

Keywords: Complex Parts Manufacturing · Quotation · Manufacturing Planning · Aerospace Industry · Semantic Web Technologies · 3D Feature Recognition

© IFIP International Federation for Information Processing 2024
Published by Springer Nature Switzerland AG 2024
C. Danjou et al. (Eds.): PLM 2023, IFIP AICT 702, pp. 163–172, 2024.
https://doi.org/10.1007/978-3-031-62582-4_15

1 Introduction

The complexity of airplane parts has increased significantly in recent years, requiring advanced manufacturing parks and innovative approaches to address the challenges of reducing costs and production time while ensuring high quality. In addition, the creation and production of an aircraft require the collaboration of engineers from various countries and backgrounds who must exchange information and expertise regarding multiple components throughout the different stages of product development or production planning [1].

As airplane technology advances, so does the complexity of its parts. For example, the fuselage of a modern airplane can consist of over 700,000 different parts, each with its unique dimensions, geometries, tolerances, and materials [1]. This level of complexity presents a significant challenge for aerospace suppliers, who must ensure that these parts are manufactured with high quality, at a reduced cost, and within a shorter production time.

Semantic Web and 3D feature recognition have been successfully applied in the quotation of complex parts manufacturing, especially in the aerospace industry [2]. The semantic web allows for the representation of information in a structured and standardized format, which facilitates the automation of processes and data-based decision-making [3].

To address these challenges, advanced manufacturing technologies and ontologies are required. These technologies include conventional and non-conventional machining, rapid manufacturing, 3D measuring machines, and others.

Industrial ontologies can be used to define and share common concepts between different systems and organizations in the manufacturing industry. Increased interoperability can lead to better collaboration and coordination, improve production agility, and make the industry more resilient to changes and disruptions. The research team also presents a methodology for developing industrial ontologies and discusses examples of their application in the manufacturing industry [4].

The manufacturing process of different airplane parts involves multiple domains and knowledge (see Fig. 1). However, it takes a long time to budget new aeronautical parts and demands extremely qualified labor for such a task. Additionally, with the amount of information involved, the probability of an error occurring in this quotation process is very high.

Due to the technological rise in the world, Industry 4.0 demands that companies suit the existence of a new market that is increasingly competitive and demands increasingly customized and high-performance products. The growing demand for these products and low cost directly conflicts with the degree of rigidity presented by most manufacturing companies [5]. Thus, companies are required to shape the paradigm of Industry 4.0, in which intelligent systems can quickly adapt to the constant and necessary changes in the smart, integrated, and customized manufacturing process.

According to [6], intelligent processes in Industry 4.0 will reduce equipment maintenance costs by 10% and 40%, reduce energy costs by 10% to 20%, and increase work efficiency by 10% and 25% until 2025. Given this scenario, business organizations that adopt intelligent systems may gain a more significant advantage over companies that mass produce standardized products [7]. Furthermore, those who choose the strategy of

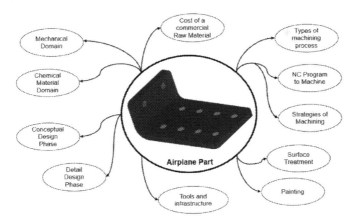

Fig. 1. Multiples domains and concepts in a single airplane part.

differentiating their products through customization tend to become more competitive and have a higher expectation of making a profit, increasing the chance of growth [8].

In addition, it is verified that conventional quotation processes have higher expenses and consume engineering time. In this context, *is it possible to automate the process of quoting aerospace parts based on knowledge representation using ontologies and semantic web-based data provided by the 3D model?*

The main objective of this research concerns the identification of the contributions and limitations of the related works in this domain and the research opportunity to cover this research gap.

Section 2 of this article presents the technological backgrounds, followed by the conceptual approach of the research in Sect. 3, and Sect. 4 presents the conclusions and ideas for future work.

2   Technological Background

2.1   Semantic Web Technologies

Semantic web technologies present information and insights in an organized manner, creating a comprehensive understanding within one or numerous fields. Furthermore, semantic reasoning systems and software agents facilitate the seamless exchange, utilization, and processing of data. To impart meaning to the data, technologies such as the Resource Description Framework (RDF) [9] and the Web Ontology Language (OWL) [10] are employed. Furthermore, these technologies are used to formally depict metadata, following the idea of ontology.

Ontologies serve as a means of enabling various functions for both humans and machines, such as information systems and cyber-physical systems. They capture knowledge within a particular domain, promote interoperability and inter-agent communication, and facilitate more adaptive automation while reducing risks [11]. This has led to the growing significance of the ontology concept in fields such as Intelligent Information Integration, Internet Information Retrieval, Knowledge Management, and the

Semantic Web [12]. The growing popularity of ontologies is attributed to the potential for providing a shared and common understanding across different domains [13]. They have been designed to impart machine-readable semantics to information sources that can be shared among systems or human entities [13]. Additionally, they are utilized by intelligent systems to promote interoperation between different systems.

A Semantic Reasoner or Rules Engine, aligned with ontologies, is a piece of software that can deduce logical consequences from a set of facts or axioms. This software enables the detection of inconsistencies in information across the product or production process. In recent studies [14–16], ontology mapping and semantic reasoning have been recognized as crucial technologies for addressing the problem of semantic interoperability. Mapping is critical in traditional applications such as information integration, query answering, and data transformation [17]. The mapping process solves the heterogeneity problem between ontologies, as it aims to find relationships between semantically connected entities from different ontologies. The process involves inputting two ontologies, each composed of various components (classes, instances, properties, rules, axioms, etc.), and outputs a similarity match [17].

Semantic web technologies make data on the web machine-readable and accessible to automated processes, improving data representation and data-driven decision-making. For example, in the aerospace industry, semantic web technologies can represent the complex relationships between airplane parts, materials, and manufacturing processes, helping to support decision-making and improve the accuracy and efficiency of cost estimation.

In the aerospace industry, using semantic web technologies can provide real-time information about manufacturing operations and quality control, helping to reduce costs and production time while ensuring high quality. Therefore, the potential of this approach in the aerospace industry should be explored further, as it can significantly enhance the aerospace manufacturing process.

Semantic web technologies and 3D feature recognition offer promising solutions, and this paper provides a preliminary discussion of how they can be used in the aerospace industry.

2.2 3D Feature Recognition

The aerospace industry has progressed and developed specialized tools, such as specialized modeling tools for aerospace sheet metal, to assist at various stages of the product's lifespan. However, despite structural sheet metal parts playing a significant role in airplanes, no automated feature recognition (AFR) method is specifically designed for them.

A feature has three main attributes: geometry, relationships with other features, and parameters. For example, the geometry of a hole is defined by its faces, while its relationship to the parent feature is represented by connecting edges. Parameters of a hole include its location (determined by an axis) and diameter (derived from its geometry). Geometry links feature to their B-rep, while feature relationships reveal the structure of geometric models. This study demonstrates that feature relationships are based on topological adjacency. Feature parameters capture design intent or engineering semantics and can be numerical or non-numerical information extracted from the geometry model.

Certain features can have different types, such as curved or planar flanges, and these types are considered additional parameters. AFR is an essential tool for various tasks in product lifecycle management, such as computer-aided process planning, data retrieval, and model difference identification. Although AFR methods exist for sheet metal parts, none are specifically tailored for the aerospace industry [18].

This method elevates the level of abstraction of information from 3D STEP models. The approach involves preprocessing the 3D STEP model to categorize the topological elements of the boundary model (B-rep model) and creating new sets of faces, face boundaries, and edges. Rule-based steps are then used to identify aerospace sheet metal features, which are described by their geometry, relationship with other elements, and pertinent parameters. Upon reading B-rep elements from STEP files, they are stored in memory as C++ objects possessing the same member attributes as the original B-rep elements. These C++ objects find their place within lists, which are implemented using C++ Vectors sourced from the Standard Template Library (STL). Vectors, serving as sequence containers, prove invaluable for managing dynamic data, as they possess the ability to expand their size based on the number of elements they encapsulate. This characteristic sets them apart from fixed-size arrays. Moreover, C++ vectors exhibit automatic storage management capabilities and demonstrate efficiency when confronted with frequent data addition and deletion operations [18].

According to [19] an AFR method for detecting shear features using geometric and topological considerations. Shear features are characteristics of sheet metal components produced through shearing procedures such as blanking, notching, piercing, and cutting. The method explicitly targets the recognition of shear features, but it also recognizes features formed through shearing and deformation processes, such as bridges. Based on [20] an AFR technique for shear features uses profile offsetting to specify the layout for punching tool paths. However, the offsetting approach exposes the areas of the parts that cannot be punched out, which requires specialized tools to manufacture. In addition, variations in the punch diameter may result in changes to the identified features, leading to inconsistent results. Recently proposed by [21] a comprehensive solution for recognizing generic deformation features. Their study treats cylindrical, conical, spherical, and toroidal faces as transitional entities to define deformation. This approach is more geometrically encompassing than any other previously published works.

Figure 2 shows the feature recognition of a 3D part. With this, it is possible to put information in the ontology model.

The suggested automated feature recognition approach involves two primary steps: categorizing and grouping elements in a 3D B-rep model and identifying aerospace sheet metal features.

3   Conceptual Approach

Manufacturing an aerospace sheet metal component consists of two primary stages: cutting the blank from a metal sheet and shaping the blank as required. In this paper, the generic features of ASM parts are classified into web features, trim features, and deformation features. Figure 3 depicts the proposed categorization of ASM model features. The web is a unique feature among ASM parts, shaped by cutting the blank and shaping operations that produce other features.

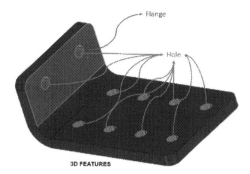

Fig. 2. Feature Recognition of 3D airplane part

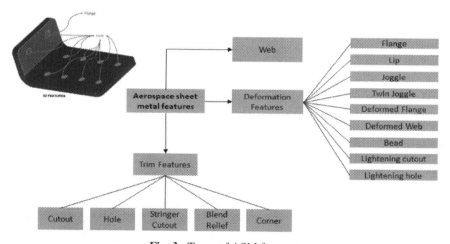

Fig. 3. Types of ASM features

The trim features include a cutout, hole, stringer cutout, bend relief, and corner. Cutouts are created by removing a section of the parent feature as long as the boundary of the parent feature remains unchanged. Holes are a specific type of cutout that has a circular shape. Stringer cutouts, on the other hand, are formed by altering the boundary of the parent feature, such as the web, and dividing the flanges or the twin joggle-induced deformed flanges.

Stringer cutouts are carved out to accommodate the placement of a stringer. Bend reliefs are cutouts that prevent sharp proximity between flanges, which can result in cracking. Since bend reliefs are designed following design guidelines, the length of the relief cuts can be determined. Corners are created by rounding off sharp, convex edges, usually near holes, and are concentric with the corresponding hole. Although both corner reliefs and corners could be considered part of the parent feature's boundary, they convey design intentions, making it essential to distinguish them as separate features.

The deformation features encompass a lightning cutout, lightning hole, flange, lip, joggle, twin joggle, deformed flange, deformed web, and bead. The deformation features are generated by deforming a section of the parent feature. This study assumes that all the bends are formed using a consistent bend radius. The lightening cutouts and holes are fashioned by removing a portion of the parent feature and constructing stiffening lips at the boundary of the removed section.

The flange is formed on the outer edge of its parent feature and is always derived from a web or another flange. It can be categorized as flat or curved, as an assembly or for stiffening, direct or bent, single or multiple, and perpendicular, open-ended, or closed.

Joggles and twin joggles are deformations that create indented areas on the web or flanges. The indented sections of the parent feature are recognized as separate features, known as either a deformed web or a deformed flange, depending on the joggled parent feature.

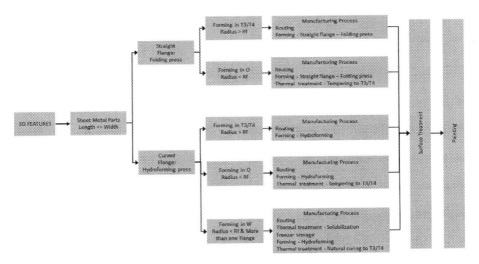

Fig. 4. Flowchart of the ASM manufacturing process

After extracting the information from the 3D model, the data goes through the rules described in the flowchart of Fig. 4. In the case of ASM, if it has a straight flange, the manufacturing process should be the folding press. The manufacturing process should be the hydroforming press if it's a curved flange.

Table 1 and Table 2 show, according to the aeronautical standard, the minimum bending operations radius and limitations for each material status and sheet metal thickness.

If the process follows the folding press and the state is between T3 and T4 with a radius greater than Rf, the manufacturing process to be followed is routing, forming, straight flange, and folding press, and then sent to the operation of surface treatment and painting. If the process follows the folding press and the state is in O with a radius smaller than Rf, the manufacturing process to be followed is routing, forming, straight flange, folding press, thermal treatment, and tempering to T3/T4, and then sent to the process of surface treatment and painting. If the process follows the hydroforming press

Table 1. The aeronautical standard is the minimum bending operations radius.

Material	State Thickness (mm)	Bending Radius (mm)											
		0.3	0.4	0.5	0.63	0.8	1.0	1.2	1.3	1.4	1.5	1.6	1.8
Sheet Metal	0	0.6	0.6	0.6	0.6	1.0	1.2	1.6	1.8	2.0	2.0	2.0	3.0
	AQ	0.6	0.6	0.8	1.0	1.6	2.0	2.5	2.8	3.0	3.0	3.0	4.0
	T3 e T4	1.0	1.0	1.6	2.0	2.5	3.0	4.0	4.5	5.0	5.0	5.0	6.0
	T6 e T81	2.4	2.4	3.2	4	4.8	6.4	–	8	–	–	10.4	11.2

Table 2. The aeronautical standard is the minimum bending operations radius.

Material	State Thickness (mm)	Bending Radius (mm)											
		2	2.3	2.5	3.0	3.2	3.5	4	4.8	5	6	6.4	9.5
Sheet Metal	0	3.0	4	4	5	5	5.6	7.2	9.5	10	12	16	35
	AQ	4	4.5	5	6	6	6.4	9.5	12.7	15	15	18	37
	T3 e T4	8	9	10	12	12	12.7	17	21.4	25	30	32	48
	T6 e T81	12.7	14.3	16.7	–	21.4	–	27	34.9	–	–	50.8	–

and the T3 and T4 state with a radius greater than Rf, the manufacturing process to be followed is routing, forming, and hydroforming, and then sent to the operation of surface treatment and painting. If the process follows the hydroforming press and the state with a radius smaller than Rf, the manufacturing process to be followed is routing, forming, hydroforming, thermal treatment, and tempering to T3/T4, and then sent to the operation of surface treatment and painting. If the process follows the hydroforming press and the W state with a radius less than Rf and more than one flange, the manufacturing process to be followed is routing, thermal treatment, solubilization, freezer storage, forming, hydroforming, thermal treatment, and natural curing to T3/T4, and then sent to the process of surface treatment and painting.

The ontology enables the representation of diverse knowledge in the parts pricing process. For instance, it allows determining the appropriate tool for bending based on the radius and thickness of the component or identifying the most suitable machine for manufacturing based on the data obtained from 3D feature recognition.

With this information extracted from the 3D model and identified by ontology and semantic web, the flow of processes to be followed, it is possible to measure the cost of each process involved and then autonomously price an aeronautical component to be manufactured.

4   Discussion and Conclusion

Many challenges have been faced with the arrival of the fourth industrial revolution, both in the academic field and industry. His arrival provided a great technological leap opportunity but suiting and mastering it is not a quick task. Due to the technological rise in the world, Industry 4.0 demands that companies suit the existence of a new market that is increasingly competitive and demands increasingly customized and high-performance products.

From the problematization, it is possible to conclude that conventional cost estimation methods, based on rule-based algorithms and manual data input, are inadequate for the complex and dynamic nature of the aerospace manufacturing industry. Instead, semantic web technologies allow information to be represented in a structured and standardized format, facilitating automation and data-driven decision-making.

The application of ontologies and semantic web technologies has the potential to enhance the aerospace manufacturing process significantly. By improving data representation, data-driven decision-making, knowledge sharing, and combining with 3D feature recognition, these technologies can help address the challenges of increasing complexity and reducing costs and pricing time in the aerospace industry. Therefore, future research should focus on exploring the full potential of this approach in the aerospace manufacturing industry and other domains.

This work allows analyzing the problem of conventional cost estimation methods, generating higher costs for the company. Therefore, propose a solution by developing a system capable of collecting data 3D model and using data crossing to a semantic web and ontology to get knowledge of the system.

The next step in the research is to validate the proposed method by applying the ontology method to improve turning machining parameters. Then, it is believed that it will be possible to implement and validate this system in an industry case, bringing benefits to the entire chain involved, mainly reducing quoting time and making the process more sustainable, both in saving environmental and financial resources.

Acknowledgements. The authors would like to thank the Pontifical Catholic University of Parana (PUCPR), the National Council for Scientific and Technological Development (CNPq) and Coordination for the Improvement of Higher Education Personnel (CAPES) for the financial support of this research.

References

1. Szejka, A.L., Mas, F., Junior, O.C.: Towards knowledge-based system to support smart manufacturing processes in aerospace industry based on models for manufacturing (MfM). In: Canciglieri Junior, O., Noël, F., Rivest, L., Bouras, A. (eds.) PLM 2021. IFIP AICT, vol. 640, pp. 425–437. Springer, Cham (2022). https://doi.org/10.1007/978-3-030-94399-8_31

2. Pereira, R.M., Szejka, A.L., Junior, O.C.: Towards an information semantic interoperability in smart manufacturing systems: contributions, limitations, and applications. Int. J. Comput. Integr. Manuf. **34**, 422–439 (2021). https://doi.org/10.1080/0951192X.2021.1891571

3. Seeliger, A., Pfaff, M., Krcmar, H.: Semantic web technologies for explainable machine learning models: a literature review. PROFILES/SEMEX@ ISWC, vol. 2465, pp. 1–16 (2019)

4. Ameri, F., Sormaz, D., Psarommatis, F., Kiritsis, D.: Industrial ontologies for interoperability in agile and resilient manufacturing. Int. J. Prod. Res. **60**, 420–441 (2022). https://doi.org/10.1080/00207543.2021.1987553
5. Royer, R.: Implantação da customização em massa na estratégia da manufatura (2007)
6. Bassi, W., Picchi, M., Gasparotto, A.: Um estudo sobre a customização de produtos. Revista Interface Tecnológica **17**(1), 292–302 (2020)
7. Maia, L.H.A.: Avaliação de Desempenho de Recobrimentos em Ferramentas de Metal Duro no Torneamento do Aço ABNT 4340 Temperado por Meio de Sinais de Emissão Acústica, pp. 10–99 (2015)
8. Machado, A., Moraes, W.: Da produção em massa à customização em massa: sustentando a liderança na fabricação de motores elétricos. Cad. EBAPE.BR, Rio de Janeiro (2012)
9. W3C: RDF - Semantic Web Standards. https://www.w3.org/RDF/. Accessed 09 Jan 2023
10. W3C: OWL - Semantic Web Standards. https://www.w3.org/OWL/. Accessed 09 Jan 2023
11. Industrial Ontology Foundry (IOF): Technical Principles – IOF Website. https://www.industrialontologies.org/technical-principles/. Accessed 09 Jan 2023
12. Imran, M., Young, R.I.M.: Reference ontologies for interoperability across multiple assembly systems. Int. J. Prod. Res. **54**, 5381–5403 (2016). https://doi.org/10.1080/00207543.2015.1087654
13. Fensel, D.: Ontologies. Springer, Heidelberg (2004). https://doi.org/10.1007/978-3-662-09083-1
14. Adamczyk, B.S., Szejka, A.L., Canciglieri, O.: Knowledge-based expert system to support the semantic interoperability in smart manufacturing. Comput. Ind. **115**, 103161 (2020). https://doi.org/10.1016/j.compind.2019.103161
15. Palmer, C., Usman, Z., Junior, O.C., Malucelli, A., Young, R.I.M.: Interoperable manufacturing knowledge systems. Int. J. Prod. Res. **56**, 2733–2752 (2018). https://doi.org/10.1080/00207543.2017.1391416
16. Chungoora, N., et al.: A model-driven ontology approach for manufacturing system interoperability and knowledge sharing. Comput. Ind. **64**, 392–401 (2013). https://doi.org/10.1016/j.compind.2013.01.003
17. Alaya, M.B., Monteil, T.: FRAMESELF: an ontology-based framework for the selfmanagement of machine-to-machine systems. Concurr. Comput. Pract. Exp. **27**, 1412–1426 (2015). https://doi.org/10.1002/cpe.3168
18. Ghaffarishahri, S., Rivest, L.: Feature recognition for structural aerospace sheet metal parts. CAD&A **17**, 16–43 (2019). https://doi.org/10.14733/cadaps.2020.16-43
19. Jagirdar, R., Jain, V.K., Batra, J.L., Dhande, S.G.: Feature recognition methodology for shearing operations for sheet metal components. Comput. Integr. Manuf. Syst. **8**(1), 51–62 (1995). https://doi.org/10.1016/0951-5240(95)92813-A
20. Devarajan, M., Kamran, M., Nnaji, B.O.: Profile offsetting for feature extraction and feature tool mapping in sheet metal. Int. J. Prod. Res. **35**(6), 1593–1608 (1997). https://doi.org/10.1080/002075497195146
21. Gupta, R.K., Gurumoorthy, B.: Classification, representation, and automatic extraction of deformation features in sheet metal parts. Comput. Aided Des. **45**(11), 1469–1484 (2013). https://doi.org/10.1016/j.cad.2013.06.010

An Approach to Model Lifecycle Management for Supporting Collaborative Ontology-Based Engineering

Manuel Oliva[1] , Rebeca Arista[2,3] , Domingo Morales-Palma[3] ,
Anderson Luis Szejka[4] , and Fernando Mas[3,5(✉)]

[1] Airbus Defence and Space, Sevilla, Spain
[2] Airbus SAS, Blagnac, France
[3] University of Sevilla, Sevilla, Spain
fmas@us.es
[4] Pontifical Catholic University of Paraná (PUCPR), Curitiba, Brazil
[5] M&M Group, Cadiz, Spain

Abstract. To reduce costs and increase production capacity, and meet market demand in a sustainable manner, the aerospace industry is placing a growing emphasis on designing the entire product life cycle. This necessitates a novel approach to make a significant advancement and ensure competitiveness in the 21st century. One solution that has emerged is the utilization of Ontology-Based Engineering (OBE) methods, processes, and tools. However, implementing OBE has presented challenges related to modeling, such as effectively managing lifecycles, workflows, and the sharing and reuse of models. To address these challenges, the authors have proposed the Models for Manufacturing (MfM) methodology, which offers a novel way to model manufacturing systems with collaborative, extensible, and reusable characteristics. These characteristics align with the concept of Model Lifecycle Management (MLM). This article highlights the difficulties faced by the aerospace industry when adopting models based on the entire product lifecycle, drawing parallels to the adoption of 3D modeling in the nineties. Furthermore, it explores how the MLM system proposed by the authors can effectively address these issues.

Keywords: Model Lifecycle Management · Ontology-Based Engineering · Models for Manufacturing · Product Lifecycle Management

1 Introduction

Rapid advancement of information and communication systems has facilitated the rapid spread and evolution of globalization. Collaborative business environments, where companies share and work together, have become commonplace. However, this new paradigm of real-time collaboration and communication presents challenges such as information security, the evolution of shared work environments, and coordination between different work environments.

Published by Springer Nature Switzerland AG 2024
C. Danjou et al. (Eds.): PLM 2023, IFIP AICT 702, pp. 173–183, 2024.
https://doi.org/10.1007/978-3-031-62582-4_16

Globalization encompasses not only the manufacturing of products, but also extends to the entire product lifecycle. Engineering systems modeling has relied on various modeling techniques, often based on the expertise of the individuals involved. This creates difficulties in collaborative processes when modeling techniques evolve or differ among teams and companies. To address this issue, two areas of focus have emerged.

The first area involves standardizing modeling techniques, leading to the emergence of model standardization through OBE [1]. A model is an abstraction or representation of a system or reality created to facilitate understanding and analyze its behavior. Modeling is employed in various disciplines, environments, and testing laboratories to analyze existing, past, or future systems.

Working in collaborative environments requires managing the lifecycle of models, enabling analysis and evaluation of various stages, defining revisions and configurations, and monitoring model evolution. This role is fulfilled by an MLM system, like Product Lifecycle Management (PLM) systems fulfill the role for CAD models.

The objective of the paper highlights the difficulties faced by the aerospace industry when adopting models based on the entire product lifecycle and explore how the MLM system proposed by the authors can effectively address these issues. The paper is structured as follows. Section 2 provides a brief review of the literature and overview of previous work. Section 3 examines the status of the European Aerospace Industry, while Sect. 4 describes the current state of model development. Section 5 presents the research and prototypes implementing a MLM system, and the final Sect. 6 concludes the paper and outlines future research work.

2   Literature Review and Previous Work

A model is a representation of a system comprising objects with attributes and relationships [2]. In engineering, multiple modeling techniques are employed, including state-diagram based system analysis [3], Object-Process Methodology (OPM) [4], Model Driven Architecture (MDA) [5], and System Engineering Model Driven (SEMD) [6] among others. These methodologies serve the purpose of organizing specifications and generating models for diverse system types. Significant efforts have been undertaken to standardize and integrate these models, with notable examples being Arcadia [7] and SysML [8].

Research efforts have also been dedicated to the development of novel languages and the integration of ontologies and semantics into modeling. Lifecycle Modelling Language (LML) [9] and Graph Object Point Property Role and Relationship (GOPPRRE) [10] are notable examples of such approaches. Ontology development has emerged as a global research topic, and OBE encompasses a range of activities involved in creating ontologies, methodologies, and supporting tools [11]. Ontology, in this context, refers to a shared understanding within a specific domain [12]. The National Institute of Standards and Technology (NIST) and projects within the European Union (EU), such as the Ontocommons project [13], funded by Horizon 2020, have explored the interoperability between different engineering disciplines and the notion of commonality in ontologies. In line with this collaboration, the Industrial Ontologies Foundry (IOF) [14] seeks to establish a comprehensive and open set of core reference ontologies for digital manufacturing. The key objectives include achieving interoperability, establishing information linkage, formalizing requirements, and ensuring quality and traceability.

The initial proposition of the concept of a MLM was introduced in 2014 at the INCOSE conference [15], where it elucidated various indispensable functionalities essential for the comprehensive management of model lifecycles. MLM bears resemblances to PLM, a well-established approach employed for the management of product lifecycle information. Metamodels play a crucial role in delineating the language used to describe models and enabling their manipulation [16]. The study of MLM has also been undertaken in conjunction with the MfM methodology [17]. MLM represents a relatively nascent challenge that has garnered limited research attention thus far. Nonetheless, it exhibits notable parallels with PLM. Considering their conceptual resemblances, techniques and tools employed in PLM can be effectively leveraged to provide support for MLM. Certain functionalities related to MLM have been subject to more extensive investigation as expounded in the reference [11].

Within the aerospace industry, the comprehensive design of the product lifecycle plays a pivotal role in achieving cost reduction, increasing production capacity, and effectively responding to market demands. The authors have introduced the concept of the industrial Digital Mock-Up (iDMU) [18], which is currently advocated as the prevailing design paradigm in the manufacturing domain. The iDMU paradigm aims to facilitate knowledge integration, enable reuse and traceability, minimize costs, improve quality, expedite time-to-market, and automate the generation of manufacturing documents. Europe is actively engaged in the development of aerospace programs, with support from the European Union (EU) and international consortia, aiming to sustain its leadership in the industry. Prominent initiatives like Clean Sky and Clean Sky 2 are geared toward enhancing the sustainability of air transport by replacing existing aircraft with fuel-efficient models that emit reduced noise and emissions while utilizing clean fuels such as hydrogen. Defense projects such as Eurodrone and New Generation Weapon System (NGWS) further enhance defense capabilities. These programs necessitate the adoption of new technologies and tools, presenting challenges in terms of cost-effectiveness, timely delivery, and certification.

The intricate nature of the aerospace industry demands the establishment of novel working models, collaborative efforts, and the utilization of modeling and simulation techniques to expedite development cycles, validate designs, facilitate certification, achieve cost reductions, and minimize environmental impact. The integration of OBE solutions, PLM systems, simulations, and production/service models is of paramount importance. Digital platforms that foster open environments, allowing for seamless model exchange and access to design data, are indispensable. To ensure, the definition of methodologies and frameworks for model development and integration through ontologies and meta-models is crucial for ensuring effective implementation.

3   Status of Model Development in Aerospace Industry

The transition from 2D drawings to 3D representation, facilitated by information systems, has yielded significant advancements in product design quality and efficiency. Nevertheless, this shift has presented challenges stemming from limited expertise and familiarity in effectively managing 3D models. In the initial stages, the engineering department independently generated and oversaw their respective 3D models stored within designated folders. The need to share models across various design areas required the establishment of user access protocols to prevent conflicting modifications. The management of model maturity and versioning emerged as critical considerations, effectively addressed through the implementation of information systems.

Subsequently, software manufacturers responded by offering model management solutions that incorporated features such as versioning and workflows, giving rise to Product Data Management (PDM). PDM mainly focused on product design aspects but lacked seamless integration with other functional areas. Over time, increasing industry requirements led to the development of collaborative PLM systems. These comprehensive PLM systems encompass the entire product lifecycle and encompass a broader scope by integrating industrial processes and resources, including the industrial Digital Mock-Up (iDMU) [18]. In general, the evolution involved transitioning from 2D to 3D, managing models between departments, and progressing from PDM to PLM for comprehensive product management.

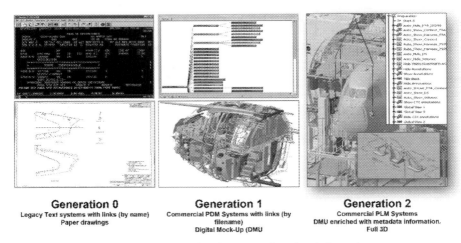

Generation 0
Legacy Text systems with links (by name)
Paper drawings

Generation 1
Commercial PDM Systems with links (by filename)
Digital Mock-Up (DMU

Generation 2
Commercial PLM Systems
DMU enriched with metadata information.
Full 3D

Fig. 1. Generations 0 to 2 showing the status of product information management.

In 2015, Mas et al. [19] conducted a comprehensive review of PLM within the aerospace industry, focusing on its impact. Their study highlighted key PLM advancements over the past 50 years, with a particular emphasis on topics related to product information management. The authors delineated three distinct generations of PLM, spanning the period from 1960 to 2015, representing an evolutionary progression from Generation 0 to Generation 2. Generation 0 entailed the utilization of legacy PDM systems coupled with attached drawings. This was followed by Generation 1, characterized by the adoption of commercial PDM systems that incorporated attached 3D Computer-Aided Design (CAD) models, leading to the establishment of a Digital Mock-Up (DMU). Finally, Generation 2 marked the integration of commercial PLM systems capable of encompassing all metadata related to the product. The evolution of product information management across these generations is succinctly summarized and presented in Fig. 1.

In 2021, an updated review was published, focusing on major players in the aerospace industry, namely Airbus and Boeing, and their projects related to Generation 3. These strategic initiatives are focused on the utilization of modeling and simulation techniques. Generation 3 is distinguished by the adoption of OBE as the foundation for product information management, accompanied by the implementation of Continuous Engineering and Digital Twin methodologies and processes [20]. Figure 2 provides a visual representation of the current Generation 3, where the central focus revolves around the model as the primary object. This perspective underscores the integration of various components, such as the 3D model, metadata, and characteristics, all of which are considered integral parts of the model itself.

 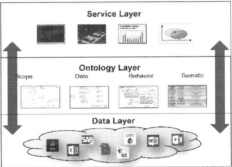

Fig. 2. Generation 3 shows the status of product information management.

In recent years, the benefits of Modeling and Simulation (M&S) have expanded beyond product design to industrial and operational systems. This has sparked interest and efforts in OBE methodologies and brought attention to the need for a more connected approach. However, the industry has struggled to learn from past mistakes in the 3D geometric realm, resulting in similar problems arising with M&S technologies in the product lifecycle. The challenges in OBE include the following challenges:

- Challenges considering models development include the lack of methodologies for the entire lifecycle, the need for comprehensive metamodels, and the absence of common development tools.
- Challenges considering framework perspective include a lack of model sharing and reusability, narrow solution focus limiting their applicability, localized optimization missing out on company-wide benefits, inadequate control over model maturity, and a lack of defined workflow for design, verification, validation, and acceptance, causing collaboration challenges.

Overall, there is a demand for a strategic vision and connected models that provide holistic solutions throughout the company, optimize local environments, and enable effective requirements management, verification, and validation.

4 Model Lifecycle Management for Supporting Collaborative Ontology-Based Engineering

The authors proposed in [21] how the MfM methodology, an OBE methodology, can be managed based on the MLM concept and how PLM tools offer the necessary functionalities to support OBE. As described above, the evolution of models follows a path like the one followed by 3D geometric models that were eventually managed from PLM [22]. Applying MfM methodology and MLM to the problems described in the previous section, practically most of them can be supported by the functionalities of an MLM. Solutions can be group following the main MLM functionalities:

- Vaulting: The need for visibility of existing models, their scope, and the solutions they provide. This would avoid the existence of redundant models and would help in the decision process of launching new projects knowing what has already been developed. It is also useful to detect which models are necessary and have not yet been developed (requirement models). The rationalization of existing models and the decisions of new ones to be developed will be supported by the visualization and queries necessary to filter them by different attributes such as scope or solutions they provide with respect to the existing ones.
- Configuration: Establish a procedure to manage model configuration. A procedure to manage the relationships between different models and to be able to detect which other models are affected by a change and evaluate the impact.
- Workflow: Model statuses and lifecycle management under workflows. Implement a procedure for managing model status (in-work, verified, validated, released, etc.).
- Visualization: Visualization of models and their properties (metadata). Definition of projects and access or operations for the different roles involved in the life cycle of a model, people, and organization.
- Interface: Models import and export capabilities to external modelers.

The proposed model is illustrated in Fig. 3. The Scope model, which establishes the boundaries within which the model operates, is represented using IDEF0 diagrams and created using Ramus software. The subsequent stage involves creating the ontology through the Data model, which defines the information managed within the selected scope. Currently, concept maps are utilized for modeling the Data model, employing software tools such as CmapTools by IHMC or GraphViz. The Behavior model is being developed using a customized diagram type in the GraphViz software. This model specifies the simulation requirements for the company processes, as previously defined in the Scope model. Lastly, the Semantic model is being constructed in the form of spreadsheets using software such as Excel. The purpose of the Semantic Model is to prevent ambiguities in database usage, ensure consistency in connections with the models, and maintain continuity in the ontologies throughout their lifecycle.

The interfaces depicted in Fig. 3 were specifically developed to interface with the modeler applications employed by the authors in their work on various industrial use cases [24, 25, 26, 27]. The MLM system accommodates the objects originating from these modelers but encounters two significant challenges: (1) all the object information is required to pass through the interface; and (2) decoration information such as positions,

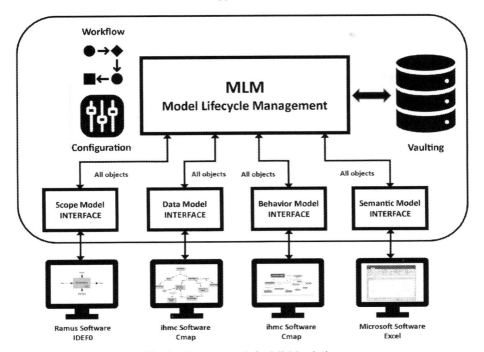

Fig. 3. First approach for MLM solution.

colors, shapes, etc., is lost in the process. Among these challenges, the first issue is identified as the primary concern, as the passed object information often includes elements specific to the modeler application that are not part of the MfM model. Consequently, this limitation restricts the utilization of any modeler within the MLM environment. Efforts are thus focused on addressing this issue to ensure seamless integration and effective model management within the MLM framework.

To address this challenge, a comprehensive set of metamodels was developed to establish the specifications that any authoring tool within the MfM methodology must adhere to. These metamodels encompass all types of objects, properties, and relationships associated with the MfM model types (Scope, Data, Behaviour, and Semantic). Metamodels were implemented both in the interfaces us let translate just the required objects as in the MLM to adapt its capabilities to them. Consequently, MLM becomes capable of managing models created by various modelers operating under different methodologies. The revised scenario is illustrated in Fig. 4, showcasing the interoperability and flexibility achieved through the integration of the metamodels into the MLM solution.

Furthermore, the proposed MLM solution incorporates several notable enhancements:

- Tool agnosticism: The MLM solution is designed to be tool-agnostic, without favoring any specific software. The authors advocate for the use of Free Open Source Software (FOSS) tools, which enable seamless reading, writing, understanding, sharing, and discussion of models among proficient engineers. This comprehensive toolset covers

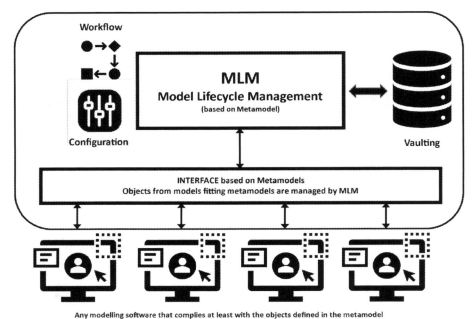

Fig. 4. New approach for MLM solution.

all aspects of modeling, from definition to simulation and trade-off processes, thereby facilitating the optimization of solutions.

- Enhanced Decision-making Environment: By providing a shared and common environment within the company, the MLM solution contributes to more informed decision-making processes. It helps align projects with the overarching company strategy, allowing for strategic decisions rather than relying solely relying on tactical considerations.
- Improved Reusability and Sharing: The MLM solution simplifies the process of reusing and sharing models within the organization by establishing clear and standardized definitions of various concepts or entities, along with their corresponding relationships. This promotes effective management of these entities and facilitates seamless collaboration among stakeholders.

The authors have introduced a prototype system architecture design and interfaces, which have been implemented using open-source software. This system serves as a collaborative PLM solution, specifically addressing the creation, management, enrichment, and reutilization of manufacturing models [19]. The efficacy of this approach has been validated through a preliminary study focused on Incremental Sheet Forming technology, specifically for the production of 3-axis Numerical Control (NC) machining metallic parts [27].

5   Conclusions and Future Research Work

This paper elucidates the imperative for the aerospace industry to embrace a disruptive leap in order to effectively confront future challenges and achieve success with a positive outlook. The current disruptive approach, centered on the application of Modeling and Simulation and specifically Ontology-Based Engineering, has led to the identification of various challenges that must be addressed to ensure the successful implementation of this approach and its resilience in the future.

One crucial aspect in the successful adoption of OBE methodologies is the establishment of a robust model management system, which can be effectively addressed through MLM which encompasses key characteristics that offer potential solutions to many of the challenges faced by the aerospace industry in managing their models. The definition of metamodels, vital for accurate design and management of models, contributes to improved coordination and integration among stakeholders.

Furthermore, the ultimate validation of the concepts, along with the inclusion of open interfaces to facilitate interoperability with diverse modeling environments across different domains, remains contingent on real-world implementation within the aerospace industry. This implementation is anticipated to support the transformative leap, enabling earlier and more sustainable impacts in product design, industrial systems, and cost reduction throughout the product lifecycle.

Acknowledgment. The authors would like to recognize colleagues from University of Sevilla, Airbus in Spain and France, and M&M Group for the support and contributions during the development of this work.

References

1. Arista, R., Zheng, X., Lu, J., Mas, F.: An ontology-based engineering system to support aircraft manufacturing system design. J. Manuf. Syst. **68**, 270–288 (2023). https://doi.org/10. 1016/j.jmsy.2023.02.012
2. Dickerson, C.E., Mavris, D.: A brief history of models and model based systems engineering and the case for relational orientation. IEEE Syst. J. **7**(4), 581–592 (2013). https://doi.org/10. 1109/JSYST.2013.2253034
3. Kordon, M., et al.: Model-based engineering design pilots at JPL. In: 2007 IEEE Aerospace Conference, Big Sky, MT, USA, pp. 1–20 (2007). https://doi.org/10.1109/AERO.2007. 353021
4. Dori, D.: Modeling Knowledge with Object-Process Methodology (2011)
5. Object Management Group: Model Driven Architecture (MDA)–MDA Guide Rev. 2.0 (2014). http://www.omg.org. Accessed 25 May 2023
6. Morel, G., Pereira, C.E., Nof, S.Y.: Historical survey and emerging challenges of manufacturing automation modeling and control: a systems architecting perspective. Annu. Rev. Control. **47**, 21–34 (2019). https://doi.org/10.1016/j.arcontrol.2019.01.002
7. Voirin, J.L.: Model-Based System and Architecture Engineering with the Arcadia Method. Elsevier (2017)

8. Object Management Group: Systems Modeling Language (OMG SysML), Version 1.3 (2012). https://www.omg.org/. Accessed 25 May 2023

9. Giammarco, K.: A formal method for assessing architecture model and design maturity using domain-independent patterns. Procedia Comput. Sci. **28**, 555–564 (2014). https://doi.org/10.1016/j.procs.2014.03.068

10. Lu, J., Ma, J., Zheng, X., Wang, G., Li, H., Kiritsis, D.: Design ontology supporting model-based systems engineering formalisms. IEEE Syst. J. **16**(4), 5465–5476 (2022). https://doi.org/10.1109/JSYST.2021.3106195

11. Arista, R., Mas, F., Morales-Palma, D., Ernadote, D., Oliva, M., Vallellano, C.: Evaluation of a commercial model lifecycle management (MLM) tool to support models for manufacturing (MfM) methodology. In: Noël, F., Nyffenegger, F., Rivest, L., Bouras, A. (eds.) PLM 2022. IFIP Advances in Information and Communication Technology, vol. 667, pp. 673–682. Springer, Cham (2023). https://doi.org/10.1007/978-3-031-25182-5_65

12. Govindan, K., Soleimani, H., Kannan, D.: Reverse logistics and closed-loop supply chain: a comprehensive review to explore the future. Eur. J. Oper. Res. **240**(3), 603–626 (2015). https://doi.org/10.1016/j.ejor.2014.07.012

13. Fisher, A., et al.: 3.1.1 model lifecycle management for MBSE. In: INCOSE International Symposium, vol. 24, no. 1, pp. 207–229 (2014). https://doi.org/10.1002/j.2334-5837.2014.tb03145.x

14. Sprinkle, J., Rumpe, B., Vangheluwe, H., Karsai, G.: 3 metamodelling. In: Giese, H., Karsai, G., Lee, E., Rumpe, B., Schätz, B. (eds.) Model-Based Engineering of Embedded Real-Time Systems. LNCS, vol. 6100, pp. 57–76. Springer, Heidelberg (2010). https://doi.org/10.1007/978-3-642-16277-0_3

15. Magas, M., Kiritsis, D.: Industry commons: an ecosystem approach to horizontal enablers for sustainable cross-domain industrial innovation (a positioning paper). Int. J. Prod. Res. **60**(2), 479–492 (2022). https://doi.org/10.1080/00207543.2021.1989514

16. Karray, M.: The industrial ontologies foundry (IOF) perspectives, industrial ontology foundry (IOF) - achieving data interoperability workshop. In: International Conference on Interoperability for Enterprise Systems and Applications, Tarbes (2021)

17. Morales-Palma, D., Oliva, M., Arista, R., Vallellano, C., Mas, F.: Enhanced metamodels approach supporting models for manufacturing (MfM) methodology. In: Proceedings, 0073:13. Tarbes, France (2022). http://Ceur-Ws.Org

18. Mas, F., Arista, R., Oliva, M., Hiebert, B., Gilkerson, I.: An updated review of PLM impact on US and EU aerospace industry. In: 2021 IEEE International Conference on Engineering, Technology and Innovation (ICE/ITMC), pp. 1–5 (2021). https://doi.org/10.1109/ICE/ITMC52061.2021.9570205

19. Arista, R., Mas, F., Oliva, M., Racero, J., Morales-Palma, D.: Framework to support models for manufacturing (MfM) methodology. IFAC-PapersOnLine **52**(13), 1584–1589 (2019). https://doi.org/10.1016/j.ifacol.2019.11.426

20. Arista, R., Mas, F., Vallellano, C., Morales-Palma, D., Oliva, M.: Toward manufacturing ontologies for resources management in the aerospace industry. In: Archimède, B., Ducq, Y., Young, B., Karray, H. (eds.) Enterprise Interoperability IX, Proceedings of the I-ESA Conferences, pp. 3–14. Springer, Cham (2023). https://doi.org/10.1007/978-3-030-90387-9_1

21. Arista, R., Mas, F., Oliva, M., Morales-Palma, D.: Applied ontologies for assembly system design and management within the aerospace industry. In: FOMI - 10th International Workshop on Formal Ontologies Meet Industry, vol. 8 (2019)

22. Morales-Palma, D., Mas, F., Racero, J., Vallellano, C.: A preliminary study of models for manufacturing (MfM) applied to incremental sheet forming. In: Chiabert, P., Bouras, A., Noël, F., Ríos, J. (eds.) Product Lifecycle Management to Support Industry 4.0. IAICT, vol. 540, pp. 284–293. Springer, Cham (2018). https://doi.org/10.1007/978-3-030-01614-2_26
23. CIMDATA Product Data Management: the definition, an introduction to concepts, benefits, and terminology (1998). https://www.cimdata.com/. Accessed 25 May 2023
24. Mas, F., Racero, J., Oliva, M., Morales-Palma, D.: Preliminary ontology definition for aerospace assembly lines in airbus using models for manufacturing methodology. Procedia Manuf. **28**, 207–213 (2019). https://doi.org/10.1016/j.promfg.2018.12.034

Learning and Training: From AI to a Human-Centric Approach

Investigation of an Integrated Synthetic Dataset Generation Workflow for Computer Vision Applications

Julian Rolf[✉] ⓘ, Mario Wolf ⓘ, and Detlef Gerhard ⓘ

Ruhr-Universität Bochum, Bochum, Germany
`julian.rolf@ruhr-uni-bochum.de`

Abstract. Object detection and other machine learning technology applications play an important role in various areas of computer vision (CV) applications within the product lifecycle, especially in quality assurance or general assembly assistance. While the implemented CV-based systems provide great benefits, training and implementing deep learning models is often a tedious and time-consuming task, especially in the field of object detection. To accomplish good results, large datasets with a high quantity of object instances in a bright variety of poses are required These are generally created manually and are therefore very time consuming to create.

To improve the training process, synthetic training data can be used. It is generated within a virtual environment using a product's geometry model. In this paper, the authors propose a synthetic dataset generator for object detection, that is integrated into a PLM system to automate the process of collecting and processing the CAD data for creating the synthetic machine learning training dataset. Domain randomization is used to eliminate the effort of creating a virtual environment, to fully automate the dataset generation, and to increase the generalization of the model. The trained detector is tested on an object detection demonstrator set-up to evaluate its performance in a real-world use case. For evaluation purposes, the authors also provide a comparison of the test results to an object detection model that is trained without domain randomization, using a very close-to-reality virtual environment.

Keywords: deep learning · synthetic data generation · PLM integration · object detection · computer vision

1 Introduction

The demand for increasing degrees of automation in different parts of the product life cycle inevitably leads to an increase in the complexity of the required assistance systems. In many areas machine learning (ML) can fulfill this need completely or partially. ML has an enormous potential for the support of assembly and disassembly processes, quality assurance, or in the improvement of product traceability, especially when using computer

© IFIP International Federation for Information Processing 2024
Published by Springer Nature Switzerland AG 2024
C. Danjou et al. (Eds.): PLM 2023, IFIP AICT 702, pp. 187–196, 2024.
https://doi.org/10.1007/978-3-031-62582-4_17

vision (CV) techniques. However, the implementation of ML systems is associated with high initial effort, particularly for the ML training process. A large amount of suitable training data is necessary for CV applications to work reliably. It is therefore necessary to create new data sets for the detection of each new individual product. This includes not only the creation of images of the respective products, but also the correct and precise labeling of the collected images. This process is not only associated with an immense amount of work but is also error-prone and literally impossible to obtain precise results manually, as is the case with the segmentation of image objects. To solve these problems, the approach is to use synthetic data. The term 'synthetic data' refers to information that does not originate from real sources. In CV, for example, a renderer is used to generate images of the desired objects. Information about position and orientation of the objects is known to the rendering engine and can therefore be added automatically to a generated data set for labelling purposes in supervised machine learning approaches. In this way, in a comparatively short time frame, any amount of data can be generated and used for training a neural network. The only prerequisite for generating the data is the availability of 3D representations of the objects of interest and their accuracy in depicting the (physical) objects that are to be detected later in the process. PLM systems are therefore ideally suited as a data source for the synthetic data generation process.

This is the first paper of a total of three. As presented in Fig. 1 the overall vision is to build an assistant assembly workstation, using deep learning algorithms to assist the assembly process. The workstation will be set up automatically, by generating the needed training data, based on the 3D representations of the objects to be detected. Further analysis on the parts will be conducted to increase the performance of the system. This includes grouping similar and ignoring smaller objects. In this paper, the authors focus on the first aspect and present a data generation pipeline to fully automate the process of generating synthetic data for object detection. Additionally, the data generator is connected to a PLM system to demonstrate the integration capabilities of the proposed solution with existing systems and workflows.

Fig. 1. Overview of the three aspects to be covered.

2   Related Work

This chapter covers use cases for CV applications in various parts of the product lifecycle. Additionally, we will cover modern synthetic data generation tools.

2.1   Computer Vision

In (dis)assembly processes, the correct identification and localization of components is a necessary requirement for the automation of work steps. For the removal of screws in a

PCB with a robotic arm, Mangold et al. used an object detector, which made it possible to determine the correct tool for different types of screws [1]. Therefore, performing the required work steps could be fully automated. Another screw detection detector for robot disassembly was implemented by Brogan et al. [2]. CV algorithms also play a major role in quality control. Basamakis et al. used an object detector to determine missing, correctly or incorrectly assembled rivets and achieved a total accuracy of 83% [3].

2.2 Synthetic Data

A variety of options exist for generating synthetic data for CV use cases, such as the Perception tool for the Unity 3D Engine [4]. Another generator is Nvidia's Omniverse Replicator, based on Nvidia's Omniverse platform [5]. It can create graphically detailed data through the native use of ray tracing. In 2022, Google Research introduced the Kubric data generator [6]. It is based on the Blender render engine and uses the Bullet physics simulation. With SynthAI [7], Siemens offers a cloud service for the generation of synthetic data based on CAD files.

Synthetic data is particularly useful when not enough data can be collected through classical methods to ensure sufficiently good performance of the trained AI model. In addition, it often requires less effort to create a suitable data set and is therefore more cost-effective. In consequence, some datasets have synthetic twins to increase the total amount of training data, like the virtual KITTI dataset [8]. Another way of creating synthetic data is by using a Generative Adversarial Network (GAN). This particular type of deep learning network can be trained on smaller real datasets and can then be used to augment the dataset with synthetic data to produce a sufficiently large amount of training samples [9].

3 Aims and Requirements

In the context of the presented research work, it is investigated how to integrate the synthetic data generation process in a PLM system for full automation. This enables the use of ML while keeping the required expert knowledge in the field of ML very low. The PLM system takes the role of the data source, workflow engine, and, after completion of the data generation, the role of the data provider for the ML application. The generated data set is used for training an object detector to check the quality of the data set. For this purpose, the object detector is tested with a dataset containing real images, which were labeled manually. In addition, a second synthetic data set is generated manually, which represents the properties of the test environment as accurately as possible. This allows a quality comparison of the two detectors and a classification of the practicability of using the general data set.

The goal of this work is the development and validation of a concept for the automatic generation of a synthetic data set for CV problems, in connection with a PLM system for the management of the detected objects. The data set generation process is to be started and monitored from within the PLM system. The generated data set should support the most common CV problems, such as 2D/3D object detection, object segmentation or depth estimation. In addition, there may be little to no knowledge about the real-world application domain of the ML model at the time of training data generation. Therefore, there is a requirement for general applicability of the dataset, which will involve domain randomization.

4 Concept

In this chapter, the concept is presented and explained in detail Firstly, the data set generator to create the CV data set based on a general service interface. Secondly, we elaborate on a PLM system integration that manages the work process as well as the parts and components involved. The last section of this chapter deals with a connector for the communication of the PLM tenant with the dataset generator.

4.1 Dataset Generator

The dataset generator is responsible for generating and creating the CV dataset independently from other components. The core of the generator is the render engine, which is used to create images from a virtual 3D scene. The properties of the scene, such as background, lighting and shadows, as well as the orientation and position of the objects in the scene, depend on the parametrization. The choice of these properties plays a decisive role in the performance of the model trained on this data set. If crucial features of the test environment or of the objects to be recognized are missing, the model is limited in its ability to correctly interpret the test inputs. However, the development of a data set adapted to the test environment requires previous domain knowledge that then influences the test environment setup. To solve this problem, the concept of domain randomization is used [10]. Here, the training environment is randomized in as many properties as possible and feasible. This increases the variance of the training data and provides a better generalization of the model. In this way, a generally usable data set can be generated, which can additionally be extended with a lesser amount of training data with high domain knowledge about the test environment to further increase the performance of the model in said environment. In the *Outcomes* chapter, the performance of an object detector trained with a synthetic dataset based on the presented concept is discussed in more detail, as well as the general usability of such a dataset.

The structure of the virtual 3D scene consists of a dome, a hemisphere with a bottom plate. A High Dynamic Range Image (HDRI) texture is placed on the inside of the geometric body. This texture contains not only a texture, but also information about the light properties of the scene. Inside the dome, an arbitrary number of objects, which are to be identified later in the process, are placed, with random orientation and position. Finally, a virtual camera is placed and aimed at the center of the floor at a random height inside a defined spectrum. To further increase randomization, the color of the objects can also be varied. Since the position and orientation of the objects are precisely known in the virtual space, annotation information, such as bounding boxes or segmentation, can be automatically generated.

In order to be able to interact with other services, such as PLM systems, the generator provides a RESTful webservice interface. An endpoint is used to place a request for data set generation. The jobs are processed according to the 'First In - First Out' (FIFO) principle. A job consists of several settings or parameters, that are stated in Table 1.

The Webhook mail address defines an optional URL. If the dataset generation status changes, a POST request with the corresponding status is sent as payload to the defined URL. This way the generator can inform about the start and the completion of the generation. Also, errors that occur during the generation can be forwarded. The return of the POST request is a UID for the unique identification of the order.

The finished data record can be downloaded as an archive via a second endpoint. In addition, the current status of the order can also be queried directly via a third endpoint. For the last two endpoints, the order UID is required.

4.2 Connector

One challenge in designing the interface of the generator is the very different approach between current PLM systems regarding interface connectivity. These vary greatly in their range of functions, such as handling with Webhooks and documents or 3D representations. Depending on the system, interaction with the interface of the data set generator is therefore not possible without restrictions. To solve this problem, a connector is designed. The connector is a micro service and is located as the middle man of the communication between the generator and the PLM system. Its task is to override or adapt the communication, so that a correct and successful data exchange can be guaranteed and the use of the full functionality of the PLM system can be ensured. Since the connectors are adapted to one system at a time, a connector must also be developed for each supported PLM system. The scope of the connector and thus also the implementation effort depends on the particular PLM system used.

Table 1. Dataset creation parameters with default, min and max values

Parameter	Default Value	Min. Value	Max. Value
Dataset Size	5	1	1000
Resolution	256×256	64×64	512×512
Scale	1.0	0.01	100.0
Color	Uniform Sample		
Obj. Min. Occurrences	1	0	Obj. Max. Occurrences
Obj. Max Occurrences	5	Obj. Min. Occurrences	
Webhook URL	Not set		

4.3 PLM-System

A PLM system is used to manage product relevant data, in particular the parts to be detected, including their 3D representations and all other. In addition, the data generation workflow of the respective item is started and monitored by the PLM system via the connector. The workflow is linked to a single object, so that a generated data set always represents exactly one object. The workflow consists of five states and is displayed in Fig. 2.

The initial state is the start state of the workflow and describes that a data generation job can be created. If a data generation job is created, the availability of the required 3D representation of the selected object as an attachment or document is checked. If the availability is given, the adapter is informed about the existence of the new order, which forwards it to the generator. In addition, the connector registers with the generator via Webhook to be informed about events. In this way, the connector can trigger the correct transition to the current state in the PLM system after receiving the information. If the generation is completed, a mail including the download link of the generator is created and sent to the owner of the respective object in the PLM system.

For the transitions to be triggered, a user with the appropriate rights profile is created and managed by the connector or used by it.

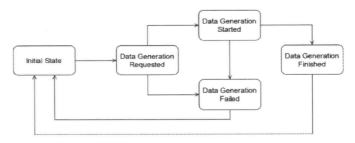

Fig. 2. Data generation workflow

5   Implementation

The implementation of the data generator is based on the data generation pipeline Kubric [6] from Google Research. It is able to generate a variety of different types of CV data sets, like segmentation, bounding box or depth estimation. Kubric uses Blender as the render engine. Also, PyBullet is used as physics engine to simulate physically correct collisions of objects, if desired. This way, more realistic data can be generated, but it also requires more complex processing of the objects, due to the generation of a collision mesh, and is therefore not used in this implementation. The objects are instead randomly placed and oriented in the scene, within an area of 0.3 m × 0.3 m around the origin at a height of 0.01 m. This ensures that the objects are captured by the camera. HDRI textures for the background and lighting of the scene are provided by Kubric via an API from HDRI-Heaven and placed on the inside of the dome. A total of up to five shots are taken per scene. The positions and orientations of the objects vary in each shot. However, the texture of the background remains the same for each scene.

The data generation takes place in a worker sub-process. The main process hosts the REST API. Jobs are managed in a FIFO data structure and processed accordingly. The Kubric process uses the webhook URL to inform if a job finished successfully or with errors. The IDs of successful jobs are managed in a separate data structure, so that the API can make these records available for download. Figure 3 shows this process sequence graphically. In addition, records older than 24 h are deleted by a third process to optimize the application's memory consumption.

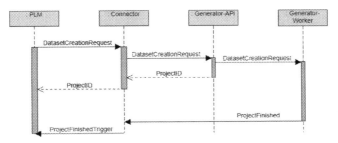

Fig. 3. Sequence Diagram of the dataset generation workflow

For the prototype implementation and evaluation purposes, Autodesk Fusion Manage 360 is used as the PLM system. A workspace in Fusion Mange is created, including a workflow as described in the concept. During the transition from the "Initial state" to the "Data Generation Requested" state, a script is called that searches for an OBJ file with the same name as the object in the attached documents in the Workspace. Thereupon a POST request is sent to the connector. This contains the parameters described in the concept for the data set generation and additionally the WorkspaceID, ItemID and TransitionsIDs of the PLM system, so that the transitions to be switched can be identified and triggered by the connector. The comment of a transition is read in by the script. This way the user is able to change the parameters for the dataset generation. The ClientID and the Client-Secret of a user with the corresponding rights profile for switching the transitions are stored in the connector.

6 Outcomes

In order to validate the general usability of a generated data set, a test procedure is developed and carried out. In total, three reality-based data sets are created. The first one consisting of 50 photos and the second one consisting of 20 photos, for 2D bounding Box detection, are created with a single object class. The third dataset also contains 50 samples but has five different object classes, instead of just one. The real scene used has good lighting conditions and a monochrome gray tabletop as background, to mimic a typical assembly workstation, but without any distracting objects. Light and background remain unchanged during the acquisition of the datasets. Only the number, orientation and position of the objects differ. The first and third one do not contain any overlapping bounding boxes and have a sparser object arrangement. The second reality-based data set contains a very tight arrangement of the objects, including overlapping bounding boxes. All three data sets are used to test three different object detectors trained with synthetic data. Detector A is trained on a data set created with the help of the data generator.

Accordingly, many properties of the virtual scene are randomized. This also applies to the color of the objects. Detector B is trained on a synthetic dataset, which is strongly adapted to the test environment, using Unity3D. In this dataset, the background, the lighting conditions, and the color of the objects match the real environment. Likewise, the objects were placed physically correctly on the background. Both training datasets only have a single object class. For the third test dataset with multiple classes, a third detector is trained using a third train dataset with synthetic images generated by the proposed data generator.

The first two training data sets each consists of 1000 images. The third one consists of 5000 images. 1000 images for each object class. Currently the data-generator is only able to handle one object at a time. To be able to train multiple classes, one dataset for each object is generated. Afterwards the datasets are combined. Figure 4 shows a comparison of images from the three synthetic training datasets and the three real test datasets.

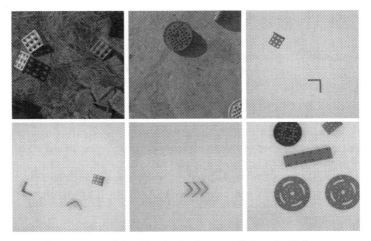

Fig. 4. Samples of the synth. training data in the first row (left and middle one generated by the data-generator, right one with Unity3D) and the real test data in the second row (left sparse, middle tight and right mul. class dataset)

The time needed to generate a single synthetic dataset for one object class with the implemented generator takes approximately 45 min for 1000 samples. The generation time is not only depended on the available hardware and the applied resolution, but it also depends on the images rendered per scene. With five images rendered each scene, half of the total generation time is conducted for creating the virtual blender environment. Increasing this value would therefore lower the generation time. All models were trained for a total of 300 epochs, using the small version of the yolov5 object detection model [11].

Precision is a percentage value that relates the amount of correctly identified objects to the amount of all objects supposedly found by the model. Recall is the amount of correctly identified objects relative to the number of objects that should have been detected.

Table 2. Test results for all three detectors

Model	Dataset	Precision	Recall	mAP50
Detector A	sparse	0.980	0.974	0.982
Detector B	sparse	1.000	0.993	0.995
Detector A	tight	0.211	0.085	0.077
Detector B	tight	0.918	0.785	0.851
Detector C	mul. classes	0.800	0.937	0.911

Mean Average Precision (mAP) also considers the Intersection of Union of the bounding boxes. For mAP50, the IoU ranges from 50% to 95%, with a step size of 5%.

As shown in Table 2 detectors A and B achieved near perfect results on the first dataset and can identify and localize the objects within the image with only a hand full of false detections. The better performance of model B is expected, due to the training data adapted to the test environment but is only slightly pronounced. The tests on the second dataset reveal a huge performance gap between the two detectors. The mean average performance of detector A is eleven times better than that of detector B.

The third detector, trained and tested on multiple classes does not perform as perfect as detector A and B on the sparse dataset, but is still able to correctly detect the objects in most cases.

7   Conclusion

In conclusion, the test procedure validates the usability of the generated data set for training an object detector for simple structured environments. This includes environments without distractions, such as objects not to be detected, with optimal lighting conditions and a rather sparse arrangement of the objects. In such a case, the comparison between model A and B shows, that dealing with the effort of adapting the training dataset to the test domain, or even creating a training dataset manually, is not worth the effort and can be automated by using domain randomization. While the data-generation pipeline currently only supports one object per dataset, combining the generated datasets for multiple class problems is a feasible approach for the number of classes tested.

In a more complex environment, the detector A trained on the randomized synthetic data fails to reliably identify and localize the individual objects, while detector B still performs reasonably well. Accordingly, the cause of the breakdown in the performance of detector A is to be found in the randomized synthetic data set and the randomized properties of the virtual scene. In order to increase the robustness of a detector trained on a randomized data set, tight arrangements of the objects in the virtual scene can be made. A subdivision of the data set into different groups based on edge cases to be mapped is conceivable.

The authors were able to demonstrate that an end-to-end process for the generation of synthetic datasets for CV problems can be fully automated and linked to a PLM system. The required expertise in the field of machine learning for the creation of such a dataset is very low and can therefore be well integrated into the PLM environment.

References

1. Mangold, S., Steiner, C., Friedmann, M., Fleischer, J.: Vision-based screw head detection for automated disassembly for remanufacturing. Procedia CIRP **105**, 1–6 (2022). https://doi.org/10.1016/j.procir.2022.02.001
2. Brogan, D.P., DiFilippo, N.M., Jouaneh, M.K.: Deep learning computer vision for robotic disassembly and servicing applications. Array **12**, 100094 (2021). https://doi.org/10.1016/j.array.2021.100094
3. Basamakis, F.P., Bavelos, A.C., Dimosthenopoulos, D., Papavasileiou, A., Makris, S.: Deep object detection framework for automated quality inspection in assembly operations. Procedia CIRP **115**, 166–171 (2022). https://doi.org/10.1016/j.procir.2022.10.068
4. Borkman, S., et al.: Unity Perception: Generate Synthetic Data for Computer Vision (2021). http://arxiv.org/pdf/2107.04259v2
5. NVIDIA Omniverse Replicator. https://developer.nvidia.com/nvidia-omniverse-platform/replicator. Accessed 25 Jan 2023
6. Greff, K., et al.: Kubric: a scalable dataset generator (2022)
7. Siemens SynthAI. https://synth.ai.sws.siemens.com/. Accessed 25 Jan 2023
8. Cabon, Y., Murray, N., Humenberger, M.: Virtual KITTI 2 (2020). https://arxiv.org/pdf/2001.10773
9. Frid-Adar, M., Klang, E., Amitai, M., Goldberger, J., Greenspan, H.: Synthetic data augmentation using GAN for improved liver lesion classification. In: 2018 IEEE 15th International Symposium on Biomedical Imaging (ISBI 2018), pp. 289–293 (2018)
10. Tobin, J., Fong, R., Ray, A., Schneider, J., Zaremba, W., Abbeel, P.: Domain randomization for transferring deep neural networks from simulation to the real world (2017). https://arxiv.org/pdf/1703.06907
11. Jocher, G., et al.: ultralytics/ YOLOv 5: v7.0 - YOLOv5 SOTA Realtime Instance Segmentation, Zenodo (2022)

Digital Technologies and Emotions: Spectrum of Worker Decision Behavior Analysis

Ambre Dupuis$^{(\boxtimes)}$, Camélia Dadouchi, and Bruno Agard

Laboratoire en intelligence des données (LID), Département de mathématiques et de génie industriel, Polytechnique Montréal, 2500 Chem. de Polytechnique, Montreal, QC, Canada
ambre-manon.dupuis@polymtl.ca

Abstract. Digital technologies enables industries to transform their processes to gain competitive advantage. Industry 5.0 puts the operator at the center of a digital and connected industry, but what about workers' emotions? To what extent do Industry 4.0 technologies in the industrial domain allows for a better understanding of the impact of emotions on workers' decision-making behavior? The overall objective of this systematic literature review is to explore the literature to assess the breadth of possibilities for analyzing emotions to understand workers' decision-making behaviors, based on data collected in industrial settings.

The analysis of 29 articles extracted from the Compendex and Web of Science search engines allowed us to define the emotional factors measured in the analysis of human decision-making behavior, the tools used, and the sectors of application. The subject is still in its infancy for the scientific community and is a source of excitement.

The results of the qualitative analysis of the articles show the predominance of text analysis (social networks and/or online reviews) for sentiment analysis. The tools used within the technology are very diverse (deep learning, machine learning, mathematical models). The same is true for the sectors of activity, although there is a particular interest in customer emotions for marketing purposes in the service industries. Finally, future research avenues are proposed, such as the analysis of the impact of emotions on the decision-making process in manufacturing is practically absent from the study.

Keywords: Industry 5.0 · Decision-making · Emotions · Human behavior · Industry

1 Introduction

To meet the challenges of increasing competition and globalization, manufacturing industries need to find competitive advantages that will lead them towards the digitalization of their processes, leading to the 4th Industrial Revolution [1]. Industry 4.0's emphasis on digital technologies, is its main purpose but also

© IFIP International Federation for Information Processing 2024
Published by Springer Nature Switzerland AG 2024
C. Danjou et al. (Eds.): PLM 2023, IFIP AICT 702, pp. 197–209, 2024.
https://doi.org/10.1007/978-3-031-62582-4_18

its biggest criticism. In doing so, it omits the impact of this transformation on organizations and workers [2], but the operator is at the heart of the production process. That's why Industry 5.0 is putting the operator at the center of the digital transformation. Since emotions play an important role in human decision-making behavior [3], the question is whether and to what extent digital technologies allow us to better understand the impact of emotions of workers' on the decision-making behavior.

To this end, this paper proposes a synthesis of the literature related to the analysis of the impact of emotions on human decision-making behavior based on digital technologies in industry. The general objective is to explore the literature in order to assess the extent of the possibilities of emotion analysis for the understanding of the decision-making behavior of workers, based on data collected in an industrial context.

In the following, the methodology used to carry out the study is presented in Sect. 2, while the results obtained are proposed in Sect. 3. These results will be discussed in the Sect. 4, which will lead to the conclusion, in Sect. 5, on the impact of emotions on the understanding of human decision-making behavior in industry through digital technologies as well as on possible research directions to be explored.

2   Methodology

2.1   Extraction Method

We followed the systematic literature review methodology developed by [4]. First, the subject and limitations of the study must be defined in order to limit its scope. This study aims to determine the extent to which Industry 4.0 technologies can be used to analyze the impact of emotions on human decision-making behavior in industry. Then, the following research questions are posed:

Whether and to what extent digital technologies allow us to better understand the impact of emotions of workers' on the decision-making behavior?

- **RC1:** What do we measure when analyzing emotions to understand human decision-making behavior?
- **RC2:** What are the tools used for the analysis of emotions impacting human decision-making behavior?
- **RC3:** What are the sectors of activity in which the analysis of emotions for the understanding of human decision-making behavior is used?

A set of keywords were iteratively defined to conduct this literature review and answer the proposed research questions.

The identification of the research need was developed from the addressed problem, namely the use of Industry 4.0 technologies for the analysis of the impact of emotions on the decision-making behavior of human in the industrial environment. Thus the expression of the research need was translated by the sentence *"analysis of emotional impact on the decision-making behavior of human in industry using industry 4.0 technologies"*.

In each iteration, the most relevant articles were analyzed to extract the relevant keywords and add them to the initial query. Each iteration targets articles similar to the articles found previously.

The final query used consists of 6 subsets of words that represent the 6 topics of the research need, as shown in Fig. 1.

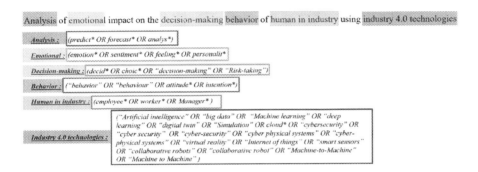

Fig. 1. Research needs analysis

The first two subsets of keywords in Fig. 1 are used to target the area of emotion analysis while the next two subsets narrow the scope of the research results to studies dealing with decision-making behavior. Finally, the last two subsets of keywords focus the study on humans in an industrial context and Industry 4.0 technologies as defined by [5].

Each subset is linked by the operator *"AND"* while the terms within the same subset are linked by the operator *"OR"*. And, finally, the query is completed by defining exclusion criteria to refine the search.

2.2   Articles Selected

Due to the important evolution observed in data analysis and artificial intelligence since the 2010 s, only articles published in English between 2010 and 2023 are selected. Journal and conference papers are considered in the search engine **Engineering Village** (with databases *Compendex, Inspec, GEOBASE, GeoRef et Knovel*) while the journal articles, the proceedings papers and the reviews will be considered in the search engine **Web of Science**.

The search with the current query was performed on January 18, 2023. Figure 2 represents the evolution of the number of articles obtained using the initial query, with the exclusion criteria defined.

A total of 44 different articles were retrieved from the two search engines. A reading of the abstracts allowed us to define 3 exclusion criteria aimed at refining the search results obtained:

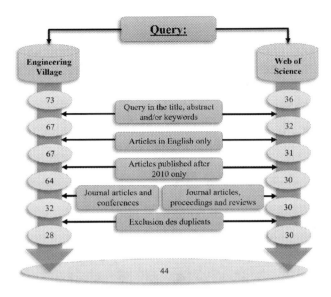

Fig. 2. Conduct of the literature review

1. Inadequacy with the industry topic: Two articles were excerpts from medical conferences. A third medical article was on the assessment of mental illness. Another was on the design of hypertext applications. These papers did not fit the purpose of supporting decision-making in an industrial context and were removed from the selected dataset.
2. Lack of decision making: One paper dealt with sentiment analysis on a social media such as Twitter, but no decision making was explored. Two articles dealing with dashboards in the education sector, with a macroscopic view of the perception of data use, did not fit the needs. One article studied the awareness of the digital footprint and the public perception of Big Data. Again, this article did not fit the research needs. All 4 articles were removed from the analysis set.
3. Misuse of terms describing Industry 4.0 technologies: Some articles do not deal with simulation technology as described by [5], but rather with serious games described by the term "simulation game" or the use of the keyword "mental simulation". Other articles deal with psychological models of technology acceptance (TAM), especially applied to "big data" technology, but not to Industry 4.0 technologies. One article uses the term "cloud" to represent the disruption of the model by the phenomenon of regret, while another uses the term "cloud model" to define the model widely used in security risk assessment. In both of these articles, the term "cloud" did not represent cloud technology. Finally, one article was misclassified with the automatic keyword "simulation" by the search engine **Web of Science** despite using a theoreti-

cal architectural model. Therefore, the term "simulation" was misused. There
were 7 articles removed.

Thus, after the exclusion criteria detailed above, a total of 29 articles remain.
This limited number of articles already allows us to conclude that the research
theme is still underdeveloped.

3   Results

First, a brief descriptive analysis will present the general characteristics of the
selected articles, followed by a more detailed content analysis to answer the
research questions.

3.1   Descriptive Analysis

Although the total number of articles is still low, the annual evolution of the
number of relevant articles published up to January 18, 2023 shows the growing
interest of the scientific community in this still nascent topic, although the year
2020 was heavily influenced by the COVID-19 pandemic. From a thematic point
of view, the overwhelming majority of the selected articles are related to **Big
Data & Analytics**, mainly using machine learning tools.

3.2   Qualitative Analysis

The results extraction phase is guided by the previously defined research questions. Data extraction and statistical follow-up are performed after reading the
abstracts of each of the selected articles. The results obtained will be synthesized
in order to identify possible research areas by answering the three previously
defined research questions.

In the field of construction, [6] analyze the relationship between personality
and safety behavior based on a personality taxonomy. [22] determine the relationship between social factors and individuals' perception of danger. [31] explains
the mechanisms of occurrence of dangerous behaviors in construction by the
theory of the double attitude. In this field, [6] uses data analysis technologies
(machine learning) with the comparison of different tools. The performance of
the linear regression model is compared to that of a feed-forward neural network
using back-propagation and Levenberg-Marquardt optimization. We also find
two studies using **simulation**. Indeed, [22] uses a dynamic system simulation
model just like [31]. However, the data source differs. [31] parameters the simulation model by combining the correlation coefficients observed in the literature.
[22] adds to the use of the literature, the data obtained by survey.

In the field of transport, [11] determine the role of driver characteristics
in modeling driving behavior. Here, the authors use the technologies of the **Big
Data & Analytics** with machine learning tools such as Hierarchical Clustering,
linear regression and Principal Component Analysis (PCA).

In the field of organizations and HR, [25] propose to understand the reactions of workers to procedural and interactional distributive (in)justice. [15] predict the performance of individuals and work teams based on the analysis of individual values and personalities. [17] evaluate the characteristics of a person on the basis of a digital trace. This research can be linked to [21] which proposes a literature review on the prediction of personality traits relevant to the human resources department based on the use of social media. [8] use adaptive interfaces based on language and social network analysis to help managers make decisions. Finally, [13] identifies research trends for sentiment analysis and opinion extraction through a knowledge management approach in organizations. In this area the most represented technology is again **Big Data & Analytics**. [15] use machine learning with tools such as SVM, XGBoost, SDG and linear regression while [25] use deep learning with a prediction model based on an artificial neural network. [17] proposes to build an ontology to be used in machine learning models. The articles [8,13,21] do not specify particular tools since they are literature reviews.

The service industries are strongly represented. In the field of e-commerce, [9] proposes a classification model of products based on their characteristics and the feelings of consumers. [20] evaluate which elements have the greatest impact on a consumer's purchase decision based on community feedback. [28] Examine how consumers' emotional and thematic preferences for products affect their purchase decision from the perspective of psychological perception and linguistic analysis. In the field of hotelling, [18] identifies key indicators perceived by travelers related to environmental management and sustainability of hotels, while [23] explains the decision-making process of travelers for the organization of their trips by analyzing feelings. In the field of tourism, [27] analyzes the impact of virtual reality during the COVID-19 pandemic on consumers' travel intentions. [26] Determine the factors that influence (positively or negatively) the opinions left by tourists about a destination. Finally, in the field of retail, there is a significant amount of literature review. [10] propose a literature review on the attitudes of retail managers towards new tools for capturing customer emotions for decision making. [14] Synthesize research on sensing consumer emotions from social media data to tailor firms' marketing strategies. [29] Address the influence of individual factors on perceptions of store type (online or offline) and consumer shopping behavior. [32] Address merchandising trends and key related issues such as virtual and augmented reality and shopper behavior. Finally, [33] compares the explanatory and predictive power of sentiment extraction methods (marketing) over several years for daily measures of customer sentiment obtained from survey data. The tools used vary widely. Some studies use commercial analysis models [9,18,33], while others are based on mathematical analysis. For example, [23] proposes to compare the performance of tools such as Choquet integrals, weighted arithmetic mean, ordered weighted averaging, and a linear model. [27] proposes to use a partial least squares structural equation model for his analysis. It is interesting to note that [27] discusses the use and application of **virtual reality** technology in data analysis. [20] proposes an alegbric SVD decomposition, then an exploratory factor analysis (EFA) and a Tobit regression. [26] uses logistic and linear regression tools. [28] uses tools closer to machine learning such as Bayesian networks, Support Vector Machines (SVM), Latent Dirichlet Allocation (LDA) and Importance Performance Analysis (IPA). The SVM algorithm is also used in the study [33] where it is compared with Sentiment Extraction Tool (SET) or commercial software solutions such as Linguistic Inquiry and Word Count (LIWC). [9] proposes its own segmentation algorithm that allows to group the comments with the closest sentiments. The article [10] does not specify the tools used, but states that it is a machine learning analysis, and the articles [14,29,32] do not specify particular tools as they are literature reviews.

In the field of finance, [16] evaluates asset prices in the stock market based on sentiment analysis and [19] presents a model that measures and analyzes the intangible assets related to the reputation of digital ecosystems and their impact on tangible assets. **Big Data & Analytics** is presented. Deep learning is used in the article [16] with a model based on LSTM (RNN). The article [19] does not detail the tools used, but mentions the use of data mining.

In the area of risk management, [30] defines a new framework for cyber risk assessment and mitigation based on exploration and sentiment analysis of text in hacker forums. [34] propose a model for predicting the risk of an attack by a particular individual based on various personality traits and contexts. In this field, the technologies of **Big Data & Analysis** and **Simulation** are represented. Thus, [30] uses the multiclass classification algorithms CART, k-Nearest Neighbor (k-NN), Ensemble Boosted Tree, Multinomial Logit and Hierarchical Logit. [34] uses dynamic system simulation (Monte Carlo simulation and sensitivity analysis) to build the Bayesian Belief Network (BBN) prediction model for attack prediction.

In the field of crisis management, [24] Although this article does not mention any of the tools used in its abstract, the analysis of the keywords allows highlighting the notion of sentiment analysis. This term is defined in the abstract of [13] and can be classified in the technology **Big Data & Analytics** as it mainly uses the artificial algorithm on textual data.

In the field of logistics, there are only manufacturing case studies with very similar research topics on reverse logistics through social media analysis. [7] Analyze positive and negative customer feedback to make strategic reverse logistics decisions, while [12] develop reverse logistics strategies based on analysis of positive/negative consumer reactions. These two articles are interesting because they deal with similar issues but use different tools. In fact, both articles propose sentiment analysis based on social networks (Twitter). [7] uses a deep learning approach with the development of a hybrid CNN-LSTM model. [12], uses a machine learning approach with the combination of Naïve Bayes, Support Vector Machine (SVM) and Maximum Entropy algorithms, combined with a voting mechanism.

We note that sentiment analysis based on social network analysis [7,12,14] and online reviews [9,18,20,23] is particularly used.

This research highlights the importance of open access databases and APIs for the advancement of research in a particular sector. Indeed, three of the four studies related to tourism and hospitality use data from the API *TripAdvisor* [18, 23,26]. The APIs of e-commerce platforms like Amazon [9,20] or their Chinese equivalent [28] are also used. The API of Twitter is also widely present in the selected articles. But, contrary to the others, this database is transversal to several domains [7,12,17,24].

Synthesis

As a synthesis, Fig. 3 represents the distribution of the selected articles according to the research topic addressed and the field of application studied.

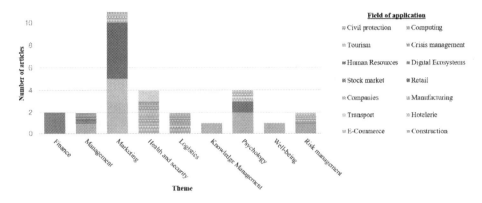

Fig. 3. Distribution of the selected articles by theme and field of application

The notions of Health and Safety are observed in the articles dealing with the construction industry but also in the transport sector. The notion of "wellness" is more related to organizations. However, [25] can be considered as a study relating to the **psychology** of the workers. This research theme is moreover the most widely represented in the application sector of organizations [15,17]. We also find the psychological theme applied to human resources [21] and to tourism [26]. Finally, the thematic of **knowledge management** and **management** are also studied in the organizations with the articles [8,13] respectively. In the thematic of **management** we also find a study applied to the management of crisis [24].

The theme of **risk management** is applied to the sector of the civil protection [34] but also to the data-processing sector with the concept of cybersecurity [30].

It is interesting to note that the manufacturing industries are particularly interested in **logistics** [7,12] while the service companies are more interested in **marketing**. Moreover, it is this topic that is the most represented in the whole of the extracted articles. The researches on marketing are mainly related to concepts applied to e-commerce, retailing, hotels and tourism. The main objective of these studies is to increase the sales by redefining the commercial and marketing strategies according to the feedbacks of the customers in order to meet their requirements. The feedbacks are collected on social networks, but also and especially on the online reviews of the different sites of service providers (TripAdvisor or Amazon). The notion of **feeling analysis** is also very present in the studies related to marketing.

Some articles determine the tendency of individuals to take risks in order to be used as input for models for the prediction of individual and team performance or of the safety behaviour of construction workers. Only [11] considers the decision-making process in its model.

4   Discussions

Previous analyses allow for answering the various research questions proposed in Sect. 2.

RC1: What Do We Measure When Analyzing Emotions for Understanding Human Decision-Making Behavior?. Language is the most analyzed element in the proposed studies. Sentiment analysis of online reviews or social media makes it possible to classify consumers' emotions. This information is then used for managerial decision-making in the definition of marketing strategies, human resource management policies or reverse logistics management strategies. However, although decision-makers use emotions to make decisions, the user decision process itself remains unexplained. Thus, the true value of language analysis for understanding the decision process remains to be determined.

RC2: What Tools Are Used for the Analysis of Emotions Impacting Human Decision-Making Behavior?. As previously mentioned, the decision-making process is still poorly explored in the literature. However, the systematic literature review allowed us to highlight the predominance and variety of tools used in **big data and analytics** technology. We find mainly segmentation tools such as SVM, CART, k-Nearest Neighbor (k-NN), Ensemble Boosted Tree, Multinomial Logit, Hierarchical Logit or Hierarchical Clustering. These tools are especially associated with sentiment analysis. Mathematical and regression tools are also well represented. We compare the performance of tools such as Choquet integrals, weighted arithmetic mean, ordered weighted averaging, and linear models. We also find the use of structural equation of partial least squares model, but also algebraic models of SVD decomposition and factor analysis (EFA). different regressions are also used, such as logistic and linear regression and Tobit. Deep learning is also represented, but to a lesser extent. Feedforward neural networks (FNN), convolutional neural networks (CNN), and recurrent neural networks (RNN and LSTM) are used. These networks can also be combined to form hybrid models (CNN-LSTM).

On the contrary, the **Simulation** technology is characterized by the uniqueness of the tool used. In fact, the only studies dealing with simulation use the simulation of dynamic systems (SD).

RC3: What Are the Industries in Which Emotion Analysis for Understanding Human Decision-Making Behavior Is Used?. The sectors of activity are varied: Construction, e-commerce, hotelerie, transportation, manufacturing, business, retail, stock market, digital ecosystems, human resources, crisis management, tourism, computing, and civil protection. However, the service industries are the most represented and are mainly associated with marketing research. Health and safety research is also well represented, but is largely dominated by studies in the construction sector. Finally, it is worth noting that the only two studies in the manufacturing sector deal exclusively with reverse logistics.

It is important to note the wide variety of results obtained. The literature review initially focused on the study of human behavior in industry. However, fields such as Marketing or Well-Being have emerged within the corpus of articles analyzed. This variety of topics and their distance from the initial industrial theme suggests a possible area for improvement. A refinement of the query to circumscribe the study and to be more efficient in reaching the set objective should be considered in the future.

5   Conclusion

Industry 5.0 puts the operator at the center of a digital and connected industry. Taking emotions into account is essential for understanding human decision-making behavior. Therefore, a systematic literature review was conducted to evaluate the impact of emotions on human decision-making behavior in Industry 4.0 technologies.

Language is by far the most analyzed element in the literature. Especially through sentiment analysis in social networks and online reviews. Marketing in service industries such as tourism, hospitality or e-commerce is the main user. But the digital technology of big data analytics is strongly represented in a variety of sectors such as psychology, construction or risk management. The variety of tools used is also interesting. Data analysis can be performed using purely mathematical tools, machine learning approaches or even deep learning.

The answers to the research questions show that emotions are still very little used to explain the human decision-making process. However, various tools allow us to use emotions to understand behaviors and guide users towards choices. Thus, the measurement of emotions does exist in the literature but it is still difficult to clearly make the link between emotions and the human decision making process.

However, it is interesting to note that none of the studies use manufacturing data. This is surprising given the high volume of decisions made in this environment. Understanding the impact on the decision-making process of operators in manufacturing remains an open area of research.

References

1. Brodeur, J., Pellerin, R., Deschamps, I.: Collaborative approach to digital transformation (CADT) model for manufacturing SMEs. J. Manufact. Technol. Manag. **33**(1), 61–83 (2022)
2. Zizic, M.C., Mladineo, M., Gjeldum, N., Celent, L.: From Industry 4.0 towards industry 5.0: a review and analysis of paradigm shift for the people. organization and technology. Energies **15**(14), 5221 (2022)
3. Abdel-Ghaffar, E.A., Wu, Y., Daoudi, M.: Subject-dependent emotion recognition system based on multidimensional electroencephalographic signals: a riemannian geometry approach. IEEE Access **10**, 14993–15006 (2022)
4. Tranfield, D., Denyer, D., Smart, P.: Towards a methodology for developing evidence-informed management knowledge by means of systematic review. Br. J. Manag. **14**(3), 207–222 (2003)
5. Moeuf, A., Pellerin, R., Lamouri, S., TamayoGiraldo, S., Barbaray, R.: The industrial management of SMEs in the era of Industry 4.0. Int. J. Product. Res. **56**(3), 1118-1136 (2018)
6. Gao, Y., Gonzalez, V.A., Wing Yiu, T., Cabrera-Guerrerod, G.: The Use of Machine Learning and Big Five Personality Taxonomy to Predict Construction Workers' Safety Behaviour. arXiv. pp.34 (2019)
7. Shahidzadeh, M.H., Shokouhyar, S., Javadi, F., Shokoohyar, S.: Unscramble social media power for waste management: a multilayer deep learning approach. J. Clean. Prod. **377**, 134250 (2022)
8. See, S.: Big data applications: adaptive user interfaces to enhance managerial decision making. In: 17th International Conference on Electronic Commerce 2015 (ICEC 2015). Association for Computing Machinery, New York, USA (03-05 August 2022)
9. Kauffmann, E., Peral, J., Gil, D., Ferrandez, A., Sellers, R., Mora, H.: Managing marketing decision-making with sentiment analysis: an evaluation of the main product features using text data mining. Sustainability **11**(15), 4235–4254 (2022)
10. Pantano, E., Dennis, C., Alamanos, E.: Retail managers' preparedness to capture customers' emotions: a new synergistic framework to exploit unstructured data with new analytics. Br. J. Manag. **33**(3), 1179–1199 (2022)
11. Witt, M., Kompaß, K., Wang, L., Kates, R., Mai, M., Prokop, G.: Driver profiling - Data-based identification of driver behavior dimensions and affecting driver characteristics for multi-agent traffic simulation. Transport. Res. F: Traffic Psychol. Behav. **64**, 361–376 (2019)
12. Ahmadi, S., et al.: The bright side of consumers' opinions of improving reverse logistics decisions: a social media analytic framework. Int. J. Logistics Res. Appli. **25**(6), 977-1010 (2022)
13. Casas-Valadez, M.A., et al.: Research trends in sentiment analysis and opinion mining from knowledge management approach: a science mapping from 2007 to 2020. In: 2020 International Conference on Innovation and Intelligence for Informatics, Computing and Technologies (3ICT). IEEE, Sakheer, Bahrain (20-21 December 2020)
14. Madhala, P., et al.: Systematic literature review on customer emotions in social media. In: 5th European Conference on Social Media ECSM 2018, Academic Conferences and Publishing International, pp. 154 – 162 (21 - 22 June 2018)
15. Altuntas, E., Gloor, P.A., Budner, P.: Measuring ethical values with AI for better teamwork. Future Internet **15**(5), 133–161 (2022)

16. Eachempati, P., Srivastava, P.R., Kumar, A., Tan, K.H., Gupta, S.: Validating the impact of accounting disclosures on stock market: a deep neural network approach. Technol. Forecast. Soc. Chang. **170**, 556–569 (2021)
17. Alamsyah, A., et al.: Ontology modelling approach for personality measurement based on social media activity. In: 6th International Conference on Information and Communication Technology, pp. 507-513. IEEE, Bandung, Indonesia (2018)
18. Saura, J.R., Reyes-Mendes, A., Alvarez-Alonso, C.: Do online comments affect environmental management? identifying factors related to environmental management and sustainability of hotels. Sustainability **10**(9), 3016 (2018)
19. Casado-Molina, A.M., Ramos, C.M.Q., Rojas-de-Garcia, M.M., Sanchez, J.I.P.: Reputational intelligence: innovating brand management through social media data. Indust. Manag. Data Syst. **120**(1), 40–56 (2020)
20. Ahmad, S.N., Laroche, M.: Analyzing electronic word of mouth: a social commerce construct. Int. J. of Inform. Manag. **37**(3), 202–213 (2017)
21. Subramanian, K.S., Sinha, V., Bhattacharya, S., Chaudhuri, K., Kulkarni, R.: A literature review on human behavioral pattern through social media use: a HR perspective. Int. J. of Cyber Behav. Psychol. Learn **3**(2), 56–81 (2013)
22. Ma, H., Wu, Z.G., Chang, P.: Social impacts on hazard perception of construction workers: a system dynamics model analysis. Saf. Sci. **138**, 105240 (2021)
23. Vu, H.Q., Li, G., Beliakov, G.: A fuzzy decision support method for customer preferences analysis based on Choquet Integral. In: 2012 IEEE international conference on Fussy systems, pp. 1-8. IEEE, Brisbane, Australia (10-15 June 2012)
24. Attanasio, A., Jallet, L., Lotito, A., Osella, M., Rua, F.: Fast and effective decision support for crisis management by the analysis of people's reactions collected from twitter. Commun. Comput. Inform. Sci. **539**, 229–234 (2015)
25. Abubakar, A.M., Behravesh, E., Rezapouraghdam, H., Yildiz, S.B.: Applying artificial intelligence technique to predict knowledge hiding behavior. Int. J. of Inform. Manag. **49**, 45–57 (2019)
26. Bigne, E., at al.: What drives the helpfulness of online reviews? a deep learning study of sentiment analysis, pictorial content and reviewer expertise for mature destinations. J. Destination Market. Manag. **20**, 100570 (2021)
27. Tan, K.L., Hii, I.S.H., Zhu, W.Q., Leong, C.M., Lin, E.: The borders are re-opening! Has virtual reality been a friend or a foe to the tourism industry so far?. Asia Pacific J. Marketing Logist. (2022)
28. Luo, Y.Y., Yang, Z., Liang, Y., Zhang, X.X., Xiao, H.: Exploring energy-saving refrigerators through online e-commerce reviews: an augmented mining model based on machine learning methods. Kybernetes **51**(9), 2768–2794 (2021)
29. Hermes, A., Riedl, R.: Influence of personality traits on choice of retail purchasing channel: literature review and research agenda. J. Theor. Appl. Electron. Commer. Res. **16**(7), 3299–3320 (2022)
30. Biswas, B., Mukhopadhyay, A., Bhattacharjee, S., Kumar, A., Delen, D.: A text-mining based cyber-risk assessment and mitigation framework for critical analysis of online hacker forums. Decis. Support Syst. **152**, 113651 (2020)
31. Zhou, M., Chen, X.C., He, L., Ouedraogo, F.A.K.: Dual-attitude decision-making processes of construction worker safety behaviors: a simulation-based approach. Int. J. of Environ. Res. Public Health **19**(21), 14413 (2022)
32. Munoz-Leiva, F., Lopez, M.E.R., Liebana-Cabanillas, F., Moro, S.: Past, present, and future research on self-service merchandising: a co-word and text mining approach. Eur. J. Mark. **55**(8), 2269–2307 (2021)

33. Kubler, R.V., Colicev, A., Pauwels, K.H.: Social media's impact on the consumer mindset: when to use which sentiment extraction tool? J. of Interactive Market. **50**, 136–155 (2020)
34. Sticha, P.J., Axelrad, E.T.: Using dynamic models to support inferences of insider threat risk. Comput. Math. Organ. Theory **22**(3), 350–381 (2016)

Prediction of Next Events in Business Processes: A Deep Learning Approach

Tahani Hussein Abu Musa$^{(\boxtimes)}$ (iD) and Abdelaziz Bouras (iD)

Department of Computer Science and Engineering, College of Engineering (Qatar University),
Doha, Qatar
ta090001@student.qu.edu.qa, abdelaziz.bouras@qu.edu.qa

Abstract. Business Process Mining is considered one of the merging fields that focusses on analyzing Business Process Models (BPM), by extracting knowledge from event logs generated by various information systems, for the sake of auditing, monitoring, and analysis of business activities for future improvement and optimization throughout the entire lifecycle of such processes, from creation to conclusion. In this work, Long Short-Term Memory (LSTM) Neural Network was utilized for the prediction of the execution of cases, through training and testing the model on event traces extracted from event logs related to a given business process model. From the initial results we obtained, our model was able to predict the next activity in the sequence with high accuracy. The approach consisted of three phases: preprocessing the logs, classification, and categorization and all the activities related to implementing the LSTM model, including network design, training, and model selection. The predictive analysis achieved in this work can be extended to include anomaly detection capabilities, to detect any anomalous events or activities captured in the event logs.

Keywords: Business Process Mining · Event log · LSTM · Business Process

1 Introduction

Business process mining or process mining is an emerging research area that gained increasing attention in both academic and industry. It aims at analyzing Business Process Models (BPM), by extracting knowledge from event logs generated by various information systems, for the sake of auditing, monitoring and analysis of business activities for future improvement and optimization throughout the entire lifecycle of such processes, from creation to conclusion [1]. It is also useful when detecting potential exceptions or anomalies in the logged workflow, and hence help identify the source of such exceptions and modify them accordingly.

Typically, a business process may be defined as a group of linked activities produced for a particular purpose [2]. A business process has a number of attributes, has a particular goal, needs a certain input, produces a certain output, has a series of events or activities carried out in a predefined order, might need some resources and involve a

© IFIP International Federation for Information Processing 2024
Published by Springer Nature Switzerland AG 2024
C. Danjou et al. (Eds.): PLM 2023, IFIP AICT 702, pp. 210–220, 2024.
https://doi.org/10.1007/978-3-031-62582-4_19

number of owners [3]. There exist various representations and notations of business processes, including Unified Modeling Language (UML), Business Process Model Notation (BPMN), Event-driven process chain (EPC) and Petri Net Markup Language (PNML) [2].

Due to the availability of event logs produced from different information systems, typical process mining techniques are utilized in process discovery, conformance checking, and model enhancement. Hence, providing valuable business perceptions and insights [4].

In this context, event logs can be considered important means for extracting knowledge about the sequence of activities, making it possible to discover, monitor and improve such processes in different application domains of interest.

However, those typical systems do not have the ability of making predictions about the next activity in business processes.

Recently, the growth and spread of Machine Learning (ML) algorithms in the field of process mining, and predictive analysis has become a motivation of using them in performing accurate predictions of future events and activities in business processes, by analyzing logged historical activities and events, extracted from event logs. Such methods include parametric regression [7], Naive Bayes classifier [5], and predictive clustering tree inducer [6].

Besides, and due to the large volumes of data presented by the extracted event logs, deep learning approaches can outperform those typical process mining approaches, bringing more accurate predictions about future events in each event log trace, through analyzing data from previous events. Deep learning approaches including deep neural networks have been widely used in predictive analysis and process mining [8].

In this paper, a Long Short-Term Memory (LSTM) Deep learning model is used to predict the next activity of a given business process. LSTM is one type of Recurrent Neural Networks (RNNs) that based on the literature, has been used extensively in time series analysis and sequential problems [9–11]. The model presented in this work has been trained on event logs that represent procurement scenario, allowing for predicting the next activity on a given trace of events that follow another activity or a series of activities as an input. To validate our approach, we obtained our preliminary results based on a dataset comprising of 300 traces. The test performed on the trained model showed that it was able to predict the next activity in our business process model presented.

The structure of this research is organized as follows: Section 2 survey some related work in literature, Sect. 3 describes our proposed approach, Sect. 4 demonstrates results and analysis, and Sect. 5 presents a discussion of our research findings and future work to be done in this area.

2   Literature Review

2.1   Process Mining

Process mining may be defined as the area that resides between data mining and Business Process Modelling (BPM) and analysis. The main goal of process mining is analyzing Business Processes through the extraction of knowledge from event logs generated by different information systems, to support the activities related to auditing, monitoring and analysis of business activities for future improvement and optimization throughout the entire lifecycle of such processes, from creation to conclusion. Several process mining algorithms exist in literature to support the activities of conformance checking, performance analysis, decision mining, organizational mining, predictions, and recommendations [12].

In process mining, data related to events and activities of a given business process are recorded in event logs, which may be defined as a "hierarchically structured file with data on the executions of business processes" [13]. Event logs are considered the input to process mining, and typically contain traces of several executions of the same business process. A *case* or *trace* is a collection of related events within a single instance of the business process. An *event* or *activity* is a single atomic part of the trace, that has several attributes describing it, beside other information related to updating the state of such events, for example (Notify Delivery of the parts). Other attributes of events may include resources allocated and timestamp [14]. The following figure illustrates the relation between and event log, a case and event.

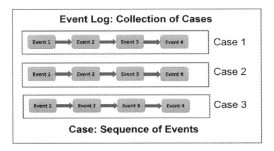

Fig. 1. Relation between Log, case, and event

According to [13], the three main tasks of process mining are: Process Discovery, Process Conformance and Process Improvement. *Process discovery* involves the use of event log as an input and the output will be the corresponding business process model without any previous knowledge of it [13]. *Process Conformance Checking* involves the comparison between a business process model and the event log generated via executing the same process model [13]. Such comparison helps evaluates whether the log information and the model are equivalent. *Process Improvement* involves applying any enhancements on the current model using the current information from the log.

Although event logs can be in either structured or unstructured, but nowadays, the eXtensible Event Stream (XES) format is the standard format for event log specification [15], considered as the 'de facto' standard for storing event logs by the IEEE task force, and it was preceded by the format Mining eXtensible Markup Language (MXML), enabling the exchange of event data between diverse systems and employing process mining techniques [15].

2.2 LSTM Neural Networks

Recurrent Neural Networks (RNNs) and its variant the Long-Term-Short Memory (LSTM) have been used extensively in time series analysis and sequential problems.

RNNs have two types of input, present and recent past. RNN use both types of input to verify how they behave in case of new data. In other words, at any given time, the output of a RNN at time step$_{t-1}$ affects its output at time step$_t$. LSTMs are considered more efficient at obtaining long-term temporal dependencies [16]. The information contained in LSTMs are beyond the normal flow of the recurrent network in a gated cell. Such Information can be stored, written, or be read from a cell, like data in a computer's memory. The cell is able to make decisions about what should be stored and when it should be allowed to read, write, and delete, through gates that open and close. Such gates are implemented with the multiplication of elements by sigmoid, resulting in having all in the range of 0–1 [16].

In the area of process mining, there exists several research works been conducted related to utilizing LSTM in predicting next event in business processes. In [17], the authors proposed the question of how to use Deep Learning techniques to train accurate models of Business Processes behavior from event logs. The proposed approach trained a neural network with LSTM architecture, in order to predict the sequence of next events, their timestamp and associated resources. An experimental evaluation on real life event logs has been performed and showed that the proposed approach outperforms previously proposed LSTM architectures targeted at this problem [17].

In [18], authors extended the body of research in this area by testing four different variants of Graph Neural Networks (GNN) and a fully connected Multi-layer Perceptron (MLP) with dropout for the tasks of predicting the nature and timestamp of the next process activity.

On the other hand, [19] proposed a model that is composed of Convolutional Neural Network (CNN) to forecast the next business activity. The use of CNN in process mining was justified by guaranteeing high accuracy in predictions of next activity, and according to the authors, CNN has achieved faster training and inference than LSTM, even when the processes tend to have longer traces.

3 Proposed Approach

3.1 Example Business Process

For demonstration purpose in this paper, we focus on a generic business process for a procurement scenario illustrated in Fig. 1, (The real work is also performed on similar scenarios provided by our project industrial stakeholders).

3.2 Simulate Business Process and Generate Event Logs

For this purpose, we used ProM 6.10[1], that is a well-known Process Mining tool developed by Developed by Process Mining Group, Math and CS department, Eindhoven

---

[1] http://www.promtools.org/doku.php?id=prom610.

University of Technology, and has several capabilities related to simulation of business processes, as well as log generation capabilities. Table 1 provides a snapshot of a generated event log resulting from the simulation.

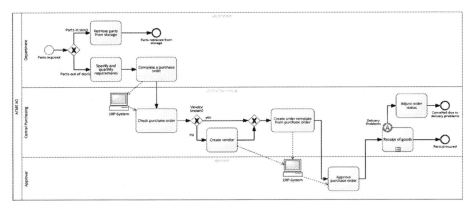

Fig. 2. Procurement Business Process

3.3   Architecture

Figure 2 illustrates the architecture of our approach to predict the next activity in the procurement business process using LSTM, adopted from [20]. The approach is composed of three phases: 1) Preprocessing of Event log, 2) Classification and 3) Prediction.

Phase 1: Event Log Preprocessing
This phase is divided into three steps

Data Extraction
In which we decide on which attributes we will select from the event log needed for the prediction, in our case the "Activity" attribute.

Trace Identification
In this step we identify the traces representing different cases in the event log, each case is represented by the corresponding trace. Those traces are extracted and appended to a text file, preserving their order in the original log.

Segmentations
After identifying the traces in the previous step, we *divide* each trace into a set of activities/events. Each event is then converted to a sequence of integers, comprising a two sequence lists of "integers", the fist contains is the list of activities (X) and the second of output activities (Y). Then, the sequence of input activities (X) is converted into a 2-dimensional matrix composed of both number of sequences and the maximum length of sequences.

Table 1. Snapshot of a generated Event Log

Process ID	Activity	Timestamp
1	Parts Required	2010-12-30T11:02:00.000+01:00
1	Retrieve Parts from Storage	2010-12-31T10:06:00.000+01:00
1	Parts Retrieved from Storage	2011-01-05T15:12:00.000+01:00
2	Parts Required	2021-01-26T21:06:18.849+03:00
2	Specify Quantify Requirements	2021-01-26T21:12:18.849+03:00
2	Specify Quantify Requirements	2021-01-26T21:13:18.849+03:00
2	Complete PO	2021-01-27T21:23:03.707+03:00
2	Check PO	2021-01-27T21:25:03.707+03:00
2	…	….
3	Parts Required	2021-01-30T22:18:32.555+03:00
3	Specify Quantify Requirements	2021-01-30T22:19:32.555+03:00
3	Complete PO	2021-01-31T22:28:32.555+03:00
3	Check PO	2021-02-01T22:14:41.262+03:00
3	Specify Quantify Requirements	2021-02-02T22:27:41.262+03:00
3	……	……
4	Retrieve Parts from Storage	2021-02-22T23:03:01.460+03:00
4	Parts Retrieved from Storage	2021-02-22T23:04:01.460+03:00
4	……	……
5	Parts Required	2021-03-26T23:00:32.002+03:00
5	Retrieve Parts from Storage	2021-03-26T23:00:38.002+03:00
5	……	……

Phase 2: Classification

In this phase we convert the sequence of integers of output activities (Y) obtained in the previous step into one hot encoding representation, identifying that the number of classes will be equal to the size of the vocabulary.

Phase 3: Prediction

This step is furtherly divided into three sub-steps

Network Design

In which we defined the design of the LSTM network in terms of input, hidden and output layers, along with specification of needed parameters.

Network Training

In this step we train the LSTM network using dataset composed of the sequence list of integers represented by the activities in matrix (X), and the one hot representation of matrix (Y).

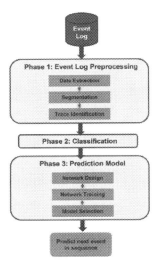

Fig. 3. Proposed Architecture

Model Selection

Proceeding the training phase, in this step the model of LSTM with best results will be identified as the final model. Such 'best results' including high accuracy would be the best one to make accurate predictions. Otherwise, another iteration on training and adjustments of the parameters will be needed.

Prediction

This is the output generated by the LSTM model, which is the prediction of the next activity in the in the business process stated earlier, from a single input activity or a series of activities.

3.4 Implementation

Table 2 lists the parameters configured for the LSTM network. For training the network, we used an event log of with 300 traces and 60 different activities in the log. During the training phase, the number of sequences that were identified was 4321. To elaborate more on the ways LSTM network works, the network accepts as input a single activity or a sequence of activities, and based on a value specified by the user, a number of predicted activities will be performed. Table 3 lists the names of the activities captured in the business process, along with their acronyms for simplifying the representation of the activity name instead of the full name. For example, SAQR represents the activity "Specify and Quantify Requirements", and CPO is the acronym for "Complete Purchase Order". As part of the implementation, Keras [21], which is a python library that enables us to build deep learning models.

Table 2. LSTM Model Parameters with their configured values

Parameter	Value
Epochs	200
Optimizer	Adam
Loss	Categorial Cross entropy
LSTM Units	100
Batch size	20

Table 3. Activity Names with their acronyms

Activity Name	Acronym
Part Required	PR
Retrieve Parts from Storage	RPFS
Specify and Quantify Requirements	SAQR
Complete Purchase Order	CAPO
Check Purchase Order	CPO
Create Order Template from PO	COTFPO
Approve Purchase Order	APO
Receipt of Goods	ROG
Adjust Order Status	AOS

4 Results

Table 4 exhibits some of the obtained results from our proposed model. The first column in the table shows the input activity, which is the one that was passed to the LSTM model as an input to the prediction phase. The target activity is the anticipated activity (or activities), i.e., those activities with highest probabilities of the targeted prediction, according to the weight of each activity. Results showed that the model proposed was able to predict the next target activity with high precision, with 85% accuracy. Figure 3 depicts the training versus validation loss for the model, and Fig. 4 illustrates the model accuracy (Fig. 5).

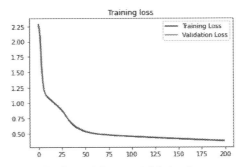

Fig. 4. Training vs. Validation Loss

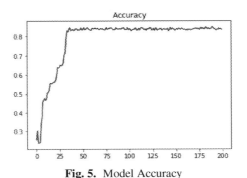

Fig. 5. Model Accuracy

Table 4. Sample results of Prediction

No.	Input Activity	Target Activity	Output Activity 1	Output Activity 2
1	**PR**	**SAQR|RPFS**	**SAQR**	**RPFS**
2	**CAPO**	**CPO|CV|**	**CPO**	
3	**COTFPO**	**APO|ROG|AOS**	**APO**	**ROG**
4	**CAPO**	**CPO|CV|COTPFO**	**CPO**	**CV**
5	**PR**	**RPFS**	**RPFS**	

5   Conclusion

Deep learning methods including LSTM can be used in predicting the next activity in a given business process model, with 85% accuracy. In this work, we utilized LSTM neural network model for such a task, by training it on traces of events extracted from event log of a simulated procurement business process. Such an approach can be used for process improvement and auditing business process models. As future work, we may test our proposed model on real world event data, and we may extend it to involve anomaly detection component, to detect anomalous events in the sequence of traces, namely, contextual anomalies using LSTM neural networks.

Acknowledgement. This research is supported by Qatar University.

References

1. Saylam, R., Sahingoz, O.K.: Process mining in business process management: concepts and challenges. In: International Conference on Electronics, Computer and Computation (ICECCO), Ankara, pp. 131–134 (2013). https://doi.org/10.1109/ICECCO.2013.6718246
2. Sarno, R., Wibowo, W.A., Kartini, F.H., Effendi, Y., Sungkono, K.: Determining model using non-linear heuristics miner and control-flow pattern. TELKOMNIKA Telecommun. Comput. Electron. Control J. **14**(1) (2016)
3. Sarno, R., Pamungkas, E.W., Sunaryono, D., Sarwosri: Business process composition based on meta models. In: International Seminar on Intelligent Technology and Its Applications (ISITIA) (2015)
4. Aalst, W.M.P.: Process Mining: Discovery, Conformance and Enhancement of Business Processes. Springer, Heidelberg (2011). https://doi.org/10.1007/978-3-642-19345-3
5. Ceci, M., Spagnoletta, M., Lanotte, P.F., Malerba, D.: Distributed learning of process models for next activity prediction. In: Desai, V.B.C., Fresca, S., Zumpano, E., Masciari, E., Caroprese, L. (eds.) Proceedings of the 22nd International Database Engineering & Applications Symposium (IDEAS 2018), Villa San Giovanni, Italy, 18–20 June 2018, pp. 278–282. ACM, New York, NY, USA (2018)
6. Pravilovic, S., Appice, A., Malerba, D.: Process mining to forecast the future of running cases. In: Appice, A., Ceci, M., Loglisci, C., Manco, G., Masciari, E., Ras, Z.W. (eds.) New Frontiers in Mining Complex Patterns. LNCS (LNAI), vol. 8399, pp. 67–81. Springer, Cham (2014). https://doi.org/10.1007/978-3-319-08407-7_5
7. Aalst, W.M.P., Schonenberg, M.H., Song, M.: Time prediction based on process mining. Inf. Syst. **36**, 450–475 (2011)
8. Evermann, J., Rehse, J.-R., Fettke, P.: Predicting process behavior using deep learning. Decision Support Syst. **100**, 129–140 (2017)
9. Li, Y., Cao, H.: Prediction for tourism flow based on LSTM neural network. Procedia Comput. Sci. **129**, 277–283 (2018)
10. Cortez, B., Carrera, B., Kim, Y.-J., Jung, J.-Y.: An architecture for emergency event prediction using LSTM recurrent neural networks. Expert Syst. Appl. (2017)
11. Liu, F., Chen, Z., Wang, J.: Video image target monitoring based on RNN-LSTM. Multimedia Tools Appl., 1–18 (2018)
12. Aalst, W.: Process discovery from event data: relating models and logs through abstractions. Wiley Interdisc. Rev. Data Min. Knowl. Discovery **8**, e1244 (2018). https://doi.org/10.1002/widm.1244
13. Van der Aalst, W.M.: Process Mining: Data Science in Action. Springer, Heidelberg (2016). https://doi.org/10.1007/978-3-662-49851-4
14. Baier, T., Mendling, J., Weske, M.: Bridging abstraction layers in process mining. Inf. Syst. **46**, 123–139 (2014)
15. IEEE Task Force on Process Mining: IEEE 1849-2016 XES Standard Definition (2016). https://www.xes-standard.org/. Accessed 9 Feb 2023
16. Greff, K., Srivastava, R.K., Koutník, J., Steunebrink, B.R., Schmidhuber, J.: LSTM: a search space odyssey. IEEE Trans. Neural Netw. Learn. Syst. **28**(10), 2222–2232 (2017)

17. Camargo, M., Dumas, M., González-Rojas, O.: Learning Accurate LSTM Models of Business Processes (2019). https://doi.org/10.1007/978-3-030-26619-6_19
18. Venugopal, I., Töllich, J., Fairbank, M., Scherp, A.: A comparison of deep-learning methods for analysing and predicting business processes. In: Proceedings of International Joint Conference on Neural Networks, IJCNN. IEEE Press (2021)
19. Obodoekwe, E., Fang, X., Lu, K.: Convolutional neural networks in process mining and data analytics for prediction accuracy. Electronics **11**, 2128 (2022). https://doi.org/10.3390/electronics11142128
20. Tello-Leal, E., Roa, J., Rubiolo, M., Ramirez-Alcocer, U.M.: Predicting activities in business processes with LSTM recurrent neural networks. In: 2018 ITU Kaleidoscope: Machine Learning for a 5G Future (ITU K), Santa Fe, Argentina, pp. 1–7 (2018). https://doi.org/10.23919/ITU-WT.2018.8598069
21. Chollet, F.K.: Keras: The Python deep learning library (2018). https://github.com/fchollet/keras

Machine Learning Algorithms for Process Optimization and Quality Prediction of Spinning in Textile Industries

Hye Kyung Choi[1] ⬭, Whan Lee[1] ⬭, Seyed Mohammad Mehdi Sajadieh[1] ⬭, Sang Do Noh[1(✉)] ⬭, Hyun Sik Son[2] ⬭, and Seung Bum Sim[3] ⬭

[1] Sungkyunkwan University, Suwon-si, Gyeonggi-do 16419, Korea
sdnoh@skku.edu
[2] Textile Material Solution Group, DYETEC Institute, Daegu, Korea
[3] Corporate Growth Support Headquarter, Korea Textile Development Institute, Daegu, Korea

Abstract. In smart manufacturing, data-driven artificial intelligence algorithms are becoming increasingly important in improving decision-making by monitoring the control, analysis, and prediction of manufacturing processes in a production system. In the textile industry, there is a strong need for smart manufacturing technologies because various parameters could affect the quality dynamically. This study aims to optimize the parameters of spinning processes by developing machine learning algorithms and models which can predict the toughness and elasticity of threads. At first, meaningful variables are extracted from the shop floor data, and then a defect classification learning model is developed to predict defects in advance. In addition, a regression model is implemented for the prediction of toughness and elasticity of the textile. By transitioning from the traditional trial and error method to the data-based method for the spinning process, production costs and time can be reduced through optimal settings of the production parameters for the spinning of the desired threads.

Keywords: smart manufacturing · data-driven prediction · textile industry · spinning process

1 Introduction

The 4th Industrial Revolution resulted in industrial changes that have accelerated the transition into smart manufacturing. Smart manufacturing (SM) controls and manages production, helps the manufacturers with business decisions, and optimizes the manufacturing process [1]. A combination of the state-of-the-art technologies, such as Industrial Internet of Things (IoT), big data, cloud computing, and Artificial Intelligence (AI) drives the development of highly sophisticated manufacturing intelligence. It plays a pivotal role in SM by optimizing the manufacturing process. Recently, there has been an increase in applications of anomaly detection technology for predicting system failure and AI prediction models for quality control in the manufacturing industry [2].

© IFIP International Federation for Information Processing 2024
Published by Springer Nature Switzerland AG 2024
C. Danjou et al. (Eds.): PLM 2023, IFIP AICT 702, pp. 221–232, 2024.
https://doi.org/10.1007/978-3-031-62582-4_20

The spinning process is a complicated manufacturing operation that involves a complex production plan. Figure 1 show an equipment of spinning process, which is the main topic of this research. The production plan needs to specify the number of yarns, the spinning process system, and the preparation method, with a variety of options available for all, which introduces a lot of complexity into the production planning stage [3].

Fig. 1. An equipment of spinning process in textile industry

Moreover, in the case of fiber industry, an appropriate production sequence is planned and various production parameters are defined, once the specific category of the final product is decided upon [3]. As a labor-intensive industry, on-site decision-making for the spinning process, heavily depends on the knowledge and the experience of the workers, which gives rise to the following problems.

i. Since the on-site decision making heavily depends on the experience of the workers, it is prone to human error and could potentially lead to increased cost, resulting in inconsistencies in quality control from the trial-and-error process.
ii. However, the defects and quality of the products cannot be examined after each sub manufacturing process and can only be examined after the entire process is completed.

To address these problems, this research attempts to predict the quality of the final products before the manufacturing process, using AI technology, by proposing a model to assist workers with decision making in setting configurations for sub manufacturing processes. Deep learning methodologies have been widely used for quality prediction tasks. In this research, among the many deep learning methodologies, the long short-term memory (LSTM) method has been applied to overcome the gradient vanishing problem. Section 3 presents the process of forecasting methods. Section 4, applications of the proposed methodology in real-life situations are presented.

2 Research Background

2.1 Spinning Process

The spinning process involves a series of sub processes, including polymerizing to create high molecular fibers, drying high-molecular polymer chips, melt spinning, forming fiber structure, cooling and drawing, and finally the process of taking up, to ultimately produce fibers that satisfy the required property values [4]. Figure 2 illustrates a flowchart of a typical spinning process. There is a lack of standardization on the extent to which variables affect the property values of the spinning process, so the variables are often arbitrarily set, invariably being affected by the worker's subjectivity. In addition, due to the long production hours, the process can only be executed once a day and defects can only be examined after all the processes have been completed, as mentioned earlier.

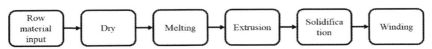

Fig. 2. Steps of a Spinning Process

2.2 LSTM: Long Short-Term Memory Network

In recent years, the neural network (NN) has emerged as a powerful tool in the fields of data analysis and time series forecasting [5]. Deep learning technologies have been applied in load forecasting, with most depending on the recurrent neural network (RNN) or LSTM [6–10]. Farooq *et al.* [11] have conducted research on artificial neural networks (ANNs) for quality characteristics estimation. The ANN was fed with parameters that affect the quality and was trained, using a combination of the Marquardt-Levenberg algorithm and Bayesian regularization. The research revealed that NNs are effective in estimating the quality characteristics of yarns. Although RNN is one of the more powerful neural networks that can internally maintain input memory, there exist the challenges of a long training process and the Levenberg vanishing gradient problem [12]. LSTM is one variation of RNN that is widely used to resolve the vanishing gradient problem [13]. LSTM is a neural network with an architecture to store sequential short-term memory, which is later processed with a secondary RNN and is frequently used in the field of AI and deep learning [14]. Hu *et al.* [15], evaluated the prediction accuracy of an optimization algorithm, using CNN-LSTM, to develop a prediction model that can accurately predict yarn quality parameters. CNN was utilized to optimize the input values, and LSTM was used to figure out the correlation between the performance index and the machining parameters and to predict the yarn quality parameters. The research indicated that the prediction accuracy of CNN-LSTM model is heavily influenced by the process parameters and the optimization algorithm. LSTM leverages the input gate, output gate, and forget gate in the memory cell of the hidden layer to discard insignificant memories, thus retaining only significant ones [16]. The vanishing gradient problem is also prevented, considerably improving the performance in the processing of long sequence inputs compared to the conventional RNN [6].

3   Methodology

This section discusses the AI algorithm development methodology that predicts the strength and elongation of products, manufactured by the spinning process. Figure 3 illustrates each step of the methodology, in detail.

3.1 Data Extraction and Processing

For this research, the data were provided by the Korea Fiber Research Institute and obtained from manufacturing processes. The data consisted of values of the process parameters and the measurements of strength and elongation. In total, 50 types of manufacturing data variables, were obtained during the spinning process. Among the 50 process variables, all the values with fixed constant value were excluded from the training of the model. Mathematical-statistical methods were utilized to derive the key factors of the process. In general, the t-test is one of the most commonly used mathematical-statistical methods to determine the correlation. However, due to the relatively large number of process variables in the spinning process, ANalysis Of VAriance (ANOVA) was utilized, which has been proven useful for analyzing multiple experimental groups. ANOVA analyzes the significant effects of a specific variable on the target variable when more than two groups exist in a study. ANOVA is an extension of the t-test for comparing means of two groups to situations where there are two or more levels of a categorical independent variable [17]. In experiments with two or more treatment groups and where the same individuals are subjected to multiple treatments or measured at different time points, the independence assumption between samples is violated. Therefore, such designs are analyzed using two-way repeated measures ANOVA (2-way ANOVA). Researchers in optometric research should carefully consider experimental design aspects and seek statistical advice to ensure the appropriate application of ANOVA [18]. To ensure the correct application of ANOVA, Armstrong [18] design and match the appropriate analysis that considered about single factor, multiple factor, sequential treatment, factor combination. In this case, there are various factors that influence the two response variables (strength and elongation), Multivariate ANOVA(MANOVA) is more effective than two-way ANOVA. Figure 4 describes two-way ANOVA and MANOVA in detail. As such, ANOVA was utilized to analyze un-fixed variables of the spinning process and 'n' number of significant variable data were extracted as a result.

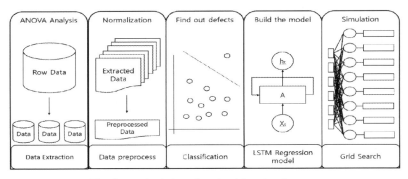

Fig. 3. Methodology for algorithm development

Based on the result of the variance analysis, data with low significance were excluded and the remaining key variables were preprocessed in the form of a standard normal distribution with mean 0 and variance 1 with anomalies excluded.

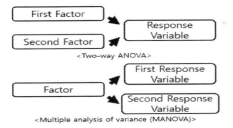

Fig. 4. ANOVA and MANOVA

3.2 Classification Model for Defect Regression Model Implementation

Removing anomalous data points is the preliminary step in implementing regression model. To accurately predict the strength and elongation of yarns, defective yarn data were regarded as outliers. After eliminating all data that were classified as outliers using a classification model, the remaining data were used for the training of the regression model. Several different models, proven to be useful for the classification task, were compared to build a defective data filtering model. The four methods used to implement the classification model were: eXtreme Gradient Boosting (XGBoost), Multi-Layer Perceptron (MLP), Random Forest (RF), and K-Nearest Neighbor (KNN). As this research distinguishes the models for predicting the strength and the elongation of the yarn, the classification model for each target value of these variables was accordingly chosen to show the best results for each. As the primary evaluation metrics for the classification model, accuracy, precision, recall, and F1 score were utilized.

3.3 Proposed Algorithm

This research built a model, using a Keras-based LSTM algorithm, in the Tensorflow 2.0 and Python 3.9.1 environment. Mean squared error was utilized to calculate the loss value for the optimization of the LSTM model and the prediction model was designed to predict two types of outputs, strength and elongation. The LSTM method used in the research stems from RNN, which is a type of deep learning algorithm, whereby dropout, batch size, epoch to store, and pass on the data were set from the previous steps for error correction in the hidden layer. The internal hidden layer of the regression prediction model consisted of a dense layer and LSTM layer; the dropout rate, to prevent overfitting; the batch size, to determine the input size for error correction; and the epoch, to state the total number of training iterations for optimization, which were all optimized through a series of trial and errors. Table 1 shows the optimal values for the parameter tuning.

Once the training for the newly implemented regression model was complete, the process simulation data were fed into the model for examination of the prediction results. The process simulation data were derived through a grid search analysis with the unique value of each significant variable.

Table 1. Parameter tuning values

Layer	Dropout	Batch size	Epoch
LSTM, Dense	0.2	16	50

4 Case Study

The selected factory for this research uses a melt spinning process and has two production lines. Each of the two hoppers consists of two extruders, A and B. Hence, there are four tanks: extruder #1 A, extruder #1 B, extruder #2 A, extruder #2 B. The above methodology was applied to a real-life site.

4.1 Data Variable Extraction

As mentioned earlier, there were 50 types of spinning data. After excluding the variables with fixed values, only 14 variables remained. Table 2 categorizes these 14 variables, based on the relevant process.

Table 2. Non-fixed variables

No	Process	Variable	UNIT	Value
1	Extrusion Process	Mesh #1 type	–	#30, #36
2		Mesh #2 type	–	#30, #36
3		Mesh #3 type	–	#30, #36
4		Mesh #4 type	–	#30, #36
5		Spinbeam Temp	°C	254, 256, 258, 260, 262
6		Mainfold A Temp	°C	254, 256, 258, 260, 262
7		Mainfold B Temp	°C	254, 256, 258, 260, 262
8		Pack front pressure	Mpa	54, 69
9		Pack end pressure	Mpa	54, 69
1	Winding Process	FR speed	m/min	4000–4400
1	Bundling Process	Migration pressure	Kg/cm^2	1.2, 1.8
1	Heat Treatment Process	GODET R/O B temperature	°C	90, 100, 105
1	Drawing Twist Process	GODET R/O A SPEED	m/min	1000–3675
2		GODET R/O B SPEED	m/min	4105–4520

If the P-value is less than 0.05, it can be said that there is significance. This means that the significance level is 95% or higher, thus hinting that the independent variable has statistically significant effects on the dependent variable. Table 3 summarizes the degrees of freedom, squared sum, mean of squared sum, and p-value. In extracting significant variables, a total of eight independent variables remained after filtering. MANOVA was utilized to verify the significance of effects of these eight independent variables on strength and elongation. As illustrated in Table 4, all variables, except drawing ratio, were confirmed as significant. However, because the drawing ratio also had an influence on strength and elongation in the two-way ANOVA analysis, it was also selected as a significant variable for the regression model.

Table 3. Suggested significant variables

Variable	Value	Sum_sq	Mean_sq	F	PR($>$F)
Spinbeam Temp	Strength	27.0066	13.5033	3.0945	$4.610038\,e^{-2}$
	Elongation	1047.7134	523.8567	392.3657	$5.912841\,e^{-2}$
Mainfold temp	Strength	4.446084	2.2230	0.5094	$6.011138\,e^{-1}$
	Elongation	1047.7134	9.0743	9.0743	$1.329997\,e^{-1}$
GODET R/O A speed	Strength	180.3400	6.9361	1.5895	$3.326776\,e^{-2}$
	Elongation	1113.1336	32.0665	32.0665	$3.487900\,e^{-2}$
GODET R/O B speed	Strength	1.0002	0.2500	0.0573	$9.938931\,e^{-1}$
	Elongation	82.9456	15.5314	15.5314	$4.909281\,e^{-1}$
GODET R/O B temp	Strength	623.6853	47.9757	10.9945	$5.06930\,e^{-21}$
	Elongation	619.3138	35.6817	35.6817	$2.52557\,e^{-21}$
Drawing ratio	Strength	21.5961	3.5993	0.8248	$5.550908\,e^{-1}$
	Elongation	25.2373	3.1504	3.1504	$4.801639\,e^{-1}$
FR speed	Strength	1.8033	0.4508	0.1033	$9.813244\,e^{-1}$
	Elongation	15.2008	2.8463	2.8463	$2.351307\,e^{-1}$
Yarn count	Strength	1.7693	0.4423	0.1013	$9.819756\,e^{-1}$
	Elongation	36.756109	6.882529	6.882529	$2.081831\,e^{-1}$
Residual	Strength	2356.3457	4.363603	NaN	NaN
	Elongation	720.9667	1.335124	NaN	NaN

4.2 Data Preprocessing

The data, used in this research were in float type. Based on the variance analysis results, insignificant variables and missing values were all removed. To improve the training process of the model, scaling techniques of normalization and standardization were utilized. During the preprocessing stage, standardization was set to transform data into a Gaussian standard distribution, where the mean is 0 and the variance is 1.

Table 4. MANOVA analysis

Variable	Pillai's trace Value	F Value	Pr < F
Spinbeam Temp	0.2009	24.1278	P < 0.001
Mainfold Temp	0.2009	24.1278	P < 0.001
GODET R/O speed A	0.2217	25.5734	P < 0.001
GODET R/O temp A	0.2315	24.8567	P < 0.001
GODET R/O speed B	0.1153	38.8318	P < 0.001
GODET R/O temp B	0.1215	41.2002	P < 0.001
FR speed	0.1163	39.2073	P < 0.001
Drawing ratio	0.0001	0.0421	P < 0.959

4.3 Classification Model for Defect Removal

Prior to implementing the regression model, the field datasets were preprocessed to remove any defects in strength and elongation, so that the model would be trained only on relevant data. To separate defective data, a classification model was built with defect set to have strength and elongation of 0 value each. As two-way ANOVA between the normal products and the defects aggravates the issue of imbalanced data, the range for a product value was divided into 0.3 units. Four classification models: XGBoost, MLP, RF, and KNN, were compared to select the best, the one with the highest accuracy in this case. The performance results are displayed in Table 5. For strength and elongation, XGBoost and MLP models were selected classification, respectively.

Table 5. Classification performance by suggested model

	Model	Accuracy	Precision	Recall	F1-score
Strength	XGBoost	0.9833	0.8333	0.8293	0.8312
	MLP	0.8750	0.5991	0.5851	0.5912
	RF	0.9833	0.8293	0.8333	0.8312
	K-NN	0.8750	0.7646	0.7486	0.7559
Elongation	XGBoost	0.3500	0.3125	0.1896	0.2350
	MLP	0.7167	0.5485	0.5556	0.5482
	RF	0.6333	0.5538	0.5747	0.5508
	K-NN	0.3500	0.2500	0.1896	0.2131

4.4 Implementation of the Regression Model with LSTM

A quality estimation regression model was first built to predict the quality of the spinning process and then trained on field data. The quality estimation model was implemented based on the LSTM method, and the parameters were tuned to obtain the most optimized results for the field data. This method has been proven to show better performance in handling the vanishing gradient problem during backpropagation. As such, the r2-scores of the regression models, trained with field data for strength and elongation were 0.7002 and 0.6992, respectively. The results of the training are illustrated in the graphs of Fig. 5 which shows that the prediction results follow approximately 70% of the actual field data.

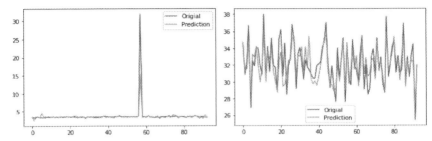

Fig. 5. Strength and elongation prediction

4.5 Process Optimization

One of the main reasons for implementing the prediction model is to facilitate on-site decision making without depending on the worker. The model attempts to set up a process that can output the desired values in advance, by predicting the final values of strength and elongation even before executing the process. As such, the process simulation data were generated to verify the usefulness of leveraging the prediction model as a tool to assist decision making for workers.

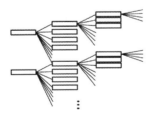

Fig. 6. Predictions of strength and elongation

The structure of the process simulation data is shown in Fig. 6. The process simulation data consist of combinations of unique values of each process variable. Therefore, all possible process scenarios that can occur on-site were transformed into input data format. In total, 238,140 datasets were generated. From these, the defective data values were removed for training and 231,050 datasets were used.

As shown in Fig. 7, measuring the similarity of the density distributions between the field dataset and the prediction results indicates that the variance is higher, and the distribution is more uniform in the simulation datasets than in the field datasets. This could be attributed to the fact that all possible scenarios of the process have been considered, thus resulting in a generally higher variance.

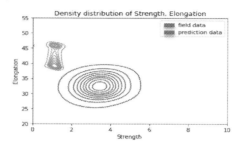

Fig. 7. Density distribution

As shown in Fig. 8, because all possible scenarios have been considered for the simulation data, the correlation analysis indicates that the correlation complexity between significant variables has been reduced. However, the effects of each variable on strength and elongation are still reflected in the field data, implying a similar correlation.

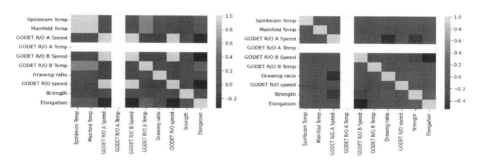

Fig. 8. Correlation analysis between field and simulation data

5   Conclusion

Today, achieving manufacturing productivity and intelligence requires the interoperability and scalability of data-driven technological systems. This has become an important issue. To optimize products and production processes, it is necessary to establish a systematic system that spans the entire lifecycle of the product. The purpose of this study was to achieve the development of the necessary data for the system and the algorithms that constitute the system to achieve systematic system construction throughout the entire lifecycle of the product. This paper proposes a methodology to optimize the fiber spinning process setting. The first contribution of this study lies in utilizing the characteristics of the spinning process data and conducting analysis to identify significant data.

In textile processes, uncertainty has traditionally been addressed by relying solely on the expertise of workers or through a trial-and-error approach to adjust optimal process settings such as temperature and speed. However, this paper identifies the variables that impact the yield and quality of textile processes through data analysis. This allows for the efficient resolution of problems that were previously dependent on workers' expertise, by leveraging data-driven approaches. Furthermore, analyzing and extracting data that influences quality can aid in the collection and accumulation of necessary data for the digital transformation of textile processes, contributing to the digitization process. The second contribution of this study involves designing various process scenarios to produce yarns with the desired quality. The strength and elongation values were then predicted based on the results obtained from these scenarios. The deep learning technique in LSTM was proven to be effective in analyzing all possible process scenarios that can occur from spinning. The proposed model had a prediction accuracy of approximately 70% and 69% for strength and elongation, respectively. In the field, knowledge about process setting variables is primarily based on the experience of production. However, the proposed predictive model in this paper allows for obtaining knowledge about process settings in unexplored environments, providing support to workers' decision-making. The algorithm can replace results that were previously obtained through the knowledge of experienced workers or through trial and error. The predictive model can support workers' decision-making by providing desired quality process settings that can be achieved, and workers can utilize the generated list of settings to configure the process. This way, the predictive model assists workers in their decision-making process and enables them to set up the process using one of the recommended configurations. Further improvements in future research, can include optimization of the prediction model so that the process would no longer need to rely solely on the experience of the workers. In addition, setting initial process configurations through predicted process results could help overcome the limitations of continuous manufacturing.

Acknowledgement. This research was supported by the MOTIE (Ministry of Trade, Industry and Energy), Korea, under the Virtual Engineering Service Platform program (P0022335) supervised by the Korea Institute for Advanced Technology (KIAT).

References

1. Váncza, J., Do Noh, S., Yoon, H.S.: Preface for the special issue of green smart manufacturing. Int. J. Precis. Eng. Manuf.-Green Tech. **7**, 545–546 (2020)
2. Kang, H.S., Lee, J.Y., Choi, S., et al.: Smart manufacturing: past research, present findings, and future directions. Int. J. Precis. Eng. Manuf.-Green Tech. **3**, 111–128 (2016)
3. Li, C., Li, J., Li, Y., He, L., Fu, X., Chen, J.: Fabric defect detection in textile manufacturing: a survey of the state of the art. Secur. Commun. Networks **2021**, 13 (2021). Article ID 9948808
4. Kim, J.: Research Group of Polymer Frontiers, Chemical Engineering and Materials Research Information Center, Korea (2019)
5. Mikolov, T., Joulin, A., Chopra, S., Mathieu, M., Ranzato, M.A.: Learning longer memory in recurrent neural networks. arXiv preprint arXiv:1412.7753 (2014)
6. Hochreiter, S., Schmidhuber, J.: Long short-term memory. Neural Comput. **9**(8), 1735–1780 (1997)

7. Abel, T., Nguyen, P.V., Barad, M., et al.: Genetic demonstration of a role for PKA in the late phase of LTP and in hippocampus-based long term memory. Cell **88**(5), 615–626 (1997)
8. Kong, W., Dong, Z.Y., Jia, Y., et al.: Short-term residential load forecasting based on LSTM recurrent neural network. IEEE Trans. Smart Grid **10**(1), 841–851 (2019)
9. Kong, W., Dong, Z.Y., Hill, D.J., et al.: Short-term residential load forecasting based on resident behaviour learning. IEEE Trans. Power Syst. **33**(1), 1087–1088 (2017)
10. Wang, J., Chen, X., Zhang, F., Chen, F., Xin, Y.: Building load forecasting using deep neural network with efficient feature fusion. J. Mod. Power Syst. Clean Energy **9**(1), 160–169 (2021)
11. Farooq, A., Cherif, C.: Development of prediction system using artificial neural networks for the optimization of spinning process. Fibers Polym. **13**(2), 253–257 (2012)
12. Yadav, A., Jha, C.K., Sharan, A.: Optimizing LSTM for time series prediction in Indian stock market. Procedia Comput. Sci. **167**, 2091–2100 (2020)
13. Salehinejad, H., Sankar, S., Barfett, J., Colak, E., Valaee, S.: Recent advances in recurrent neural networks. arXiv preprint arXiv:1801.01078 (2017)
14. Monner, D., Reggia, J.A.: A generalized LSTM-like training algorithm for second-order recurrent neural networks. Neural Netw. **25**, 70–83 (2012)
15. Hu, Z.: Prediction model of rotor yarn quality based on CNN-LSTM. J. Sens. (2022)
16. Gers, F.A., Schmidhuber, J., Cummins, F.: Learning to forget: continual prediction with LSTM. Neural Comput. **12**(10), 2451–2471 (2000)
17. Pak, S.I., Oh, T.H.: The application of analysis of variance (ANOVA). J. Vet. Clin. **27**(1), 71–78 (2010)
18. Armstrong, R.A., Eperjesi, F., Gilmartin, B.: The application of analysis of variance (ANOVA) to different experimental designs in optometry. Ophthalmic Physiol. Opt. **22**(3), 248–256 (2002)

E-Learning Content Creation for Interdisciplinary Master of Science Program in Product Lifecycle Management (PLM)

Alexandra Saliger[1]([✉]), Yannick Juresa[2], Manfred Grafinger[1], and Jens C. Göbel[2]

[1] Institute of Engineering Design and Product Development, TU Wien, Getreidemarkt 9, 1060 Vienna, Austria
alexandra.saliger@tuwien.ac.at

[2] Institute of Virtual Product Engineering (VPE), RPTU Kaiserslautern-Landau, Gottlieb Daimler Str. 44, 67663 Kaiserslautern, Germany

Abstract. This contribution introduces an educational Product Lifecycle Management (PLM) curriculum suitable for undergraduate and young professionals, which enables a graduation as Master of Science. The main goal is to increase the accessibility of valuable PLM content for educational purposes and to introduce students to the productive "hands on" features of PLM by developing an international, interdisciplinary and online-based program. This program focusses on a fundamental PLM understanding, the usability of available software systems and an industrial PLM integration and application. A highly professional cooperation between five technical universities from Turkey, Spain, Germany and Austria, international experts designed interactive online classes, which are suitable for distance learning. Sharing technology for international educational platform enables students, professors and partners from industry to share their PLM expertise with students and improve the common and different understandings. This new PLM program shows the usability of the applications, which were implemented and tested in a pilot training course. The results are very promising and further suggestions and feedback for improvement will be taken into account. Student feedback and the exceptional motivation of the whole project team shows that interdisciplinary online education has a promising future in international academic PLM education integrating practical use cases.

Keywords: product lifecycle management · online platform · curriculum

1 Introduction

Young professionals need new requirements to develop smart products in an industrial environment. The development of smart products in various disciplines, but also across a value network, creates ever greater complexity. Graduates and employees who can deal with the complex structures in companies and products could reduce existing uncertainties in industrial companies. To educate these young professionals, on the one hand, the

C. Danjou et al. (Eds.): PLM 2023, IFIP AICT 702, pp. 233–243, 2024.
https://doi.org/10.1007/978-3-031-62582-4_21

234   A. Saliger et al.

requirements and problems for graduates in the field of engineering have to be broken down, and on the other hand, in addition to the professional skills, social skills have to be addressed by a degree program.

To educate young professionals with the right technical knowledge, it is necessary to establish a curriculum that teaches how to deal with complex information systems in the product life cycle and product development processes. Today's smart products are interdisciplinary and complex, requiring various IT solutions during the development and use phases [1]. Product lifecycle management (PLM) refers to central IT solutions supporting product development and creation. The integration of tools and data in a digital environment that spans the entire product lifecycle plays an important role [2]. In this context, PLM is considered a business paradigm that companies view as a key factor for success in the engineering and technology sector [3]. Several PLM solutions have evolved over time and exist in parallel in many companies today [4]. However, the competence to work in a collaborative environment with PLM is not very common among graduates in the job market [5]. Another point is that most companies operate globally. This means that in addition to having a complex IT environment in the company, being able to work together in a large international and distributed team is a key skill [6]. Therefore, it is essential to bring the two topics together and to sensitize and introduce graduates to the topic of PLM through an international course of study. This can be covered in the location-independent collaboration by integrating different international universities into one study program.

Until 2009, there were no study programs for students that also considered the technical processes in the context of PLM [7]. The current situation at the universities involved in the research project shows that PLM has so far only been offered as a component of other courses of degree programs. PLM usually only plays a major role in the master's program and is attempted to be covered by a single course. There are no specific PLM courses in the bachelor's degree programs, but only in individual lecture units, for example, in digital engineering. Singular educational institutions quickly reach their limits with specialized courses of study, whether for bachelor's or master's degrees. This makes it more important to bring together the existing know-how from different teaching and research institutions to create and manage smart products in an industrial environment. Through the synergies of new partnerships in the field of teaching, innovative and future-oriented courses of study for specializations can be introduced. Various measures can be taken to promote internationalization and the ability to work in a team in addition to the specialized topics. According to studies, international study programs with various educational cooperation partners in two or more countries and exchange semesters in other countries greatly impact students' professional, personal, and social skills. In addition to the professional maturity that is strengthened by the acquisition of a master of science degree, study-related stays abroad also result in the acquisition of important competencies and skills, e.g., in areas such as intercultural learning, personality development, promotion of social skills, and foreign language acquisition [8]. The experience gained here also plays a not inconsiderable role in later professional life. Another study found very positive effects especially in European experiences abroad of students through the exchange program ERASMUS, among others in the areas of a biographical change, which stands for a growing self-confidence, but also the ability to deal with unforeseen events more easily, as well as that the stay abroad has a positive influence on the academic and professional career [9].

The coronavirus pandemic has forced educational institutions to move to alternative teaching formats more quickly. This change has brought digital formats into focus, allowing for different learning experiences. E-learning has several implications for the study. As a tool, it is indispensable in modern and forward-looking educational institutions, especially since the beginning of the corona crisis. To use e-learning profitably, several requirements must be met. On the one hand, the requirements for a study program and the professional acquisition of competencies in it, on the other hand, due to a new learning method, requirements for the platform, the design of it as well as the learning offers are equally important. To be able to deal with all topics such as e-learning via a platform, mapping of the individual courses in the learning platform, general learning objectives and competencies in the courses as well as the teach-the-teacher concept, the Connect4PLM project is divided into six different work packages, each with its own intellectual results.

2 PLM Master Curriculum Approach

PLM represents a specialization in engineering. Both, to be able to understand PLM from a student perspective and to be able to apply PLM inevitably require fundamental knowledge from the field of engineering. Especially in the abstract university environment, problems from the field of PLM, which arise from the challenges of companies, are difficult to convey [10]. Therefore, it is essential that a potential PLM curriculum builds on a science or engineering undergraduate or bachelor's degree. This is, among other things, one of the reasons why the Connect4PLM research project is specifically pursuing the development of a master's curriculum. To achieve the necessary learning objectives, it is therefore also necessary to design a holistic and multidisciplinary curriculum approach [11]. There are already offers, e.g., how young professionals and other employees can be educated in PLM. One example is the PLM Professional Program from Fraunhofer IPK [12]. Nevertheless, the course of studies presented here already focuses on students who want to specialize in PLM before starting their careers.

Studies from the 2000s show a strong industrial need for students well-trained in PLM, so a solid foundation exists, and additional employee training is no longer necessary [13]. Current studies continue to show that there is a need for further action in the area of teaching in educational institutions [14]. The structure of a curriculum must cover various considerations in advance [15]. Resource planning and the relationship between knowledge and skills are part of it, in addition to pedagogical practice, curricula, and methods. As a more constructive definition, a curriculum is "the multi-layered social practices, including infrastructure, pedagogy and assessment, through which education is structured, enacted and evaluated" [15, p.136]. The core of a curriculum can be based, among others, on nine terms [16]. These target resources, contents, goals, environment, and circumstances. Figure 1 shows the curriculum spider web, which describes the terms and the corresponding question.

The curriculum concept is an important part of the Connect4PLM project. However, in order to be able to deal with all topics sufficiently, the entire project is divided into six work packages, each of which is led by a different university. The exact structure of the preceding and following work packages can be found in Sect. 3. The initial structure of the curriculum was defined according to the topics and the guiding questions of the curriculum spider in work package four (Curriculum development). However, it was clear from the outset that not all topics could receive the same attention. Therefore, within the work package for building a general curriculum, assumptions were already made to be able to set a focus and postpone other topics to a later point in time. Figure 1 shows the consideration of the different topics for a general structure.

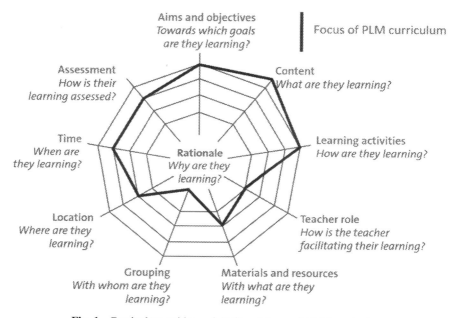

Fig. 1. Curriculum spider web [16] and focus of PLM curriculum

Work package four was initially worked on by a single university, but all partner institutions have influenced the curriculum structure in various feedback and iteration rounds.

3 Connect4PLM Project Educational Framework

Connect4PLM project aims to combine background knowledge of the academic partners from Turkey, Spain, Germany and Austria in order to generate awareness about PLM. The basic idea of the project is to develop an e-learning platform for interdisciplinary master of science program to train graduates, researchers and young professionals in PLM. The approach to be implemented at the project partners' universities is based on practice-oriented courses designed to stimulate technical learning through the development of competencies essential to the ability of individuals to perform specific functions efficiently and solve complex scientific and practical problems. The project Connect4PLM enhance accessibility, fosters learning experiences, supports scalability, encourages collaboration, aligns with industry needs, and facilitates continuous professional development. It addresses the evolving demands of the PLM field and empowers individuals to gain the skills and knowledge necessary for successful careers in PLM. It is built on the following concepts, shown in Fig. 2.

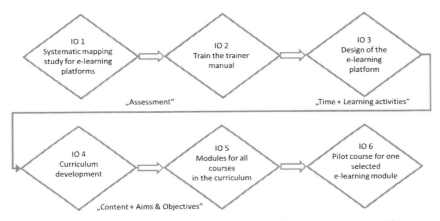

Fig. 2. Structure of the project with the associated intellectual outputs (IO)

3.1 Systematic Mapping Study for E-Learning Platforms

The objective of a systematic mapping study is to apply a systematic mapping process and create a systematic map for e-learning platforms. First, research questions are defined. The Master of Science program will be made available on the e-learning platform. Research questions are generated for each of the requirements. The keywords are selected from the abstracts of relevant studies and classified according to the requirements of the e-learning platform.

3.2 Train the Trainer Manual

The goal of the "Train the Trainer Manual" is to support PLM trainers in the methodology defined in the ADDIE model (model for the development of an instruction system) for e-learning. The ADDIE model has five phases: analysis, design, development, implementation and evaluation. PLM trainers have various roles in e-learning, such as instructional designer, subject matter expert, online moderator and tutor. The ADDIE manual itself consists of the instructions for the needs analysis, target group analysis, task and topic analysis in the analysis phase. The e-learning course combines the teaching methods of presentation, application and collaboration. When the delivery strategy is defined, the appropriate e-learning formats such as video conferencing, live webcasting, audio conferencing, application sharing and animations, slides are selected. For self-paced e-learning, practice and tests consist mainly of questions associated with answer options and feedback.

3.3 Design of the E-Learning Platform

The e-learning platform should be open source. The most important characteristic of open source platforms is that they can be extended easily and cost-effectively. This function is provided by the extensibility of the database model. The database model of the e-learning platform is expanded to include the specific requirements for the MSc title. The main features of the platform include the preparation of interactive course content, support for multi-client user management and access to course content from mobile devices. The spread of the custom e-learning platform for PLM will be possible through the spread of the open-source e-learning platform.

Appropriate configurations are supported to enable local use of the e-learning platform in the participating countries of the project. The learning analytics module is developed in the platform and the e-learning process thus becomes an essential part of a personalized master's degree.

3.4 Curriculum Development

The aim of the MSc curriculum is to develop the qualification description, the curriculum itself and the curriculum content of the master's program. For this purpose, the content of existing international master's programs is evaluated and the necessary information and reports on the requirements and needs for curriculum development are obtained. A selection of 26 bachelor's and master's degree programs with strong tendencies towards PLM topics was made for closer examination. This selection was mainly but not only based on European institutions, both universities and universities of applied sciences. Based on these reports and instructions, all project partners will create the description of the Master of Science program. After the curriculum has been developed, the qualification description is developed, which contains all the requirements for the master's degree. The tasks leading to the design and development of the master's curriculum are as follows:

– Evaluation of the content of existing international master programs
– Development of curriculum and qualification descriptions
– Development of curriculum and their content materials
– Improvement of academic staff in blended learning

3.5 Modules for All Courses in the Curriculum

This project focuses on the creation of a new curriculum and the implementation of e-learning modules for the Master of Science program on PLM. The proposed study plan will be evaluated and approved by the European partner institutions. The master's program also includes blended learning technology in PLM and e-learning in the Master of Science education system. It obviously requires academic staff to have enough knowledge, experience and skills to share blended learning technologies while also teaching the course content of PLM. Since the master's program will have a close relationship with existing PLM education and with the institutions where PLM has been used extensively, academic staff should also know how to use the e-learning tools directly in the teaching process. Members of each partner university and institution are sent to a German University to complete this training. The development of the curricula and the teaching materials is the responsibility of the five European Universities from Germany, Austria, Spain and Turkey to teach them at each partner university. The involvement of the PLM-experienced partner university with all partners will help to integrate the master's course with high quality and ensure the future viability of the master's course.

3.6 Pilot Course for One Selected E-Learning Module

The Connect4PLM project aims to create a new curriculum and e-learning platform for the PLM Master of Science program. The created curriculum is to be developed, created and implemented by the partner institutions. The concept of the pilot course will be exemplary for adapting several courses for PLM to e-learning. The prerequisite for the pilot course is a completed bachelor's degree in any engineering discipline. Interest and feedback from university teachers is very welcome! In our Pilot Course in PLM, we used Moodle. It suits our online platform perfectly, because the knowledge imparted by teachers needs to build the mind of the student, without so many books in between and, through collaborative learning. The individual approaches and strategies used by professors in different countries to impart knowledge to students and inspire them to learn are likely to build on academic training, instinct and intuition using the various teaching methods laid down in our curriculum. In May 2022, the Connect4PLM team conducted a pilot course as part of the 2nd Transnational Meeting. Each module was followed by an assessment of the learners' acquired knowledge, conducted after the hands-on activity. This finding is used to enhance learners' online experience and allow teachers to monitor their progress in real time.

4 PLM Curriculum Structure

In the following we describe the concept, the didactic approach and the content of the courses, which were set up in cooperation and coordination with all project partners. The course concept builds on the one hand the needs of the students and the challenges and needs of the industry and on the other hand the pedagogical experiences of the activity partners (Fig. 3).

Several qualifications and prerequisites should be considered when implementing the master's program:

- Bachelor's Degree in a relevant field such as engineering, computer science, business, or a related discipline. The degree should be from an accredited institution.
- relevant work experience in areas related to PLM. It demonstrates practical knowledge, real-world application, and a clear understanding of the industry context.
- proficiency in the English language for international students. Applicants may need to submit English language test scores, such as TOEFL, to demonstrate their proficiency.

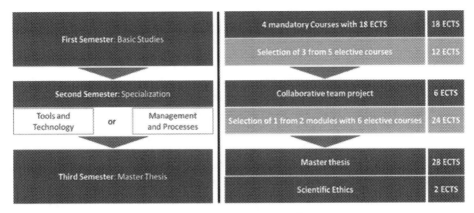

Fig. 3. Curriculum sequence and content

- technical skills to navigate online learning platforms, use communication tools, submit assignments, and engage in online discussions.
- prerequisite knowledge in relevant subjects: foundational knowledge in engineering principles, manufacturing processes, computer-aided design (CAD), supply chain management.

4.1 Basic Studies in PLM Curriculum

The concept for the first semester, which provides the basic knowledge, is shown in detail in Fig. 4. Here one can see the main focus which needs to be achieved. It consists of the four mandatory and five elective modules.

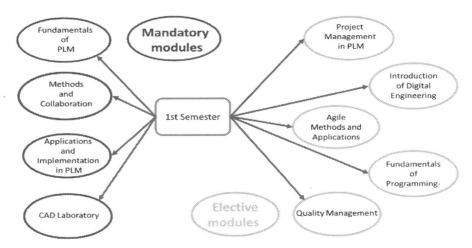

Fig. 4. Course structure overview 1st Semester

4.2 Specialization in PLM Curriculum

The courses address not only bachelor students, but also all current and future employees of manufacturing companies - technicians, maintenance managers, industrial engineers, quality managers, production managers, etc. - in general for everyone who is interested in gaining more knowledge and practical experience in the area of PLM. Accordingly, the second term is divided in two main subject areas: "Tools and Technologies" and "Management and Processes" (Fig. 5). Each of them contain six selectable courses. The participants must choose out of these twelve courses at least six to achieve the requirements.

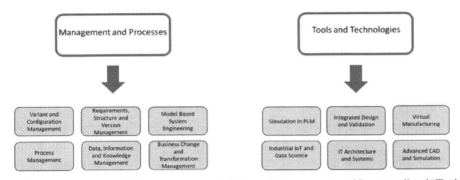

Fig. 5. Course structure 2nd Semester in specializations "Management and Processes" and "Tools and Technologies"

Furthermore a "Collaborative team project" will be mandatory for all students in the second term (Fig. 3). Many subjects are directly related to product life cycle management. The project team will assure that all the lectures have a certain amount of overlapping throughout the chosen courses so that the intersections are suitable for many other topics in the field of product development and various fields in mechanical engineering.

4.3 Graduation Semester in PLM Curriculum

In the third semester, a master thesis with 30 Credit points (28cp + 2cp) needs to be written. This completes the training and the academic degree Master of Science is awarded. The emerging trends of global multidisciplinary teams and the associated skills of cross disciplinary work in the field of PLM should be reflected in this master thesis. The graduates of the final semester are absolutely capable of dealing with the rapid development of current and emerging technologies which need to be understood and above all their potential has to be captured for the advancements of the industry.

5 Conclusion

The curriculum developed within the EU founded project Connect4PLM supports closing gap in the PLM education. This is suitable for bachelor graduates, who are aiming for a masters degree in PLM and young industrial professionals. Both meet the industry's needs, which is looking for well-trained and experienced engineers in the PLM area. On the one hand, the graduates are motivated to work in several production branches of the industry, on the other hand, young professionals are enabled to learn new technical topics that are valuable for their future leaders positions. Even if young engineers only have limited time in active professional life to take part in university courses, the present project will raise PLM training in multidisciplinary engineering areas to a new level. This makes a significant contribution to improve and expand the use of PLM solutions in industry applications. Furthermore, added value for companies that need excellently trained PLM graduates to solve complex projects is created.

The rapid developments in current technologies, both in industry and in university education, are taken in to account and the potential for further development in education is massively supported with the project. The specific PLM skills which graduates could need are addressed and represented in the proposed curriculum. The international design, considering also the emerging trends, is done by our global multidisciplinary Connect4PLM project team. The framework is currently being defined. The authors strive to create a comprehensive online platform that can be implemented in the next project.

Acknowledgment. The Connect4PLM research project has been funded by the EU. The European Commission Erasmus+ programme financial support and also the great cooperation and experienced expertise of all partner universities is gratefully acknowledged.

References

1. Eickhoff, T., Eiden, A., Göbel, J., Eigner, M.: A metadata repository for semantic product lifecycle management. Procedia CIRP **2020**(91), 249–254 (2020)
2. Eigner, M., Stelzer, R.: Produktdaten-management und product lifecycle management. In: Eigner, M., Stelzer, R. (eds.) Product Lifecycle Management, pp. 27–45. Springer, Heidelberg (2009). https://doi.org/10.1007/b93672_3
3. Gandhi, P.: Product lifecycle management in education: key to innovation in engineering and technology. In: Fukuda, S., Bernard, A., Gurumoorthy, B., Bouras, A. (eds.) Product Lifecycle Management for a Global Market. IAICT, vol. 442, pp. 121–128. Springer, Heidelberg (2014). https://doi.org/10.1007/978-3-662-45937-9_13
4. Göbel, J., Abramovici, M.: Modeling product lifecycle management (PLM) harmonization problems in industrial companies. In: Cosic, P. (ed.) Proceedings MOTSP 2015 Conference, Brela, Croatia, 10–12 June 2015, Croatian Association for PLM, Zagreb (2015). ISSN 1848-9591
5. Sauza Bedolla, J., Ricci, F., Martinez Gomez, J., Chiabert, P.: A tool to support PLM teaching in universities. In: Bernard, A., Rivest, L., Dutta, D. (eds.) Product Lifecycle Management for Society. IAICT, vol. 409, pp. 510–519. Springer, Heidelberg (2013). https://doi.org/10.1007/978-3-642-41501-2_51
6. Probst, A., Gerhard, D., Ebner, M.: PDM field study and evaluation in collaborative engineering education. In: Auer, M.E., Guralnick, D., Simonics, I. (eds.) Teaching and learning in a digital world. AISC, vol. 715, pp. 407–415. Springer, Cham (2018). https://doi.org/10.1007/978-3-319-73210-7_49
7. Kakehia, M., Yamadab, T., Watanabe, I.: PLM education in production design and engineering by e-Learning. Int. J. Prod. Econ. **122**(2009), 479–484 (2009)

8. Tham, B., Feldmann-Wojtachnia, E.: DAAD Wirkungsstudie. Nutzung der Partnerschaften und Kooperationsprojekte in Erasmus+ durch die deutschen Hochschulen. Kooperationsprojekt im Rhamen der DAAD Studie 2017. Teilstudie 2. Forschungsgruppe Jugend und Europa am Centrum für Angewandte Politikforschung LMU München (2017)

9. Tekin, U.: Auswirkungen des ERASMUS-Programms auf Universitäten und Studierende in der Türkei. In: Pusch, B. (ed.) Transnationale Migration am Beispiel Deutschland und Türkei. Springer, Wiesbaden. (2013) https://doi.org/10.1007/978-3-531-19177-5_16

10. Fradl, B., Sohrweide, A., Nyffenegger, F.: PLM in education – the escape from boredom. In: Ríos, J., Bernard, A., Bouras, A., Foufou, S. (eds.) Product Lifecycle Management and the Industry of the Future. IAICT, vol. 517, pp. 297–307. Springer, Cham (2017). https://doi.org/10.1007/978-3-319-72905-3_27

11. Fielding, E., McCardle, J.R., Eynard, B., Hartmann, N., Fraser, A.: PLM in design and engineering education: international perspectives. Concurrent Eng. Res. Appl. (2014)

12. Stark, R., Adolphy, S., Woll, R.: PLM Professional – Entstehung einer Disziplin. Produktdaten J. **23**(2), 12–15 (2016). ProSTEP Darmstadt

13. Drăghici, G., Brîndaşu, P.D., Savii, G.G., Drăghici, A.: The need for PLM education to satisfy industrial requirements. In: 4th Balkan Region Conference on Engineering Education & MSE. Sibiu, Romania (2007)

14. CIMdata: Are Students "Real-World" Ready? The Challenge in Preparing Students for Industry 4.0 (2017). https://www.cimdata.com/en/resources/complimentary-reports-research/white-papers

15. Priestley, M., Nieveen, N.: Understanding curriculum. In: Chartered College of Teaching (ed.) The Early Career Framework Handbook, pp. 135–143. SAGE Publishing, London (2020)

16. van den Akker, J.: Curricular development research as a specimen of education design research. In: Plomp, T., Nieveen, N., Enschede, N.E. (eds.) Educational Design Research, 52–71, Netherlands Institute for Curriculum Development (2013)

Enhancing Collaborative Design Through Process Feedback with Motivational Interviewing: Can AI Play a Role?

Sabah Farshad$^{(\boxtimes)}$ ⓘ, Yana Brovar ⓘ, and Clement Fortin ⓘ

Skolkovo Institute of Science and Technology, Moscow 121205, Russia
Sabah.Farshad@Skoltech.ru

Abstract. Collaborative design is a key element in Product/System development. However, delivering true collaboration in multidisciplinary teams is challenging. Feedback systems are one of the solutions to improve collaboration; although teams normally receive feedback on outcomes, the collaboration process itself is neglected. During a PBL course, 40 engineers from 22 disciplines and 12 countries were distributed in six teams. In addition to receiving outcome feedback, we used Motivational Interviewing (MI) techniques to provide process feedback for half of the design teams whereas the other half only received outcome feedback. At the same time, we employed a pre-trained Machine Learning (ML) technique to compare the teams' progress through teams' communication and sentiment analysis. Our results show that; (i) adding process feedback in the early stages of the design process enhances the collaborative design. (ii) ML algorithms can predict the progress. We suggest further research using Natural Language Processing (NLP) and supervised ML techniques for designing a new AI team-mate and mentoring assistant, as well as fostering Human-AI interaction styles via MI methods.

Keywords: Collaborative Design · Feedback · Machine Learning · Motivational Interviewing · PBL

1 Introduction

In the universal competition between companies, developing new products has become a tough challenge where innovation and collaboration in the design process are two primary keys to success, and design projects depend on controlling the collaboration among the multidisciplinary actors [1]. According to surveys [2], 40% of engineering time is directly impacted by the ability to work together. Moreover, engineering efficiency as a top goal for product development success is significantly dependent on effective collaboration. In addition, many companies struggle with poor collaboration and its cost has never been higher. However, in the process of product and system development, which is following the Systems Engineering (SE) methodology, the issue of ensuring effective team collaboration is rarely addressed even though it is widely accepted as necessary

© IFIP International Federation for Information Processing 2024
Published by Springer Nature Switzerland AG 2024
C. Danjou et al. (Eds.): PLM 2023, IFIP AICT 702, pp. 244–253, 2024.
https://doi.org/10.1007/978-3-031-62582-4_22

[3]. SE aims to enable the successful realization, use, and retirement of engineered systems. Still, in the SE models (e.g., V-model) the focus is on the baselines, documents, reviews, and audits of the technical process [5], not on the collaboration process. Based on such a procedure, reviews and feedbacks target the evaluations of the design to ensure compliance with the technical requirements (Verification), and the stakeholders' needs (Validation), while the team interactions are not reviewed. In general, in design teams and PLM systems, no feedback is provided on the collaboration process of the design team itself. At the same time, interaction issues appear to be the most fundamental arguments concerning collaborative design, particularly when computer systems such as PLM systems are used in the process [6]. One of the effective strategies to improve interaction is MI, a guiding and mentoring style of communication [7] that is not sufficiently covered in engineering and design studies. Meanwhile, Artificial Intelligence (AI) is able to identify emotions and intentions of human interactions through ML using NLP techniques [8]. Now the question is how can we improve collaborative engineering design using AI capabilities, interaction improvement techniques, and process feedback in SE and PLM systems environments? To address this question, in this study we employ MI to empower team members to facilitate the process of collaboration during an SE project. Then we compare the progress in a case study including two groups of test and control teams to examine the effectiveness of MI in a team process feedback on the engagement and the final SE project outcome. At the same time, we test the predictability of the process through pre-trained machine learning techniques that opens doors for further development of team support through intelligent systems, particularly in collaborative design learning. This is important because in large-scale engineering design projects where hundreds and sometimes thousands of collaborators are working on the same project, the use of human support interventions through human agents are, if not impossible, very costly and difficult to scale. However, an AI agent that can handle this progress can possibly turn it into a cost-effective and highly scalable approach. Based on these, the hypotheses of this research are: (1) Using MI as a method in process feedback can significantly improve collaborative design in a design project. (2) AI can predict the progress through pre-trained ML and NLP techniques by teams' sentiment analysis.

2   Literature Review

This section first summarizes evidence on the improvement of collaborative design. It then explains the use of the MI and its significance. Next, the AI capabilities in detecting interaction sentiment are described. Finally, some research on using AI for the application of MI in human-AI interaction is explored.

2.1   Improving Collaborative Design

Research on collaborative design mostly investigated the technical side of collaborative design such as design, engineering, and manufacturing through computer-aided approaches [9–11], web-based systems [12, 13], or information sharing systems and enterprise resource planning [13–15]. Management, social and cognitive aspects have also been studied [17–19]. However, evidence indicate that the non-technical dimension of collaboration is the main effective factor in engineering projects [20], it has also been documented that early in a project's life cycle, when the conceptual design is being developed, non-engineering factors are most likely to influence the system's design [21]. Wang et al. [20] showed that the team collaboration atmosphere is the most significant factor, followed by the ability of collaborators to learn in terms of team efficacy in collaborative design. Wang et al. suggest that the human interaction process is one of the most influential elements of collaborative design. Stempfle & Badke-Schaub [22] investigated collaborative design challenges, focusing on the cognitive processes of design teams during the design process. They analyzed the entire communication of three design teams and concluded that teams spent about 70% of their interaction on the content, and 30% on the group process.

2.2 Motivational Interviewing (MI)

MI is a communication strategy and mentoring style that has been investigated in various settings including leadership and management [23–25], sport and human coaching [26], healthcare [27], higher education and training [28] etc. MI is an evidence-based evolution of Rogers's person-centered counseling approach, a directive method to enhance readiness for change by helping people explore and resolve ambivalence [31]. The rapidly growing evidence for MI indicates its significant effectiveness in various systematic reviews and meta-analyses [30–34]. The high effectiveness of MI across various settings suggests a need to understand and apply this style in collaborative design and engineering. Miller & Rollnick [35] in their book "Motivational interviewing: Helping people change" define MI as follows:

"MI is a collaborative, goal-oriented style of communication with particular attention to the language of change. It is designed to strengthen personal motivation for and commitment to a specific goal by eliciting and exploring the person's own reasons for change within an atmosphere of acceptance and compassion."

The fundamental elements of MI consist of three qualities [36]: (i) MI is a guiding technique of communication that lies between active listening and guiding through giving information. (ii) MI is designed to empower people to change by discovering their own meaning, values, and capacity for change. (iii) MI encourages a natural changing process and respects individual autonomy through a respectful, curious approach. According to Miller & Rollnick, in the MI method, the mentor interacts with the person as an equal partner and avoids unrequested advice, directing, confrontation, warnings, or instructing. It is not a way to "get people to change" or techniques to push people and impose on the conversation. The principles and skills of MI can be applied in a variety of conversational contexts, but MI is especially useful in the following situations: (a) High ambivalence, where people are stuck with vague feelings about change. (b) Low confidence is when people have doubt about their ability to improve. (c) Low desire, when people are not sure if they want to make a change or not. (d) Low importance, when the advantages of change and the disadvantages of the current situation are not clear.

As cited in [37], based on the Miller and Moyers method, the acronyms OARS and EARS are used to summarize the idea for acquiring MI. OARS and EARS refer to Open-ended questions/elaborating, Affirmations, Reflective Listening, and Summaries.

Klonek & Kauffeld, [37] use a metaphor to describe the application of the OARS; the verbal skills of the MI idea can be compared to the oars on a boat, which the trainees use as dynamic micro-tools within verbal interaction (Fig. 1). The OARS of a boat help the trainee safely go across the river, similar to basic communication skills that help interaction go smoothly. Dynamic interactions with a conversational partner are represented by the river. The rock in the river represents resistance to change. "To roll with resistance", not confrontation with it, is one of the main principles of MI.

Fig. 1. The metaphoric meaning of OARS as micro-tools to navigate through a dynamic interaction (i.e. river) in a conversation (Klonek & Kauffeld, 2015)

Klonek & Kauffeld [37], showed that MI is significantly effective in verbal communication skills, increases motivation to interact, and reduces the confrontational behaviors of engineering participants.

2.3 AI Application in Sentiment Analysis

Sentiment analysis through conversations using AI is an emerging and yet challenging approach that aims to discover the emotional states and their changes in people participating in a conversation. There is a wealth of information in the interactions that affect speakers' emotions in a complex and dynamic manner, and in recent years many promising studies have been conducted on how to accurately and comprehensively model these complex interactions [38]. As chatbots are common in everyday life and their role in teamwork is becoming more important, in a study on emotion recognition through conversations sentiment, Majid & Santoso [39] developed a chatbot called Dinus Intelligent Assistant (DINA) to assist student administration services. The study used datasets of textual-based content of conversation dialogues. They preprocessed the conversations using sentiment analysis and then applied neural networks to categorize the emotions. The result showed an accuracy of 0.76, meaning that the algorithm can reliably recognize emotions from text-based conversations. Dehbozorgi [40] used sentiment analysis on verbal data from team discussions to create an indicator for individual performance. The study conducted a successful attitudinal components detection that correlates with performances through NLP algorithms.

2.4 AI and MI

One of the advantages of MI is the possibility to employ its techniques through AI and ML in chatbots. Hershberger et al. [41] developed and tested a training tool to assess dialogue in MI that uses NLP to provide MI metrics. The study objective was to implement MI more widely in a manner that can provide immediate and efficient feedback for MI training. The results of this study that was conducted in a therapy setting, showed that MI metrics can be detected by AI. The developed model produces metrics that a trainer can share with a student, or clinician for immediate feedback, while it decreases the need to rely on subjective feedback and time-consuming review procedures. Almusharraf et al. [42] trained, and tested an automated MI-based chatbot that is capable of drawing out reflection and generate MI-oriented responses in a conversation with cigarette smokers. The chatbot could stimulate reflections on the pros and cons of smoking through MI techniques.

3   Methodology

This study incorporates a multi and interdisciplinary research literature review, along with a case study conducted during an eight-week SE course, where we established a test-control group setup to compare and evaluate the effects of variables. To observe the normal distribution, the teams were equally distributed based on expertise, gender, project of interest, and nationality. The students went through an SE learning process, according to NASA and INCOSE handbooks using a Project-based Learning (PBL) approach. The test (experiment) group consisted of teams with lowest grades during the Preliminary Design Review (PDR). Pre and post-intervention results were compared using data graphs and statistical analysis by comparing the PDR with the Critical Design Review (CDR) grades. The intervention consisted of using MI techniques in the test group through brief interviews and short talks. The study was double-blind, meaning that neither the experiment/control group nor the review team was aware of the ongoing research. To compare the text-based sentiment of the teams, we used the Google Cloud Platform (GCP) to perform a sentiment analysis that evaluates a given text and identifies the overall emotional opinion within the text, particularly to determine a writer's attitude as positive, negative, or neutral [44]. We used the pre-trained Natural Language features of GCP. The document sentiment includes the overall sentiment of the text that consists of Score and Magnitude; a score of the sentiment ranges between -1.0 (negative) and 1.0 (positive) and corresponds to the overall emotional tendency of the text. The magnitude indicates the overall strength of the emotion (negative/positive) and, (Fig. 2). At the end of the course, all participating teams were asked to export all chat conversations to a single file to be used in the analysis (Only if all team members agreed). The research team undertook to respect privacy so that the names are coded, and the conversations are not to be published without permission. One of the participating teams refused to share the content and is excluded from the study, but this issue did not have a negative impact on our study as both the test and control groups had three teams at the end. For each period, the entire conversation was analyzed and the average sentiment score for the given period was extracted.

Fig. 2. An example of one sentence sentiment level (areas: Green; Positive, Yellow; Neutral, Red; Negative) (Color figure online)

4 Case Study

The Northern Sea Route (NSR) is a shipping route that crosses the seas of the Arctic Ocean. Annual cargo shipments on the NSR are up to 33 million tons mainly as an energy highway for the export of hydrocarbons and other natural resources [45]. The NSR has nine main ports, and each port has different levels of resources. With the aim of improving the shipping process on the NSR, the following SE projects were defined to the teams: (1) Spare part delivery from the main ports to ships with measurements of temperature, pressure, winds, and visibility on the route using Unmanned Aerial Vehicle (UAV). (2) Charging station, interaction with UAV, and navigation. (3) Port fuel transfer automated system in all weather conditions. (4) Ambulance system for emergency evacuation from ships based on a UAV system. (5) Emergency drug delivery to a ship or port based on a UAV system. (6) Coordination of various UAVs to carry out different tasks, to the ports and/or the ships; central coordination system. (7) Satellite communication and observation along the sea route in all weather conditions with monitoring of ice thickness and prevailing winds. Participants had the chance to select three projects of interest in order of priority. The deliverables for each review stage were according to the "V model" of SE. Totally, 46 engineers from 22 disciplines and 12 countries were distributed in seven teams after filing out a form containing their demographic information, degrees, expertise, and favorite project. Three teams were included in the test and four teams in the control group. One of the control group teams, consisting of six members, was excluded from the study. The final number of participants in each group was 20 individuals, organized into two teams of seven and one team of six.

5 Results and Analysis

While the PDR results of the test group with a mean grade of 68% were lower than in the control group which had a mean of 72%, this difference statistically is not significant (Independent T-Test with 95% confidence: $0.27 > 0.5$). However, a comparison of the CDR results indicates that both groups showed improvement, but the difference is significant (Independent T-Test with 95% confidence: $0.017 < 0.5$). Figure 3 shows the comparison of the two groups. The data-mining process for analyzing the text-based conversation reveals that both groups stayed in the Neutral area with no significant differences. The entire conversations of all teams from the start day to the PDR were analyzed and the average grade was calculated. The results show -0.08 and -0.04 respectively for test and control group. The same process was repeated over the time span from PDR to CDR. Figure 3 illustrates the results. As the graph shows, the results of text-based sentiment analysis using AI and the results of changes in scores in CDR are meaningfully correlated.

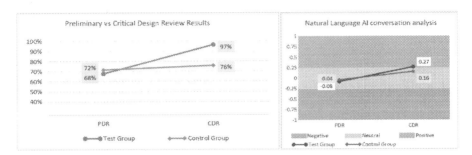

Fig. 3. Left graph: Test and Control groups' grades in PDR and CDR; Right graph: Sentiment Analysis with Google Cloud Natural Language

6 Discussion

While previous research to improve collaborative design in engineering has mainly focused on technical aspects, non-engineering factors are likely to influence system design the most, particularly in the early phases. On the other hand, positive interpersonal relationship enhances individuals' enthusiasm for collaboration. A large body of research confirms the significant positive influence of MI as a communication strategy and mentoring style. Although MI is effective, it is not easy to employ it in collaborative engineering projects or educational PBL with a large number of participants. However, AI advances using ML and NLP shows promise toward using an automated intervention through chatbots or other Human-AI interactions. However, this topic has not been studied in engineering, especially in the field of collaborative design and PBL. The case study results show a significant positive effect of MI in improving collaborative engineering design in SE and PBL outcomes. This is in line with previous systematic analysis about the positive influence of MI to improve interactions (e.g., [32, 33]); however, the effect of using MI in a process feedback on collaborative design had not been studied before. This result supports the first hypothesis of the study. In addition, the second hypothesis is also supported by the results. This is in accordance with previous studies that investigated the application of speech analysis in predicting performance [46]. Although pre-trained AI detects the overall sentiment of the team, it is not able to recognize the engagement quality and collaborative process, however, a supervised ML that has been specified to classify speech according to collaboration activities, can increase the quality of monitoring.

7 Conclusion

Improving collaboration is a widely recognized need, however, it is challenging, expensive, and faces the issue of scalability. Improving interactions is one of the most effective ways to support team collaboration, and MI has proven to be an effective strategy for enhancing interactions. We also demonstrated its effectiveness in engineering design and PBL and suggested that state-of-the-art technology has the potential to further develop it. This paper is icebreaking from different perspectives and opens the door for future studies. First, it reveals the significant effect of Motivational Interviewing as a way to improve interaction and therefore the design in engineering collaborative design and project-based learning. Second, sentiment analysis is a powerful tool to recognize the team's challenges and track the changes after interventions. Third, the literature review results show a promising capability of AI as a new member of engineering teams that can monitor interaction and start mentoring through MI techniques, which is a part of our outlook for future studies. This point is also important in the Human-AI collaboration because it provides a basis to identify an effective interaction style of an intelligent machine with its human colleagues.

References

1. Pol, G., Merlo, C., Legardeur, J., Jared, G.: Implementation of collaborative design processes into PLM systems. Int. J. Prod. Lifecycle Manag. 3(4), 279 (2008)
2. Boucher, M.: What's the cost of poor collaboration? (survey results). Tech-Clarity, Inc. (2020). https://tech-clarity.com/cost-of-poor-collaboration-in-engineering/9942
3. DeFranco, J.F., Neill, C.J., Clariana, R.B.: A cognitive collaborative model to improve performance in engineering teams-a study of team outcomes and mental model sharing. Syst. Eng. 14(3), 267–278 (2011). https://doi.org/10.1002/sys.20178
4. INCOSE: Systems Engineering. https://www.incose.org/about-systems-engineering/system-and-se-definition/systems-engineering-definition
5. Clark, J.O.: System of systems engineering and family of systems engineering from a standards, V-Model, and dual-V Model perspective. In: 2009 3rd Annual IEEE Systems Conference, pp. 381–387 (2009). https://doi.org/10.1109/SYSTEMS.2009.4815831
6. Kvan, T.: Collaborative design: what is it? Autom. Constr. 9(4), 409–415 (2000). https://doi.org/10.1016/S0926-5805(99)00025-4
7. Rollnick, S., Miller, W.R.: What is motivational interviewing? Behav. Cogn. Psychother. 23(4), 325–334 (1995). https://doi.org/10.1017/S135246580001643X
8. Prabha, M.I., Umarani Srikanth, G.: Survey of sentiment analysis using deep learning techniques. In: 2019 1st International Conference on Innovations in Information and Communication Technology (ICIICT), pp. 1–9 (2019). https://doi.org/10.1109/ICIICT1.2019.8741438
9. Li, W.D., Lu, W.F., Fuh, J.Y.H., Wong, Y.S.: Collaborative computer-aided design—research and development status. Comput. Des. 37(9), 931–940 (2005). https://doi.org/10.1016/j.cad.2004.09.020
10. Qin, S.F., Harrison, R., West, A.A., Jordanov, I.N., Wright, D.K.: A framework of web-based conceptual design. Comput. Ind. 50(2), 153–164 (2003). https://doi.org/10.1016/S0166-3615(02)00117-3
11. Zheng, Y., Shen, H., Sun, C.: Collaborative design: improving efficiency by concurrent execution of Boolean tasks. Expert Syst. Appl. 38(2), 1089–1098 (2011). https://doi.org/10.1016/j.eswa.2010.05.004

12. Shen, W., Barthès, J.: An experimental environment for exchanging engineering design knowledge by cognitive agents. In: Workshop on Knowledge Intensive CAD, pp. 19–38 (1997)

13. Zhang, S., Shen, W., Ghenniwa, H.: A review of internet-based product information sharing and visualization. Comput. Ind. **54**(1), 1–15 (2004). https://doi.org/10.1016/j.compind.2003.09.002

14. Roy, U., Bharadwaj, B., Kodkani, S.S., Cargian, M.: Product development in a collaborative design environment. Concurr. Eng. **5**(4), 347–365 (1997). https://doi.org/10.1177/1063293X9700500407

15. Numata, J.: Knowledge amplification: an information system for engineering management. Sony's Innov. Manag. Ser. **17** (1996)

16. Simpson, M., Viller, S.: Observing architectural design: improving the development of collaborative design environments. In: International Conference on Cooperative Design, Visualization and Engineering, pp. 12–20 (2004)

17. Cross, N., Clayburn Cross, A.: Observations of teamwork and social processes in design. Des. Stud. **16**(2), 143–170 (1995). https://doi.org/10.1016/0142-694X(94)00007-Z

18. Lang, S.Y.T., Dickinson, J., Buchal, R.O.: Cognitive factors in distributed design. Comput. Ind. **48**(1), 89–98 (2002). https://doi.org/10.1016/S0166-3615(02)00012-X

19. Girard, P., Robin, V.: Analysis of collaboration for project design management. Comput. Ind. **57**(8–9), 817–826 (2006). https://doi.org/10.1016/j.compind.2006.04.016

20. Wang, J., Yuan, Z., He, Z., Zhou, F., Wu, Z.: Critical factors affecting team work efficiency in BIM-based collaborative design: an empirical study in China. Buildings **11**(10), 486 (2021). https://doi.org/10.3390/buildings11100486

21. Greene, M., Papalambros, P.Y., McGowan, A.-M.: Position paper: designing complex systems to support interdisciplinary cognitive work. In: DS 84: Proceedings of the DESIGN 2016 14th International Design Conference, pp. 1487–1494 (2003)

22. Stempfle, J., Badke-Schaub, P.: Thinking in design teams - an analysis of team communication. Des. Stud. **23**(5), 473–496 (2002). https://doi.org/10.1016/S0142-694X(02)00004-2

23. Organ, J.N.: Motivational interviewing: a tool for servant-leadership. Int. J. Servant-leadersh. **15**(1), 209–234 (2021)

24. Niesen, C.R., Kraft, S.J., Meiers, S.J.: Use of motivational interviewing by nurse leaders. Health Care Manag. (Frederick) **37**(2), 183–192 (2018). https://doi.org/10.1097/HCM.0000000000000209

25. Marshall, C., Nielsen, A.S.: Motivational Interviewing for Leaders in the Helping Professions: Facilitating Change in Organizations. Guilford Publications (2020)

26. Wierts, C.M., Wilson, P.M., Mack, D.E.: Awareness and use of motivational interviewing reported by Canadian university sport coaches. Int. J. Evid. Based Coach. Mentor. **17**(1) (2019)

27. Simmons, L.A., Wolever, R.Q.: Integrative health coaching and motivational interviewing: synergistic approaches to behavior change in healthcare. Glob. Adv. Heal. Med. **2**(4), 28–35 (2013). https://doi.org/10.7453/gahmj.2013.037

28. Ogles, B.M., Wood, D.S., Weidner, R.O., Brown, S.D.: Motivational Interviewing in Higher Education: A Primer for Academic Advisors and Student Affairs Professionals. Charles C Thomas Publisher (2021)

29. Marker, I., Norton, P.J.: The efficacy of incorporating motivational interviewing to cognitive behavior therapy for anxiety disorders: a review and meta-analysis. Clin. Psychol. Rev. **62**, 1 (2018). https://doi.org/10.1016/j.cpr.2018.04.004

30. Lundahl, B.W., Kunz, C., Brownell, C., Tollefson, D., Burke, B.L.: A meta-analysis of motivational interviewing: twenty-five years of empirical studies. Res. Soc. Work. Pract. **20**(2), 137–160 (2010). https://doi.org/10.1177/1049731509347850

31. Hettema, J., Steele, J., Miller, W.R.: Motivational interviewing. Annu. Rev. Clin. Psychol. **1**(1), 91–111 (2005)
32. Magill, M., et al.: A meta-analysis of motivational interviewing process: technical, relational, and conditional process models of change. J. Consult. Clin. Psychol. **86**(2), 140–157 (2018). https://doi.org/10.1037/ccp0000250
33. Schwalbe, C.S., Oh, H.Y., Zweben, A.: Sustaining motivational interviewing: a meta-analysis of training studies. Addiction **109**(8), 1287–1294 (2014). https://doi.org/10.1111/add.12558
34. Rubak, S., Sandbæk, A., Lauritzen, T., Christensen, B.: Motivational interviewing: a systematic review and meta-analysis. Br. J. Gen. Pract. **55**(513), 305–312 (2005)
35. Miller, W.R., Rollnick, S.: Motivational Interviewing: Helping People Change. Guilford Press (2012)
36. MINT: Understanding Motivational Interviewing, Motivational Interviewing Network of Trainers (2021). https://motivationalinterviewing.org/understanding-motivational-interviewing
37. Klonek, F.E., Kauffeld, S.: Providing engineers with OARS and EARS. High. Educ. Ski. Work. Learn. **5**(2), 117–134 (2015). https://doi.org/10.1108/HESWBL-06-2014-0025
38. Zhang, Y., et al.: A quantum-like multimodal network framework for modeling interaction dynamics in multiparty conversational sentiment analysis. Inf. Fusion **62**, 14–31 (2020). https://doi.org/10.1016/j.inffus.2020.04.003
39. Majid, R., Santoso, H.A.: Conversations sentiment and intent categorization using context RNN for emotion recognition. In: 2021 7th International Conference on Advanced Computing and Communication Systems (ICACCS), pp. 46–50 (2021). https://doi.org/10.1109/ICACCS51430.2021.9441740
40. Dehbozorgi, N.: Sentiment analysis on verbal data from team discussions as an indicator of individual performance. The University of North Carolina at Charlotte (2020)
41. Hershberger, P.J., et al.: Advancing motivational interviewing training with artificial intelligence: ReadMI. Adv. Med. Educ. Pract. **12**, 613–618 (2021). https://doi.org/10.2147/AMEP.S312373
42. Almusharraf, F., Rose, J., Selby, P.: Engaging unmotivated smokers to move toward quitting: design of motivational interviewing-based chatbot through iterative interactions. J. Med. Internet Res. **22**(11), e20251 (2020)
43. IBM Corp: IBM SPSS Statistics for Windows, Version 22.0. IBM Corp, Armonk, NY (2013)
44. White, T.E., Rege, M.: Sentiment analysis on google cloud platform. Issues Inf. Syst **21**, 221–228 (2020)
45. Arcticportal: The northern sea route: from strategies to realities. Arct. J. (2021). https://arcticportal.org/ap-library/news/2487-arctic-journal-the-northern-sea-route-from-strategies-to-realities
46. Chowdary Attota, D., Dehbozorgi, N.: Towards application of speech analysis in predicting learners' performance. In: 2022 IEEE Frontiers in Education Conference (FIE), pp. 1–5 (2022). https://doi.org/10.1109/FIE56618.2022.9962701

Designing a Human-Centric Manufacturing System from a Skills-Based Perspective

Marco Dautaj[1,2](✉) ⓘ, Maira Callupe[1] ⓘ, Monica Rossi[1] ⓘ, and Sergio Terzi[1] ⓘ

[1] Politecnico di Milano, Via Lambruschini 4b, 20156 Milan, Italy
{marco.dautaj,maira.callupe,monica.rossi,sergio.terzi}@polimi.it
[2] University School for Advanced Studies IUSS Pavia, Palazzo del Broletto, Piazza Vittoria 15, 27100 Pavia, Italy

Abstract. This publication aims to delve into the concept of human-centric manufacturing systems (**or human-centered**) and the essential skills needed for their successful implementation and promotion. Through a combination of expert focus group and stakeholders' workshop, this research strives to develop a comprehensive understanding of the key elements that define a human-centric manufacturing (HCM) system and the crucial role played by human factors. By analyzing the outcomes of these workshops, the authors intend to uncover valuable insights regarding the skills and knowledge required for designing, developing, and implementing HCM systems. Main concepts and skills have been discovered across three main pillars: Empowerment, Inclusivity and Safety. This exploration will also encompass a deep comprehension of the role of the various functional units within an organization, emphasizing the significance of collaboration and communication amongst all stakeholders involved within a HCM context. The findings from this study make valuable contributions to the formulation of strategies aimed at facilitating the transition from an Industry 4.0 to Industry 5.0 manufacturing system and the emerging Society 5.0.

Keywords: Industry 4.0 · Industry 5.0 · Skills · Human-centric Manufacturing · Focus Group

1 Introduction

Along with the technological development, new ways of conceiving the link between manufacturing context and human-centric approaches have emerged. One of these ways is the so-called "Industry 5.0" [1, 2] which emerges from the idea that Industry 4.0 places less emphasis on the fundamental principles of social fairness and sustainability, instead prioritizing digitalization and AI-powered technologies to enhance production efficiency and flexibility. With the introduction of Industry 5.0, there is a shift in perspective that underscores the significance of research and innovation in enabling the industry to serve humanity in the long run while respecting the limits of our planet. Furthermore, another paradigm has been coined, in particular it is "Society 5.0" [3] that "represents

© IFIP International Federation for Information Processing 2024
Published by Springer Nature Switzerland AG 2024
C. Danjou et al. (Eds.): PLM 2023, IFIP AICT 702, pp. 254–265, 2024.
https://doi.org/10.1007/978-3-031-62582-4_23

the vision of a new human-centered society, where advanced technologies are applied in everyday life, and in different spheres of activity, to provide products and services satisfying various potential needs as well as reducing economic and social gaps, for the benefit and convenience of all citizens" not clearly focusing on the industrial aspects [2]. The core of Industry 5.0 can be defined as follows: Industry 5.0 places the wellbeing of workers at the heart of the manufacturing process (human-centric manufacturing system), making production respect the limits of our planet and the harmonious symbiotic relationship between man and machine, achieving social goals that go beyond jobs and economic growth, as well as the sustainable development goal of an uber-smart society and ecological assets, becoming a robust and resilient provider of prosperity in an industrial community of shared futures [4]. This new perspective sheds light on both how a working environment should be, and which are the paramount abilities and skills an employee should have. In order to promote a human-centric manufacturing (HCM) system vision that places fundamental human needs and interests at the center of the production process, moving away from a focus on technological advancements towards a more human-centered and society-centered approach [1] workforce needs, at least, to acquire and master those skills developed in Industry 4.0 systems [5]. Furthermore, through the introduction of Industry 5.0, businesses must pay particular attention to the well-being of their workers, on many dimensions, whose main pillars can be identified in: *safety, empowerment* and *inclusivity* [6].

The term Industry 5.0 was firstly introduced on December 1st, 2015, in an article published by Michael Rada on LinkedIn social network. For Michael Rada, Industry 5.0 is the "efficient use of machines and people labor in a synergistic environment" [7].

The European Commission for Innovation, Research, Culture, Education and Youth [8] points out that Industry 5.0 complements and extends Industry 4.0. It has a focus on aspects that are not only economic or technological in nature, but which will be decisive for the place of industry in European society in the future. This includes ecological, social and fundamental rights aspects. Industry 5.0 is neither the successor nor the replacement for the existing Industry 4.0 [9]. It is the result of a forward-looking exercise designed to help shape the way in which European industry can co-exist with emerging trends and needs in society.

Furthermore, it is important to ensure that workers have access to the necessary resources and support, such as mental health services, to enable them to cope with the changes brought about by the transition [8].

This work has been conducted in order to ease the shift from the current Industrial concept (I4.0) to the new one (I5.0) and to create the foundation for the education of companies and people that will deal with the future HCM system. The main objective of this paper is to explain and delineate the peculiarities of a HCM system, focusing on the skills a worker should possess and why we need such an industrial conception, creating awareness around the meaning of HCM and its three main constructs, Empowerment, Inclusivity and Safety, and identify which are the skills employees should have. It is also important to highlight the connection between a Human-centric manufacturing system and the digitalisation of Product Lifecycle Management, in particular the Digitalisation of Manufacturing System, indeed in the article [10] it has been stressed that a virtual validation by means of a PLM software can ameliorate and improve safety processes of a manufacturing system. Therefore, the use of digital tools for building an industry environment, where Empowerment, Inclusivity and Safety are at the centre, is paramount and essential. This paper is the result of what have been done during two workshops for the European Project DE4Human, in which academic and industrial specialists have taken part. DE4Human project has been funded by EIT Raw Materials and its mission is upskilling white collars of automation manufacturing to use design thinking methodologies to redesign human centric manufacturing process towards a safer (DE4SAFETY), more inclusive (DE4INCLUSIVITY) and empowered (DE4EMPOWER) processes and workplace.

This article is divided into five sections. First, the Introduction through which a rough introduction of Industry 5.0 concept and the HCM challenges has been provided. Secondly, a theoretical background of the main pillars characterizing Industry 5.0 will be presented. Thirdly, the research methodology based on two workshops, will be thoroughly described. A fourth section will provide the information gathered by means of the two workshops. The following part will be about the discussion of results. Finally, the authors will conclude the work by providing limitations and future research directions.

2   Theoretical Background: Industry 5.0 and HCM System

Industry 5.0 is the term used to describe the latest stage of industrial evolution, characterized by a fusion of advanced technologies such as artificial intelligence (AI), the Internet of Things (IoT), and smart automation systems with the human workforce, with the aim of creating a collaborative environment between human and machine [11]. Industry 5.0 can be seen as the answer to the demand for a new HCM system, starting from the reorganisation of production processes (in terms of structure, organisation, management, knowledge, philosophy and culture), in order to have a positive impact on the business perspective and on all the components of the innovation system [12].

This new phase of industrial development is characterized by smart, connected factories that are highly efficient, flexible, and capable of self-optimization, which is poised to foster and focus on human's creativity. Industry 5.0 also involves a deep integration of digital technologies with physical production processes, enabling real-time data collection and analysis, autonomous decision-making, and highly personalized and adaptive manufacturing [8]. This leads to increased efficiency, reduced waste, and improved quality control. Overall, Industry 5.0 represents a major shift in the way goods are produced, with a focus on creating a more sustainable, efficient, and agile manufacturing process. Furthermore, in a recent Financial Times article [13] it emerges that many industries are already in the process of embracing digitalization and AI, resulting in the automation of certain technical skills. Consequently, employers are now prioritizing human-centric abilities like problem-solving, creativity, critical thinking, cognitive agility, and empathy, so it is important to develop a set of skills which positively impact on a HCM system.

With the introduction of Industry 5.0, some concepts strictly related to the well-being of the workers have emerged, such as Safety, Empowerment and Inclusivity within a manufacturing system. Indeed, there is a strong correlation between Industry 5.0 in terms of safety. The integration of advanced technologies in Industry 5.0 has the potential to make manufacturing processes safer for workers. For example, the use of robots and automation can help to reduce human exposure to hazardous working conditions, such as in tasks that involve handling dangerous materials or working in confined spaces [14–16]. Moreover, the use of data and analytics in Industry 5.0 can help to identify potential safety hazards and prevent accidents before they occur. For example, sensors and other monitoring devices can be used to monitor the health and safety of workers in real-time, and predictive analytics can be used to identify potential safety issues and develop preventive measures. In summary, as the advanced technologies integrated in Industry 5.0 can enhance worker safety, and the focus on human well-being in human-centric manufacturing can help to ensure that workers are protected from harm in the workplace [14, 17–19].

Industry 5.0 has also a strong connection with the empowerment of workers; indeed, it has been conceived as a human-centric system in which the employees' talents are nurtured and fostered by means of "combining human beings' creativity and craftsmanship with the speed, productivity and consistency of robots" [20]. Another perspective on empowerment comes from [21] in which Industry 5.0 "emphasizes on empowering the human being specially the customers through fulfilling their personalizing and customized needs". One of the central objectives of Industry 5.0 is to reshape industrial employment in terms of improving the wellbeing and empowering workers via assistive technologies [4, 22]. As a path to the digitalised production of the future, this integration of human workers should be built on the achievement of the I4.0 technology-driven orientation [11, 21] approaches for a flexible and human-centered integration and support of employees in the digitalized and interconnected production of the future.

Even though there is paucity of literature, another important aspect is the connection between the concept of Industry 5.0 and inclusivity (both from a company and worker perspectives). The development of worker-inclusive decision-making tools and human-centric and flexible work planning models is beginning to emerge in recent academic literature. Some emphasize that in order to develop more work-inclusive solutions, workers need to be involved in both the individual data collection phase and the decision-making phase [23–26] and others say that there is the need to train and instruct workers through an ad-hoc mentoring and tutoring approach [27] in order to make the employees more included in the manufacturing process loop.

This work refers to safety, inclusivity and empowering as the three pillars of HCM, as core system to put in place to accomplish the Industry 5.0 goals.

3 Research Objectives and Methodology

The current work purses two main objectives: i) to obtain a user-centric view of what comprises a HCM system, and ii) to obtain a skills-oriented view of the requirements for the facilitation and realization of a HCM system. In order to achieve this purpose, two consecutive workshops were held wherein participants were asked to assume the role of designers of a HCM system. The choice of methodology arises from two key notions presented in the previous sections: the human-centricity that guides the Industry 5.0 paradigm as well as the importance of cooperation between academia and industry to address the emerging challenges. These are respectively embedded into the workshops by the adoption of user-centered design (UCD) and collaborative design (CD) as approaches during their design and execution phases. UCD is an iterative design approach based on the active involvement of users, focusing on their needs throughout the design process [28]. CD, on the other hand, brings together actors from different disciplines to share their knowledge about the design process as well as the artifact being designed [29]. Table 1 summarizes the main information related to the two workshops, in particular in the first workshop has been used an expert focus group approach while in the latter a stakeholder workshop approach, including the objective of each activity as well as the participants involved. The following sections cover in detail the design and execution of both workshops.

Table 1. Characteristics of the two workshops

Workshop	1. Expert Focus Group	2. Stakeholder Workshop
Objective	Define the three main pillars that comprise HCM and to discuss the skills that facilitate their realization	Identify the main functional units within an organization involved in the implementation of a HCM system and to discuss the skills that facilitate its realization from an organizational perspective
Participants		
Manager/PM	4	8
Professor/Lecturer	4	5
PhD Candidate	2	3
Company Employee	1	2
Consultant	1	1
Executive Director/ Director	0	5
Total	12	24

3.1 Workshop No. 1 – Expert Focus Group

The expert focus group approach has been adopted for the first workshop [30, 31]. The in-presence workshop was scheduled during 1.5 h and brought together 12 experts from industry and academia who worked together to discuss the concept of HCM. After a brief introduction about Industry 5.0 and HCM, participants were presented with a series of prompt questions and were asked to use post-its to give their answers (see Fig. 1). The objective was to zoom-in on the characteristics of a HCM system, focusing on the skills that facilitate the realization of its three main pillars: safety, inclusivity, and empowerment. The prompt questions were as follows:

- What is a HCM system?
- What is the meaning of *safety* within a HCM?

- What are the skills related to safety that can support the realization of a HCM?
- What is the meaning of *inclusivity* within a HCM?
- What are the skills related to inclusivity that can support the realization of a HCM?
- What is the meaning of *empowerment* within a HCM?
- What are the skills related to empowerment that can support the realization of a HCM?

Fig. 1. First Workshop

3.2   Workshop No. 2 – Stakeholder Workshop

The second workshop had a duration of 2 h and was held remotely making use of the collaborative tool MURAL, counting with the participation of 24 individuals working in industry and academia [32]. The participants were presented with a series of prompt questions and were asked to use post-it in MURAL to give their answers (see Fig. 2.). The objective was to identify the main functional roles and units within an organization involved in the implementation of a HCM system and to discuss the skills that facilitate its realization from an organization-al perspective. The prompt questions were as follows:

- What are the roles/job positions/functions/units a manufacturing company should pay attention for implementing a HCM system?
- What are the competences/skills that can support a HCM system in an existing company?
- How such competences/skills should be transferred to those actors in an existing company?

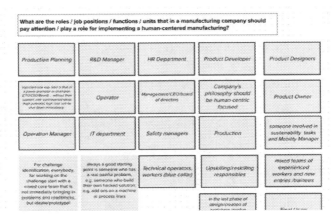

Fig. 2. Second Workshop

4   Results and Discussion

4.1   Workshop No. 1 – Expert Focus Group

As a result of the first workshop, the concept of HCM was explored in detail taking as a basis the three main pillars of safety, inclusivity, and empowerment.

Safety. The participants' responses related to safety within the larger context of a HCM indicated an understanding of the concept in three different dimensions: *emotional*, *professional*, and *physical*.

- *Emotional Safety.* Refers to whether employees feel valued and see themselves as belonging to a team. This feeling of "safety" may emerge from being treated with respect and as a valuable component of the system and as more than just an employee.
- *Professional Safety.* Related to the perception of a person's job as the source of their livelihood and the means to provide for their families. This feeling of "safety" may emerge from perceiving that one's job position is not at risk; otherwise, this sense of imminent threat may undermine an employee's performance and force them to take unnecessary risks to protect their employment.
- *Physical Safety.* Makes reference to comfort in relation to the activities conducted as part one's job. This feeling of "safety" may emerge from working in an environment with furnishings that enable comfortable body posture, keeping an adequate temperature and access to all required tools to perform one's tasks.

Inclusivity. The participants' responses related to inclusivity within the larger context of a HCM indicated an understanding of the concept in two different dimensions: personal inclusivity and work-related inclusivity.

- *Personal Inclusivity.* Refers to being accepting of personal characteristics that are inherent and do not affect a person's performance in their job. These characteristics may include age, gender, religion, ethnicity, disability, economic status, among others.

- *Work-Related Inclusivity.* Related to the different levels of skills and competences across employees within an organization. This understanding of inclusivity is of particular relevance in instances where existing employees are given larger roles within the organization or new employees enter the organization and may be perceived as unprepared.

Empowerment. The participants' responses related to empowerment within the larger context of a HCM indicated an understanding of the concept in two different dimensions: individual empowerment and structural empowerment.

- *Individual Empowerment.* Refers to a feeling of confidence over one's capabilities, actions, and decisions. This feeling of "empowerment" may emerge from perceiving the impact of one's actions within their organization.
- *Structural Empowerment.* Refers to the strategies and practices that are put in place and shape a workplace with the objective of facilitating individual empowerment. Therefore, structural empowerment refers to initiatives and practices that foster the sharing of power, decision-making, and control over resources.

Fig. 3. Dissection of the three concepts. In particular, two skills levels have been identified: Personal Skills (one person icon), Organizational Skills (three people icon)

The dissection of these three concepts together with the skills relevant to their realization as identified by the participants (see Fig. 3.) allowed for the further identification of two skill levels across all three pillars: personal skills and organizational skills.

- *Personal Skills* refer to the way in which a person interacts with other people and the surrounding environment. Examples of personal skills include autonomy, resilience, proactivity, etc.

- *Organizational Skills* refer to the achievement of an organization's objectives in relation to the performance of one's duties. Examples of organizational skills include multidisciplinarity, delegation, time management, etc.

4.2 Workshop No. 2 – Stakeholder Workshop

The second workshop adopted an organizational perspective to identify relevant skills for the realization of a HCM system. The participants' answers showed that these skills are relevant and required not only at the factory or at the shop floor level, but throughout the organization in different forms -i.e. different skills- and at different levels. The effective realization and implementation of a HCM system requires the involvement of the entire organization, especially during its design phase. As a result of participants' answers to the prompt questions, it was possible to identify three different types of skills from an organizational perspective: *Technical Skills, Transversal Soft Skills, Managerial Skills* (see Table 2).

- *Technical Skills.* Hard skills that are specific to a functional unit. Examples of hard technical skills include rapid prototyping in the R&D department and technology literacy in the Production department.
- *Transversal Soft Skills.* Soft skills that are relevant to the entire organization. Examples of transversal soft skills include creativity, communication, and resilience.
- *Managerial Skills.* Soft and hard skills that are relevant across management. Examples of managerial skills include leadership, change management, and networking.

Table 2. Organizational Skills Levels

Technical Skills	Transversal Soft Skills	Managerial Skills
Product/Process Design (**R&D**)	Creativity	Holistic thinking
Design for automated assembly (**R&D**)	Collaboration	Leadership
Design for X (**R&D**)	Critical Thinking	Change Management
Design Thinking (**R&D**)	Empathy	Task Decomposition
Rapid Prototyping (**R&D**)	Fairness	Networking
Knowledge about regulations related to hazardous materials (**R&D and Safety**)	Respectfulness	Abilities to over-come
	Research Skills	criticalities and problems
	Abilities to innovate and be intuitive	Awareness about resources consumption
Technology Literacy (**Production & Supply Chain**)	Collaborative Environment	Optimization skill
Lean Thinking (**Production & Supply Chain**)	Communication	Life Cycle Thinking
	Resilience	

5   Conclusion, Limitations and Future Research

Along with the technological development, new ways of conceiving the industrial manufacturing context development have emerged. One of these concepts is the so-called Industry 5.0 [2]. The core of Industry 5.0 can be defined as follows: Industry 5.0 places the wellbeing of workers at the heart of the manufacturing process (human-centric manufacturing system), making production respect the limits of our planet and the harmonious symbiotic relationship between man and machine, achieving social goals that go beyond jobs and economic growth, as well as the sustainable development goal of an uber-smart society and ecological assets, becoming a robust and resilient provider of prosperity in an industrial community of shared futures [4].

Thanks to the two workshops the authors have been able to delineate the main concepts of HCM system and the three main pillars linked to it: *Inclusivity*, *Empowerment* and *Safety*. The main skills that should be developed within a HCM system have been listed and grouped in three main categories: *Technical Skills*, *Transversal Soft Skills* and *Managerial Skills*. Furthermore, as a workshop result, the authors have discovered "Hard Skills" that are specific to a functional unit and paramount for the development of a HCM system (see Table 2).

From the research some limitations have emerged. One of the limitations could be industry related, in particular the skills discovered could be limited to a specific sector and therefore some skills could not fit well a company's working environment. In order to overcome this limitation, interviews with operators could be conducted. Another limitation is related to the conception of the three main pillars, indeed different pillars could be discovered in the literature for other sectors and manufacturing systems based on social and less tangible characteristics.

For future studies, researchers should validate e include practical examples, for instance if industry is considering, what have been developed by this project, a core aspect within their working environment. Furthermore, an important focus should be put on the education, in fact DE4Human project sheds light on the education side and how it could ameliorate the transition towards a safety, inclusive and empowered Industry 5.0 manufacturing system.

To conclude, research should also be carried out on the correct use of the concepts, because sometimes some paradigms have overlaps (e.g., Industry 5.0, Society 5.0 and Working World 4.0).

Acknowledgments. This project has received funding from EIT Manufacturing under grant agreement ID 23173 (correspondent to the project shortly entitled "DE4Human", "Developing design driven innovation skills for human centered manufacturing system").

This paper and related research have been conducted during and with the support of the Italian inter-university PhD course in sustainable development and climate change (link: https://www.phdsdc.it).

This work was produced while attending the PhD programme in PhD in Sustainable Development And Climate Change at the University School for Advanced Studies IUSS Pavia, Cycle XXXVIII, with the support of a scholarship financed by the Ministerial Decree no. 351 of 9th April 2022, based on the NRRP - funded by the European Union - NextGenerationEU - Mission 4 "Education and Research", Component 1 "Enhancement of the offer of educational services: from nurseries to universities" - Investment 4.1 "Extension of the number of research doctorates and innovative doctorates for public administration and cultural heritage".

References

1. Xu, X., Lu, Y., Vogel-Heuser, B., Wang, L.: Industry 4.0 and Industry 5.0—inception, conception and perception. J. Manuf. Syst. **61**, 530–535 (2021). https://doi.org/10.1016/j.jmsy.2021.10.006
2. Dautaj, M., Rossi M.: Towards a new society: solving the dilemma between society 5.0 and industry 5.0. In: Junior, O.C., Noël, F., Rivest, L., Bouras, A. (eds.) Product lifecycle management. green and blue technologies to support smart and sustainable organizations: 18th IFIP WG 5.1 international conference, PLM 2021, Curitiba, Brazil, July 11–14, 2021, Revised Selected Papers, Part I, pp. 523–536. Springer, Cham (2022). https://doi.org/10.1007/978-3-030-94335-6_37
3. Fornasiero, R., Zangiacomi, A.: Reshaping the Supply Chain for Society 5.0 BT - Advances in Production Management Systems. Artificial Intelligence for Sustainable and Resilient Production Systems. Presented at the (2021)
4. Leng, J., et al.: Industry 5.0: Prospect and retrospect. J. Manuf. Syst. **65**, 279–295 (2022). https://doi.org/10.1016/j.jmsy.2022.09.017
5. Acerbi, F., Rossi, M., Terzi, S.: Identifying and assessing the required I4.0 skills for manufacturing companies' workforce. Front. Manufac. Technol. **2**, 1–19 (2022). https://doi.org/10.3389/fmtec.2022.921445
6. Romero, D., Stahre, J.: Towards the resilient operator 5.0: the future of work in smart resilient manufacturing systems. Procedia CIRP **104**, 1089–1094 (2021). https://doi.org/10.1016/j.procir.2021.11.183
7. Rada, M.: INDUSTRY 5.0 - from virtual to physical. https://www.linkedin.com/pulse/industry-50-from-virtual-physical-michael-rada. Accessed 10 Jan 2023
8. Miraz, M.H., Hasan, M.T., Sumi, F.R., Sarkar, S., Hossain, M.A.: Industry 5.0. In: Machine vision for industry 4.0, pp. 285–300 (2022). https://doi.org/10.1201/9781003122401-14
9. Directorate-General for Research and Innovation: Industry 5.0: Towards more sustainable, resilient and human-centric industry. https://research-and-innovation.ec.europa.eu/news/all-research-and-innovation-news/industry-50-towards-more-sustainable-resilient-and-human-centric-industry-2021-01-07_en. Accessed 10 Jan 2023
10. Caputo, F., Greco, A., D'Amato, E., Notaro, I., Spada, S.: On the use of Virtual Reality for a human-centered workplace design. Procedia Struct. Integrity. **8**, 297–308 (2018). https://doi.org/10.1016/j.prostr.2017.12.031
11. Nahavandi, S.: Industry 5.0—a human-centric solution. Sustainability **11**, 4371 (2019)
12. Carayannis, E.G., Christodoulou, K., Christodoulou, P., Chatzichristofis, S.A., Zinonos, Z.: Known unknowns in an era of technological and viral disruptions—implications for theory, policy, and practice. J. Knowl. Econ. 1–24 (2021)
13. Chehidi, T.: Human skills will be vital for future jobs | Financial Times. https://www.ft.com/content/aa925f98-500f-4fe1-80fe-db1003f935c9. Accessed 15 April 2023
14. Li, S., et al.: Proactive human–robot collaboration: Mutual-cognitive, predictable, and self-organising perspectives. Robot Comput. Integr. Manuf. **81**, 102510 (2023). https://doi.org/10.1016/j.rcim.2022.102510
15. Adel, A.: Future of industry 5.0 in society: human-centric solutions, challenges and prospective research areas. J. Cloud Comput. **11** (2022). https://doi.org/10.1186/s13677-022-00314-5
16. Peruzzini, M., Grandi, F., Pellicciari, M.: Benchmarking of tools for user experience analysis in Industry 4.0. Procedia Manuf. **11**, 806–813 (2017). https://doi.org/10.1016/j.promfg.2017.07.182
17. Wang, H., et al.: A safety management approach for Industry 5.0's human-centered manufacturing based on digital twin. J. Manuf. Syst. **66**, 1–12 (2023). https://doi.org/10.1016/j.jmsy.2022.11.013

18. Khamaisi, R.K., Brunzini, A., Grandi, F., Peruzzini, M., Pellicciari, M.: UX assessment strategy to identify potential stressful conditions for workers. Robot Comput. Integr. Manuf. **78**, 102403 (2022). https://doi.org/10.1016/j.rcim.2022.102403

19. Li, C., Zheng, P., Yin, Y., Pang, Y.M., Huo, S.: An AR-assisted Deep Reinforcement Learning-based approach towards mutual-cognitive safe human-robot interaction. Robot Comput. Integr. Manuf. **80**, 102471 (2023). https://doi.org/10.1016/j.rcim.2022.102471

20. Martynov, V. V, Shavaleeva, D.N., Zaytseva, A.A.: Information Technology as the Basis for Transformation into a Digital Society and Industry 5.0. Presented at the (2019)

21. Kumar, R., Gupta, P., Singh, S., Jain, D.: Human empowerment by Industry 5.0 in digital era: analysis of enablers. Lecture Notes in Mechanical Engineering, pp. 401–410 (2021). https://doi.org/10.1007/978-981-33-4320-7_36

22. Turner, C., Oyekan, J., Garn, W., Duggan, C., Abdou, K.: Industry 5.0 and the circular economy: utilizing LCA with intelligent products. Sustainability **14**, 14847 (2022). https://doi.org/10.3390/su142214847

23. Vijayakumar, V., Sgarbossa, F., Neumann, W.P., Sobhani, A.: Framework for incorporating human factors into production and logistics systems. Int. J. Prod. Res. **60**, 402–419 (2022). https://doi.org/10.1080/00207543.2021.1983225

24. Finco, S., Abdous, M.A., Calzavara, M., Battini, D., Delorme, X.: A bi-objective model to include workers' vibration exposure in assembly line design. Int. J. Prod. Res. **59**, 4017–4032 (2021). https://doi.org/10.1080/00207543.2020.1756512

25. Katiraee, N., Calzavara, M., Finco, S., Battini, D.: Consideration of workforce differences in assembly line balancing and worker assignment problem. IFAC-PapersOnLine. **54**, 13–18 (2021). https://doi.org/10.1016/j.ifacol.2021.08.002

26. Battini, D., Berti, N., Finco, S., Zennaro, I., Das, A.: Towards industry 5.0: a multi-objective job rotation model for an inclusive workforce. Int. J. Prod. Econ. **250**, 108619 (2022). https://doi.org/10.1016/j.ijpe.2022.108619

27. Katiraee, N., Finco, S., Battaïa, O., Battini, D.: Assembly line balancing with inexperienced and trainer workers. IFIP Adv. Inf. Commun. Technol. **631** IFIP, 497–506 (2021). https://doi.org/10.1007/978-3-030-85902-2_53

28. Mao, J.Y., Vredenburg, K., Smith, P.W., Carey, T.: The state of user-centered design practice. Commun. ACM **48**, 105–109 (2005). https://doi.org/10.1145/1047671.1047677

29. Kleinsmann, M., Valkenburg, R.: Barriers and enablers for creating shared understanding in co-design projects. Des. Stud. **29**, 369–386 (2008). https://doi.org/10.1016/j.destud.2008.03.003

30. Robinson, N.: The use of focus group methodology - With selected examples from sexual health research. J. Adv. Nurs. **29**, 905–913 (1999). https://doi.org/10.1046/j.1365-2648.1999.00966.x

31. Leung, F.H., Savithiri, R.: Spotlight on focus groups. Can. Fam. Physician **55**, 218–219 (2009)

32. Coleman, S., Hurley, S., Koliba, C., Zia, A.: Crowdsourced Delphis: designing solutions to complex environmental problems with broad stakeholder participation. Glob. Environ. Chang. **45**, 111–123 (2017). https://doi.org/10.1016/j.gloenvcha.2017.05.005

Smart Processes: Prediction, Optimization and Digital Thread

Product Model for Lifecycle Support of Mechanical Parts

Hiroyuki Hiraoka[✉] and Arata Hori

Chuo University, Kasuga, Bunkyo-ku, Tokyo 112-8551, Japan
hiraoka@mech.chuo-u.ac.jp

Abstract. Appropriate maintenance support for products throughout their lifecycle is required to realize a sustainable circular society. For this purpose, we are developing a part agent system to support the lifecycle of individual parts. A part agent is a network agent provided for each part that monitors the state of the part, manages the related information on the part, and advises its user on its maintenance. The part agent performs a short-term lifecycle simulation of the corresponding part to estimate its prospective states and select preferable maintenance actions. The product information on which the lifecycle simulation is performed requires not only the shape, assembly, and kinematic information but also the information on the product's behavior, including deterioration and deviation. In this paper, a product model that represents such information for the lifecycle support of products is proposed. Example applications of the model for the part agent include the lifecycle simulation of a simple manipulator and the generation of display information to support users disassembling a product.

Keywords: Product model · lifecycle · Part agent

1 Introduction

Effective reuse of mechanical parts is one of the crucial measures for developing a sustainable society. For that purpose, not only products but also their lifecycles should be designed and controlled appropriately. There are studies on lifecycle design [1, 2], but the execution of the lifecycle is also important. We are developing a part agent system that consists of network agents and radio frequency identification (RFID) tags attached to the significant parts and modules with functionality [3, 4]. A part agent manages the state of the corresponding part and provides the user with advice on maintenance actions.

A part agent performs lifecycle simulation (LCS) of the corresponding part to predict the state of the part for the short-term future by using its lifecycle model based on the conditions of its environment and usage. Researchers have developed product life cycle simulation systems to design products taking into consideration their lifecycles as well as to design the lifecycles of products [2, 5]. LCS for a part agent is focused on a product and its parts at the stage where the product is used. The simulation is not aimed at the design phase but is intended for iterative use in the operational or management phase [6].

© IFIP International Federation for Information Processing 2024
Published by Springer Nature Switzerland AG 2024
C. Danjou et al. (Eds.): PLM 2023, IFIP AICT 702, pp. 269–278, 2024.
https://doi.org/10.1007/978-3-031-62582-4_24

The LCS needs a product model to manage the changes in the behavior of the part due to deterioration and interaction among parts. Hiraoka proposes LCS based on an assembly model and its application to a manipulator with a modularized structure [7]. This paper further discusses the requirements for the product model for LCS and proposes a product model structure not only for lifecycle simulation but also for the extensive lifecycle support activities.

2   The Model Information of Mechanical Part Required for the Lifecycle Support

Proper management of mechanical parts and products is important to promote their reuse. We assume that part agents perform the following functions and processes to manage the lifecycle of parts: A part agent collects the current state of the corresponding part using available sensory functions. Operational data to control the part could also be used for this purpose. The part agent forecasts the state of the part in the near future based on acquired sensory data and information on its lifecycle and product data. For the prediction, part agents perform the lifecycle simulation of the part. The part agent proposes to the user the appropriate maintenance actions for the part based on the result of the prediction. The proposal will include the necessary information for the replacement of parts, such as the disassembly procedure of the part.

Based on this scenario, we consider the following information to be required for the product model: First, information is required on the deterioration and the behavior affected by the deterioration. An important difference between as-designed product information and product information for lifecycle support is the information on deterioration. It includes information on how the deterioration progresses under the conditions of the part, such as the load and the environment, and information on how the deterioration causes the deviation of the part's shape.

Next, the structure of the product model should accommodate the replacement of parts. The model of a part should independently contain all the information on the part, including the deterioration level and historical data on the operational and maintenance activities on the part.

Information to support the disassembly of parts should be provided to replace them. A proper procedure for disassembling or assembling a part aids users and maintenance personnel in replacing the part. The designer of the product is assumed to provide this information, depending on the part to be replaced.

Lifecycle information on assemblies and parts is required. The designers of a product also design the lifecycle of the product [2]. A part agent manages the lifecycle of the corresponding part based on the designed lifecycle information. Considering that every part may follow a different lifecycle through its replacement and reuse, every part, as well as assemblies, should contain its own lifecycle.

We created a product model for lifecycle support, taking these capabilities into consideration, and applied it to the lifecycle simulation of a simple manipulator with 3 degrees of freedom.

3 Proposed Representation of the Product Model

3.1 Assembly structure and shape

The fundamental structure of our product representation is the assembly structure. Researchers propose general-purpose assembly models [8, 9]. We are proposing a simple hierarchical structure of the assembly model with shape information based on [10, 11]. Figure 1 shows an instance of the assembly model. In the figure, a block depicts an instance of an entity, and an arrow represents the attribution to another instance. An assembly consists of plural parts, and each part has its own shape and assembly features. An assembly feature is a portion of a shape, such as a pin, connecting to another assembly feature, such as a hole, of a different part. In addition to the relations between parts and assemblies, those between connected parts and those between connected assembly features are also defined. For the shape information, we use a B-rep solid model of polyhedrons, for simplicity. A shape feature is a group of faces that compose an assembly feature.

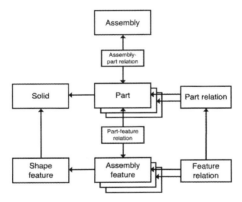

Fig. 1. Model of assembly and shape

3.2 Deterioration

Deterioration is important information for managing the lifecycle of a part. However, it is difficult to create a deterioration model that is universally applicable to predict the development of deterioration in mechanical products in various situations. We assume that force applied to the part causes the deterioration of a mechanical part, which generates its geometrical deviation. We implemented deterioration as a function that takes force as an input parameter and returns the deviation value as a result. Any deterioration phenomenon can be represented as a subclass of this function.

For the manipulator case, we assume that joint deviation is proportional to the accumulation of elastic deviation. The elastic deviation of joints is estimated as the force and torque applied to the joint multiplied by its compliance matrix [12].

3.3 Kinematics and Dynamics

The force applied to the part is derived from the analysis of its motion and force, i.e., kinematics and dynamics. For that purpose, a kinematic model of the product with its mass properties is constructed [7]. Figure 2 shows an example kinematic model of the links of a manipulator. A joint axis is an assembly feature that represents the connecting point of a part. It has a transformation matrix T_k representing its position and orientation. A joint connects two joint axes on different parts. A joint k has a transformation matrix $T_{jk}(q_k)$ between two joint axes, which changes its value according to the joint variable q_k, and an additional transformation T_{dk} representing the deviation when the joint deteriorates.

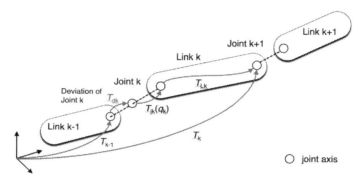

Fig. 2. Kinematic model

With this kinematic model, the motion values, i.e., position, orientation, velocity, and acceleration of the joints and the links, are calculated. This forward kinematics calculation is performed in series from the base link module towards the tip module, one by one, along the kinematic structure of the manipulator. Force and torque are calculated based on the acquired motion values, also in series, but from the tip module toward the base link module [13]. This serial nature of the calculation enables the replacement of modules with deviations.

Figure 3 shows the calculation scheme for lifecycle simulation of manipulators based on these models of kinematics and dynamics in coordination with the deterioration model described in 3.2. Based on this scheme, a prototype lifecycle simulation for a simple manipulator was implemented. The manipulator has a serial structure with two rotational joints and two links with a 0.7 m length. The simulated motion is to move the tip of the manipulator repeatedly between 2 points, (0.12, 0.16) and (0.10, 0.10). The compliance matrix of joints is derived from the catalog of their bearings. Two manipulators are employed with different tasks: Manipulator 1 moves for 1 h per day carrying 0.1 kg of weight, and Manipulator 2 moves for 30 min a day carrying 0.02 kg of weight. The deterioration value of each joint is calculated, and if the deterioration value exceeds a predefined level, the joint is replaced with another joint with the least deterioration value.

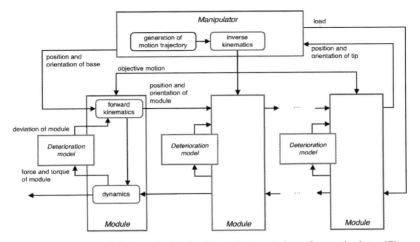

Fig. 3. Scheme of the calculation for lifecycle simulation of a manipulator [7]

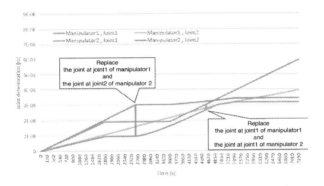

(a) With replacements of joints

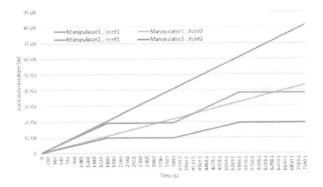

(b) Without replacements of joints

Fig. 4. Simulation of replacement of manipulator joints:

The results of the simulation are shown in Fig. 4. Two replacements occurred in this simulation. Less variance in the deterioration of joints is observed in the case with replacements compared to the case without replacements.

3.4 Disassembly Support

Disassembly information is structured at three levels, as shown in Fig. 5. The top level is a disassembly procedure, which represents a procedure to disassemble an assembly. It is assumed that parts are disassembled one by one from the assembly, and the disassembly procedure is an ordered list representing the sequence of parts to be disassembled. To disassemble a part, multiple connecting relationships between assembly features must be removed. A disassembly operation sequence represents a sequence of operations to disassemble a part, and each disassembly operation represents an operation to remove a feature relation. Additionally, a disassembly operation includes an operation display that contains graphical information, such as arrows, for depicting its disassembly motion.

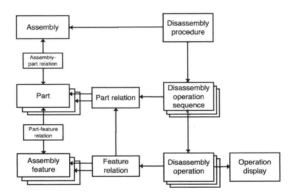

Fig. 5. disassembly procedure

We developed a system, as shown in Fig. 6, to assist users in disassembling a product using augmented reality (AR) technology based on the disassembly information [14]. Information for disassembling the product is displayed overlaid with the camera image calibrated by a marker on the product. Figure 6 shows a generated display image showing an operation to open the cover of the battery box to replace the battery of a computer.

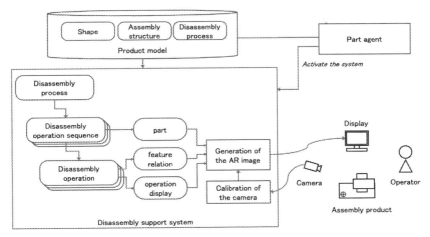

Fig. 6. Disassembly support system

3.5   Management of the Lifecycle of Assembly

A part agent provides advice to users based on the lifecycle information on the maintenance of the corresponding part, such as whether to replace the part, discard it, or continue to use it. Figure 7 shows an instance of the lifecycle model [4]. The lifecycle of a part consists of lifecycle stages, such as production, use, and disposal, and the lifecycle paths connecting them. Each part has both its own lifecycle and the lifecycle stage that it is currently at.

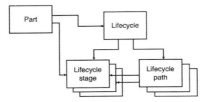

Fig. 7. Lifecycle model

Each lifecycle stage has functions for future prediction that calculate the values related to the stage, such as cost, environmental load, and benefit. These calculations are performed based on the lifecycle simulation of the part. The lifecycle path has a probability that a part may take the path in the future, which is also estimated based on the lifecycle simulation.

4   Integrated Product Model for Lifecycle Support

Figure 8 shows the overall structure of the proposed product model. Note that some details are not presented. The diagram was drawn based on the notation of Unified Modeling Language (UML) [15], i.e., blank arrows denote inheritance relations, and lines with diamonds denote aggregation relations. Blue boxes denote entities on kinematics, green ones on lifecycle, and yellow ones on disassembly support.

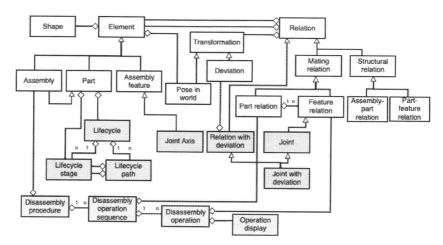

Fig. 8. Integrated product model for lifecycle support

With the information from the product model, we are developing functions for lifecycle support. Figure 9 shows a lifecycle simulation performed by the cooperation of parts. The behavior of a part in a lifecycle stage is simulated based on the objective motion assigned to the stage. As described in 3.3, a forward kinematics calculation derives the motion of a part in coordination with connected parts. Then the dynamics calculation is performed based on the motion values, also in coordination with other parts, to obtain the force and torque. The deterioration of the part is estimated with the force and torque using a function representing the deterioration phenomenon. The derived deviation of the shape or kinematic structure is used in the next cycle of the lifecycle simulation. Those derived values on the part's behavior, namely the motion, force and torque, and deviation, are used to evaluate the activity of the lifecycle stage by values such as the cost, benefit, and environmental load.

At assembly level, the calculation of kinematics and dynamics is performed similarly based on the results of part-level calculations. The objective motion derived from the planned activity is distributed to the parts. A deviation in the assembly, such as the deviation of the tip position of a manipulator, is calculated. Note that we assume the deterioration does not occur in the assembly but only in the parts. Based on the results of the activity at the assembly level and the evaluation results collected from the parts, the activity of the assembly is evaluated, and advice for the user that supports the decision on maintenance of the assembly is generated, such as whether a part should be replaced or not.

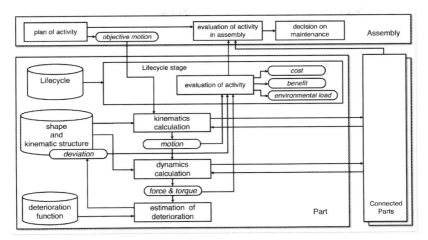

Fig. 9. A lifecycle simulation of a part

5   Discussion and the Remaining Issues

We recognize the remaining issues to be solved for lifecycle simulation for part agents. One is how to deal with the deterioration that affects the lifecycle of parts. We assume the availability of the deterioration model, which represents the quantitative relationship between the cause and the result of the deterioration. We also assume that the deterioration occurs only in parts and not in assemblies. As it is very difficult to capture deterioration phenomena, these assumptions may not hold.

The proposed scheme is currently limited to the domain of mechanical kinematics and dynamics, and the target product is assumed to be mechanical. Different types of parts, such as flexible parts, are not considered. We consider that we can apply our scheme to other domains based on their disciplines.

We used a criterion to replace a part at a predefined deterioration value in the prototype lifecycle simulation. It worked to a certain degree for averaging the deterioration of parts, but we need to clarify our objective and devise an effective strategy for the replacements. We should also develop an appropriate method for decision-making at the assembly level. We need to clarify what constitutes a proper communication and collaboration among part agents and their assembly agents.

We are developing experimental manipulators with a modularized structure to investigate the effectiveness of prospective maintenance with the replacement of parts by part agents.

6   Conclusion

An integrated product model for the lifecycle support of mechanical parts is proposed. A part agent performs a lifecycle simulation on the corresponding part based on the model to predict the future state of the part and provide advice on the maintenance of the part. We will evaluate the effectiveness of the model through the implementation of the system based on the model.

Acknowledgement. This research is supported by a Chuo University Personal Research Grant. The lifecycle simulation described in 3.2 was partially developed by Mr. Tomoya Toshida in his graduate thesis research.

References

1. Hauschild, M., Jeswiet, J., Alting, L.: From life cycle assessment to sustainable production: status and perspectives. CIRP Ann. **54**(2), 1–21 (2005)
2. Umeda, Y., et al.: Toward integrated product and process life cycle planning—an environmental perspective. CIRP Ann. **61**(2), 681–702 (2012)
3. Nakada, T., Hiraoka, H.: Network agents for supporting consumers in the lifecycle management of individual parts. Int. J. Product Lifecycle Manage. **5**(1), 4–20 (2011)
4. Hiraoka, H., et al.: Part agent advice for promoting reuse of the part based on life cycle information. In: Nee, A., Song, B., Ong, S.K. (eds.) Re-engineering Manufacturing for Sustainability, pp. 335–340. Springer, Singapore (2013)
5. Komoto, H., Tomiyama, T., Nagel, M., Silvester, S., Brezet, H.: Life cycle simulation for analyzing product service systems. In: 2005 4th international symposium on environmentally conscious design and inverse manufacturing, pp. 386–393. Tokyo (2005)
6. Hiraoka, H., Nagahata, T., Saito, H., Tanigawa, T.: Estimation of prospective states of mechanical parts for lifecycle support by part agents. In: Fortin, C., Rivest, L., Bernard, A., Bouras, A. (eds.) PLM 2019. IAICT, vol. 565, pp. 213–222. Springer, Cham (2019). https://doi.org/10.1007/978-3-030-42250-9_20
7. Hiraoka, H., Tanaka, T., Sugahara, T., Saito, H.: Assembly model of mechanical products for their life cycle simulation. Procedia CIRP **98**, 553–558 (2021)
8. Fenves, S.J., Foufou, S., Bock, C., Sriram, R.D.: CPM2: a core model for product data. J. Comput. Inform. Sci. Eng. ASME **8**(1) (2008)
9. Rachuri, S., et al.: A model for capturing product assembly information. J. Comput. Inform. Sci. Eng. ASME **6**(1), 11–21 (2006)
10. Sugimura, N., et al.: A study on product model for design and analysis of mechanical assemblies based on STEP. Trans. Inst. Syst. Control Inform. Eng. **8**(9), 466–473 (1995). (in Japanese)
11. ISO/TS 10303–109:2004(E) Product data representation and exchange: Integrated application resource: Kinematic and geometric constraints for assembly models, (2004)
12. Takano, S.: Kinematics of robots in detail, Ohmsha (in Japanese) (2004)
13. Luh, J.Y.S., Walker, M.W., Paul, R.P.C.: On-line computational scheme for mechanical manipulators. J. Dyn. Syst. Meas. Contr. **102**(2), 69–76 (1980)
14. Nagasawa, A., Fukumashi, Y., Fukunaga, Y., Hiraoka, H.: Disassembly support for reuse of mechanical products based on a part agent system. In: Hu, A., Matsumoto, M., Kuo, T., Smith, S. (eds.) Technologies and Eco-innovation Towards Sustainability 11, pp. 173–183. Springer, Singapore (2017)
15. ISO/IEC 19501:2005 Information technology — Open Distributed Processing—Unified Modeling Language (UML) Version 1.4.2 (2005)

Comparative Analysis of the Sustainability of Injection Molding and Selective Laser Sintering Technologies for Spare Part Manufacturing

Philipp Jung[✉], Klaas Tuschen, Kristin Zagatta, and Iryna Mozgova

Paderborn University, Warburger Street 100, 33098 Paderborn, Germany
philipp.jung@upb.de

Abstract. In order to achieve the climate targets, set by governments worldwide and the United Nations, measures must be taken to reduce CO2 emissions. A large proportion of CO2 is emitted during the manufacturing process, so there is a high potential for savings here. Making production more flexible can therefore help to reduce the CO2 footprint of products. In production, the combination of the injection molding process with lifecycle-depending use of the additive manufacturing process laser sintering should improve ecological sustainability. To investigate the effectiveness of combining these manufacturing processes, two scenarios are compared and analyzed. Some of the product lifecycle management factors such as part design, total number of pieces and product lifetime remain constant. In the first scenario, injection molding is used as a manufacturing process throughout all product lifecycle phases from market introduction to end-of-life to supply the demand. In the second scenario, injection molding is substituted by laser sintering depending on the lifecycle phase. In later product life phases, laser sintering replaces injection molding in order to no longer have to maintain or reproduce and store parts. The comparison of the scenarios is performed using the measured net energy consumption in combination with literature values and assumptions for all further product lifecycle management phases. Finally, product lifecycle management factors are determined that allow decision support in the demand-tailored selection of the manufacturing process to improve ecological sustainability.

Keywords: sustainability · lifecycle assessment · low carbon manufacturing · additive manufacturing

1 Introduction

Increasing economic output while maintaining quality standards and climate protection is a major task for companies. In the automotive industry, technologies such as lightweight construction help to achieve CO2 savings through weight reduction. However, in order to reduce the CO2 footprint before a vehicle has driven its first kilometer, CO2 intensive manufacturing processes should be investigated. During production approx. 10 tons of CO_2 are emitted per combustion vehicle [1]. Assuming a total mileage of 240.000 km and an average emission value of 150 g/km [2], 36 tons of CO_2 are emitted via the fuel burned. This shows that new solutions must be found to save CO_2 in production, because approx. 20% of the CO_2 emissions are emitted in production. Due to the decreased emissions by fuel, this percentage is even higher for electric vehicles [1].

Between 2015 and 2019, more than 90 million vehicles (cars, trucks, buses) were produced worldwide every year. Due to the corona pandemic, production in 2020 fell to 77 million automobiles per year, but this will increase again [3]. Statistically, an automobile has a lifespan of 18 years [4]. However, it usually only takes six to eight years for a new model in a series to come onto the market. Each of these automobiles is assembled from many different components, all of which are required in large numbers. Manufacturing processes such as injection molding (IM) for plastic components are used, which have short cycle times and thus ensure economical production [5, 6]. In order to have components available over the entire product lifecycle and beyond, components pre-manufactured with IM and kept in stock or the corresponding tools are kept in stock so that they can be continuously reproduced [5, 6] In order to save CO_2 in production over the entire product life cycle in the future, this approach should be rethought so that neither components nor tools have to be kept in stock, but the manufacturing process of components can be designed flexibly. Depending on the phase of the product lifecycle, the ecologically most sensible process must be selected, for example to quickly and easily replace IM, which makes sense in the introductory phase due to the large quantities, with Selective Laser Sintering (SLS) after the end of life for spare parts.

However, it is important to know the respective boundary conditions, such as the mechanical properties, so that the substitution is technically possible and the mechanical and optical component quality is maintained. To investigate this approach, two scenarios are set up in Sect. 3. In the first scenario IM is fixed as manufacturing process for the hole product lifecycle. In the second scenario IM is replaced by SLS after series production of the vehicle has ceased, when only spare parts are needed. To know the differences between the two manufacturing processes when a spare part is needed, a comparison of the processes is made in Sect. 2. The component used for the investigation of the scenarios is a plastic component that is used to clad the exterior of a vehicle. It will be shown in Sect. 4. The comparison of the two scenarios will be carried out in the Sect. 5.

This approach can make an important contribution to achieving the climate targets.

2   Comparison of Manufacturing Processes

To enable the complex comparison of the scope of product lifecycle between two different manufacturing processes, IM and SLS are briefly explained.

IM is one of the most important primary forming processes in polymer technology. Due to the short tact times and the high output rate, this process is of great importance for mass production. The raw material is added as a solid granulate. The granulate is plasticized by friction and pressure inside the plasticization unit as well as by externally supplied thermal energy. The plasticized material is injected through a nozzle into the provided mold (cavity), which forms the component [5].

The IM process is characterized as a fully automated discontinuous process. This means that the IM machines operate fully automatically over a period of time, while periodic interventions such as collecting the parts or refilling of raw material are carried out by operators. The forming of the components results from geometrically defined cavities in a special manufactured tool. Due of the manufacturing effort and price of the tool such as the process properties, the freedom regarding the geometry or the place of use, this limits the flexibility of manufacturing. As IM is a highly rapid and scalable manufacturing technique, it can produce high quantities of parts in short cycle times. For this reason, it is an advantageous option for centralized high-volume manufacturing of parts with simple and complex geometries. [5, 6].

SLS is a powder bed-based additive manufacturing process. The material is provided in the form of a solid dry powder. A laser is used to selectively melt the powder and bond it together layer by layer to form a structural component. In this process, no support structures are required. The non-melted powder can be used as recycling powder mixed with the new powder and then partially recirculated into the process. [7].

The SLS build process itself is fully automated. The steps involved in pre- and postprocessing, on the other hand, must be carried out by operators. These include setting up and cleaning the machine as well as removing and finishing the components [8]. The SLS process, in contrast to IM, is a tool-free process, requiring no mold. The part geometry is determined by a digital 3D model, allowing shape changes to be made easily and cost-efficiently [9]. Due to the tool-free manufacturing, a component can be manufactured on any SLS system, which increases the flexibility of the process. As the build of the components is based on melting the material layer by layer, long manufacturing times are characteristic for the SLS technology. Consequently, the SLS process is suitable for "on-demand" manufacturing of prototypes and low-volume series. [8–11].

Figure 1 summarizes a comparison of the two different manufacturing processes of IM and SLS. It shows the main process steps between product application and product use.

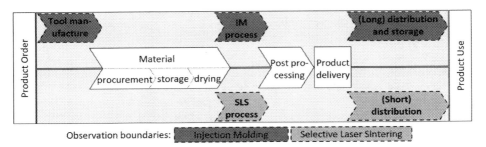

Fig. 1. Order of IM and SLS process for spare part manufacturing

In order to obtain a good expressiveness in the comparison of IM and SLS, only the explicit procedural differences in the two scenarios are compared. It is assumed that the process steps occurring in both processes are very similar and therefore have no influence. The material describes its procurement, its storage and its drying. Both manufacturing processes, whether plastic powder or plastic granules as the starting material, require these steps, which is why the consideration of the material is neglected. Post-processing, such as painting the component, is also not taken into account, since it is irrelevant for post-processing how the component was manufactured.

A significant difference between the processes is the use of a mold in IM in contrast to SLS. Material and energy must be invested before the first component is produced, so $CO2$ emissions start before the actual manufacturing. Another difference, as described, is the manufacturing process itself of IM and SLS. After the products have been manufactured, the last process difference follows. Because of the manufacturing effort, the price and the fixed component-related mold of the tool, IM is only economic with a large number of components with each tool [5, 6]. So for one component IM is located at the location of the component-related tool. This leads to long transport distances. In contrast, the corresponding SLS component can be manufactured on any SLS machine, since only the CAD file is required. Therefore, it can be produced very close to the place of application, so that the delivery distances are comparatively short. Closely related to this is the last major difference in the process, the supply of the components. Whereas with IM, manufacturing must either take place from stock or, if necessary, the manufacturing machine must be extensively retooled, and SLS enables a tool-less manufacturing on demand.

3   Setting up the Scenarios

The scenarios are based on the process differences and specific boundary conditions just described, which apply equally to both scenarios. In order to be able to compare two fundamentally different manufacturing processes, it must be ruled out that the manufacturing process itself provides important component properties that stand in the way of process substitution. Such a case would be, for example, a safety-relevant component that depends on certain strengths and rigidities. For this reason, process substitution can only be used for components that are not exposed to high mechanical loads. In addition, the processes should use a similar material. Therefore, a plastic component from the automotive sector is selected that is required in large quantities, a trim cap for a reversing light, (see Fig. 3).

The model series in question will be produced for six years, after which approx. 1,66 million cars of the model series will be on the European market [12]. The trim cap is installed once in each car. Due to the average useful life of 18 years, however, spare parts must still be available. If the statistics of annual body damage to cars in Germany (approx. 50 million registered cars [13]; 2 million vehicles with body damage per year [14]) are extrapolated to Europe, taking into account the model series and assuming that every second damage is rear damage, it can be assumed that approx. 33.000 additional trim caps will be required annually over the next twelve years.

The trim caps are produced at three different locations in Europe. This serves to shorten the transport routes to the respective vehicle assembly, which is also divided over several locations. On the other hand, three manufacturing sites serve as security in the event of a possible failure of a manufacturing line.

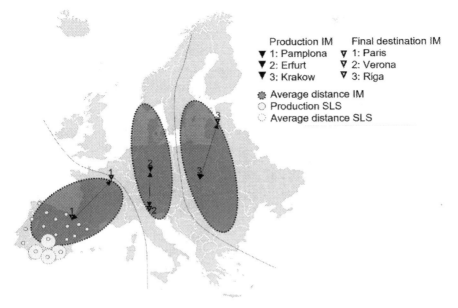

Fig. 2. Manufacturing locations and delivery routes

3.1 Scenario Injection Molding

In the first scenario, the required spare parts are still produced using the IM process. To do this, the molds should be stored in such a way that the required number of cover caps can be reproduced for each location and district. The three locations of the IM plants are in Palermo, Erfurt and Krakow. Due to the proximity to the manufacturing site, the decorative caps are manufactured and shipped from there. With 33.000 decorative caps required per year, 2.750 trim caps are required per quarter and location and shipped from there. In order to be able to map the shipping, an average shipping radius is assumed. Paris is 874 km from Pamplona, Verona from Erfurt 806 km and Riga from Kraków 967 km, the respective destination addresses for describing an average shipping route.

3.2 Scenario Selective Laser Melting

In the second scenario, the need for spare parts is served by SLS. As already described in the Process Comparison section, the significant differences to IM are used here. The manufacturing facility for SLS that is closest to the place of use can always be selected, so that the delivery routes are reduced to an average of 100 km (Fig. 2). For optimal use of the shipping route, 4 fairing caps are always sent at the same time so that the workshop concerned has a minimum stock. The manufacturing of the cap can be done together with other components within a joint construction order, so that only the absolute demand is produced.

4 Experimental Setup

4.1 Examined Part

A sample part optimized for IM manufacturing was selected for the test, as this process is the primary manufacturing process of this study due to its higher efficiency for larger quantities.

The manufactured object is part of an exterior trim of a car (see Fig. 3a). The bounding box of the part is 90 mm × 90 mm × 25 mm. The volume is 21 cm^3 and the surface is 207 cm^2.

To clarify the direct impact of the conversion to another manufacturing process combined with the chance of tool-free decentralized spare part supply the focus of this study was narrowed. Only the following directly allocable emitters were considered:

- Energy consumption manufacturing;
- Emission allocable to transport of parts.

4.2 Selective Laser Sintering (SLS)

For the SLS manufacturing an EOS P395 was used. This manufacturing device has a build chamber size of 340 × 340 × 600 mm^3 and is operated with an CO2-laser with a maximum power input of 50 W. The machine is designed for industrial applications. The manufacturing time is the sum of the time that is necessary to preheat the build chamber, the manufacturing time and the time to actively cool down the machine in a controlled manner.

The energy consumed by SLS manufacturing therefore adds up to:

$$W_{SLS} = W_{preheat} + W_{manufacturing} + W_{cooldown}, \tag{1}$$

where:

$$W = energy; \; P = power; \; t = time, $$

$$W = P * t \text{ in } W, \tag{2}$$

$$W_{preheat, \, machine} = const, \tag{3}$$

$$W_{cooldown} = P_{cooldown} * t_{manufacturing}, \tag{4}$$

$$P_{cooldown} = const. \tag{5}$$

As the cooldown is similar to the manufacturing time like [15] concluded

Baumers et al. [16] have shown the base consumption of the SLS machine is stable. Therefore, a repetition of the same part/level combination was not necessary during the experiments and the energy-consumption was multiplied according to the number of total parts needed to be manufactured for the scenario.

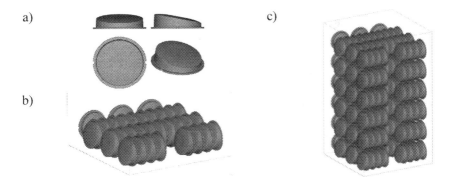

Fig. 3. Part (a), nested parts in SLS machine (31 parts (b), 186 parts (c))

For the experiment this translates to a manufacturing of 31 parts nested in the same level of the build chamber, whereas a filled chamber would output 186 parts. The consumed energy during the preheat phase is constant for each build job. The required cooldown time for the SLS machine depends on the total build time of the job and can therefore be calculated. The energy consumed during the manufacturing is the product of consumed energy by one level multiplied by the repetitions of the same.

$$W_{manufacturing} = W_{manufacturing/level} * n_{levels}, \tag{6}$$

where:

$$n_{levels} = \text{Number of repeating levels.}$$

As Fig. 3 shows, one level fits 31 parts and the whole chamber fits 6 repeated levels. The experiment was reduced to one level and the results were extrapolated for the further calculations of different scenarios.

4.3 Injection Molding (IM)

An Arburg ALLROUNDER 370 A with a EUROMAP 170 injection unit was used for manufacturing of the IM parts. This machine has a maximum closing force of 600 kN.
    The energy consumed by IM manufacturing adds up to:

$$W_{IM} = W_{preheat, machine} + W_{shot} * n_{shots}, \tag{7}$$

where:

$$n_{shots} = \text{Number of shots.}$$

For the comparability of the two manufacturing processes, 31 parts were planned to be likewise produced in IM, because one filled level in the SLS machine fits 31 parts (Fig. 3b). The calculations in the scenarios are derivatives from the experiments.

5 Results

Table 1 shows the results of the calculation of the scenarios. A distinction is made between the two manufacturing processes and the locations. The result is the CO2e value in kg per delivered component. The values for IM are closer to each other than the values for SLS. The values for SLS are further apart between the locations. Spain has the lowest and Poland the highest emissions.

Table 1. CO_2e in kg per delivered part

Manufacturing process	Location	CO_2e [kg]
Injection molding	Pamplona/Spain	0,34
	Erfurt/Germany	0,32
	Krakow/Poland	0,41
Selective laser sintering	Spain*	0,27
	Germany*	0,41
	Poland*	0,78

* The CO_2e of the power consumptions of the countries was used for the decentralized manufacturing, that was also considered for IM. A future more detailed study should consider facilitating a more precise breakdown into smaller areas with the basic values valid for CO_2 emissions for the production of electrical energy in the respective region.

In Fig. 4, the CO2e values are shown separately for transport and production to allow a more detailed examination. When comparing the different processes, it is noticeable that the ecologically better manufacturing process depends on the place of manufacture. In Poland and Germany, IM is still the preferred process for spare parts. Only in Spain SLS is more ecological for spare parts in these scenarios. Looking at the areas of transport and manufacturing, it can be seen that SLS emits more CO2 in manufacturing and IM more CO2 in transport.

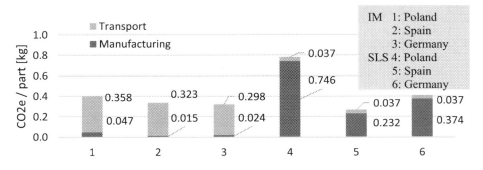

Fig. 4. CO_2e [kg]/delivered part distinguished according to Manufacturing and Transportation

IM consumes only 6.5% of the energy needed to produce an SLS part. Switching from IM to SLS, on the other hand, reduces emissions from transport by around 90%. The advantage of shortened transportation is counteracted by the high energy requirement per component in SLS. The energy requirement per component in IM production has hardly any influence.

Since the CO2e value in the production of SLS per component varies greatly between the sites, the composition of the electricity mix (Fig. 5) is taken into account.

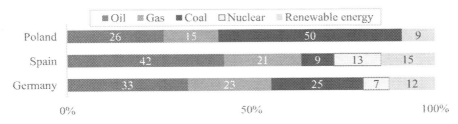

Fig. 5. Allocation of Energy Production [17]

It shows that Poland, which has the highest emissions per component for IM and SLS, gets most of its energy from fossil fuels (91%). Only 9% comes from renewable energy. Nuclear forms of energy are not used. In Spain, which has the lowest CO2 emissions per component in manufacturing, the share of renewable and nuclear energy is 28%, 11% higher than in Poland. The share of gas is also significantly higher.

6   Conclusion

In summary, it can be stated that decentralized manufacturing can significantly reduce emissions due to transport. Due to the high share of emissions due to production in SLS, SLS is particularly interesting if the share of renewable energies in the electricity mix is very high or if other forms of energy with low CO2 emissions have a high share. In IM, it is mainly the length of the transport route that is decisive. Since the share of emissions in manufacturing per component is very low in IM, the composition of the electricity mix has little influence. Thus, the decision whether IM and SLS are more ecologically sustainable in the provision of spare parts depends to a large extent on the location and the electricity mix there. To take advantage of the low emissions in transportation in SLS, the electricity mix should be ecologically better composed to make SLS more interesting in manufacturing as well. In addition, the SLS process would benefit from an increase in the energy efficiency of the manufacturing process, so that the machines themselves require less electrical energy and thus indirectly emit less CO2. In order to address the issue of sustainability in its entirety, the approach of ecological analysis will be supplemented by economic aspects in the future. To this end, a recommendation for action will be developed. The recommendation for action is intended to support the improvement of economic and ecological sustainability in production between two manufacturing processes. In addition, it necessary to examine how social sustainability is influenced, for example, by the relocation of jobs due to changed production locations through more ecological energy mixes or mor ecological transport. In this comparison, the three-dimensional optimization between the electricity mix, the transport and the manufacturing process is the challenge and must be further investigated in the future.

Acknowledgement. This work is a part of the research project BIKINI (funding code 03LB3018C), which is funded by the Federal Ministry for Economic Affairs and Climate Action.

References

1. ADAC Homepage. https://www.adac.de/verkehr/tanken-kraftstoff-antrieb/alternative-ant riebe/klimabilanz/#neu-die-interaktive-lca-plattform. Accessed 15 Feb 2023
2. VOLKSWAGEN AG: Emissionswerte & Emissionsnormen. PDF. Accessed 15 Feb 2023
3. Statista Homepage. https://de.statista.com/statistik/daten/studie/151749/umfrage/entwic klung-der-weltweiten-automobilproduktion/. Accessed 15 Feb 2023
4. Statista Homepage. https://de.statista.com/statistik/daten/studie/316498/umfrage/lebens dauer-von-autos-deutschland/. Accessed 15 Feb 2023
5. Schüle, H., Eyerer, P.: Polymer Engineering 2, 2nd edn. Springer Vieweg, Berlin (2022)
6. Groover, M.P.: Automation, production systems, and cumputer-integrated manufacturing, 4th edn. Pearson, Harlow (2016)
7. Breuninger, J., Becker, R., Wolf, A., Rommel, S., Verl, A.: Generative Fertigung mit Kunststoffen, 1st edn. Springer Vieweg, Berlin (2013)
8. Horsthotte, R., Bruning, B., Prümmer, M., Arntz, K., Bergs, T.: Determination of the level of automation for additive manufacturing process chains. In: 2nd conference on production systems and logistics, pp. 585–594. Publishing (2021)
9. Gibson, I., Rosen, D., Stucker, B., Khoransani, M.: Additive manufacturing technologies, 3rd edn. Springer, Cham (2021)
10. Thompson, M.K., et al.: Design for Additive Manufacturing: Trend, opportunities, considerations, and constraints. In: cirp annals – manufacturing technology, pp. 737–760. Elsevier (2016)
11. Schmidt, T.: Potentialbewertung generativer Fertigungsverfahren für Leichtbauteile. Springer, Berlin (2016)
12. Statista Homepage. https://de.statista.com/infografik/5140/europa-faehrt-golf/. Accessed 15 Feb 2023
13. Umweltbundesamt Homepage. https://www.umweltbundesamt.de/daten/verkehr/verkehrsi nfrastruktur-fahrzeugbestand#entwicklung-des-kraftfahrzeugbestands. Accessed 15 Feb 2023
14. Statista Homepage. https://de.statista.com/statistik/daten/studie/3396/umfrage/anzahl-der-str assenverkehrsunfaelle-mit-personenschaden/. Accessed 15 Feb 2023
15. Greiner, S., Jaksch, S., Cholewa, S., Drummer, D.: Development of material-adapted processing strategies for laser sintering of polyamide 12. Advanced industrial and engineering polymer research, pp. 251–263. KeAi Publishing (2021)
16. Baumers, M., Tuck, C., Bourell, D.L., Sreenivasan, R., Hague, R.: Sustainability of additive manufacturing: measuring the energy consumption of the laser sintering process. In: Proceedings of the institution of mechanical engineering, part b: journal of engineering manufacture, pp. 2228–2239 (2011)
17. Iea data and statistics Homepage. https://www.bpb.de/kurz-knapp/zahlen-und-fakten/europa/ 75140/energiemix-nach-staaten/. Accessed 17 Feb 2023

Hybrid Production Structures as a Solution for Flexibility and Transformability for Longer Life Cycles of Production Systems

Dorit Schumann$^{(\boxtimes)}$ (iD), Marco Bleckmann (iD), and Peter Nyhuis (iD)

Leibniz University Hannover, Institute of Production Systems and Logistics,
An der Universität 2, 30823 Garbsen, Germany
`schumann@ifa.uni-hannover.de`

Abstract. The dynamic environment of manufacturing companies not only directly impacts products and their design but also affects the configuration of production systems. Traditionally used production principles, such as the flow principle, which could previously be operated economically for (homogeneous) standard products, now no longer meet the requirements due to the increasing number of product variants and decreasing volumes. Other production principles, such as job shops or cell production, offer more flexibility to cover the diversity of product variants. However, considering the economic and logistical targets, those principles have longer throughput times and higher levels of work-in-process. This paper presents the idea of combining different production principles into hybrid structures to combine the advantages of flow production, such as high utilization and short lead times, with the flexibility of other principles. Hybrid production structures make the production system more flexible and transformable so that changes can be implemented more efficiently, and the life cycle costs of a production system can be reduced as a result. However, the configuration of a hybrid production structure requires the consideration of different influencing factors and the quantification of the economic and logistical targets. For this purpose, a research approach is presented in this paper.

Keywords: Hybrid Production Structures · Flow Production · Job Shop Production · Flexibility · Interdependencies · Transformability · Life Cycle

1 Introduction

The changing paradigm in manufacturing from mass production to personalized production has increased product variety and requires companies to adapt their technologies and production strategies to remain profitable [1]. In a recent study, companies were asked about their perspectives on potential new production technologies and systems in terms of their ability to solve the current challenges in designing production systems. The main results were that modular production cells, their flexible linking and a reconfigurable production layout would be helpful for the realization of a high number of product variants

C. Danjou et al. (Eds.): PLM 2023, IFIP AICT 702, pp. 289–299, 2024.
https://doi.org/10.1007/978-3-031-62582-4_26

and the integration of new variants [2]. Another challenge to the increasing diversity of variants is constantly changing product characteristics resulting from shorter product life cycles. In industries with high development speeds, it is not possible to forecast which technologies will be used in the long term. Flexible and transformable production systems would still be able to produce changing product characteristics without significant changes to the system so that the life cycle of the production system itself is extended [3]. For the application of flexible production systems, various approaches exist in the literature, which considers structural and organizational possibilities for gaining flexibility [4–6]. While flexibility means only adapting to changing requirements on a limited time scale without major investment or significant structural changes, a transformable production must also integrate pre-planned potentials for structural adjustments that can be activated as soon as needed [7]. The paper aims to combine production principles for a flexible and transformable production system to realize a high diversity of product variants, changing demands and the adaptation of product properties over a longer life cycle.

A production system can be characterized by its principle, which means the arrangement of the basic system components like workpieces, people and equipment. A frequently used production principle is flow production. For high volumes, this principle is particularly economical and is convincing regarding logistical performance with high utilization at balanced cycles, short throughput times and a low level of work-in-process (WIP). If more flexibility is required in terms of product, operation sequence and production capacity, job shop production is an option. However, it has comparatively lower logistical performance due to longer lead times and high WIP [8]. With today's trends of personalization, digitalization, and automation, production conditions, manufactured goods as well as production programs regarding product variants and volumes are changing, so conventional production principles must be expanded and adapted to alternative forms to continue producing economically [9]. The adaptation of production principles has been researched several times. An overview of existing approaches is given in the following Sect. 2. After presenting the idea of hybrid production structures as a combination of different production principles, the paper shows how this can reduce the life cycle costs of a production system in Sect. 3. A case study compares the life cycle costs of flow production and cell production and how these costs can be reduced by a hybrid structure of both production principles combined. However, in order to take advantage of hybrid production structures, they must first be configured, and due to the differences between production principles, it is expected that there will be interactions between numerous influencing factors. Therefore, a research approach is presented in Sect. 4 to quantify the interrelationships by combining different production principles. After the discussion of the research approach in Sect. 5, the final section will conclude the work.

2   Related Work

There are different approaches in the literature on handling the diversity of product variants with a flexible and transformable production system. These approaches focus on manufacturing, assembly systems or both in general. One idea is the production stage concept, which separates the variant neutral processes into the pre-production stage from the variant generating processes into the final production stage [8, 10]. Bley and Zenner present an integrated consideration of all required product variants in an interface between product design and variant-oriented assembly planning [3]. Hu et al. review the research of assembly system design and operations under high product variety. Besides other aspects, they considered line balancing, reconfigurable assembly planning and scheduling in mixed model assembly [4]. The need for changeable production systems due to higher product variety and shorter product life cycles is already discussed for a long time. Koren [11] is one of the drivers in this research and represents the principles and future trends of reconfigurable manufacturing systems. Compared to flexible manufacturing systems (FMS), which enable the manufacturing of different product variants on the same manufacturing system, reconfigurable manufacturing systems (RMS) are designed to react quickly to sudden changes in requirements and to adapt structure, software and hardware regarding production capacity and functionality within a part family. RMS have the advantages of both dedicated serial lines and flexible manufacturing systems and can be expanded modularly for higher production capacity on the one hand and are flexibly adaptable on the other. [11] In addition to reconfigurable systems, modular and hybrid approaches were developed. The solutions range from modular assembly systems [12] to line-less mobile assembly systems [9] and continuous to matrix structures [6, 13, 14]. Many concepts examine the effects of variant diversity, using the automotive industry as an example. Kern et al. investigate alternatives to assembly line production in the automotive industry and present principles of a modular assembly concept [15]. The developed concept has uncoupled workstations without a uniform cycle time, and the products are transported by automated guided vehicles (AGV) for more flexibility. Focusing on modularized, order-flexible and adaptable assembly processes, Küber et al. develop a method for cross-architecture assembly line planning to enable the lines to realize more than one vehicle architecture [16]. Combining a line assembly with matrix production, respectively flexible segments, is referred to as a hybrid assembly structure by Göppert et al. [17] and Kampker et al. [18]. Göppert et al. consider a holistic way to combine the advantages of both structures with the integration of planning software and digital twins, as well as a flexible layout and AGV transport routes [17]. Kampker et al. compare the conventional mixed-model assembly line with hybrid assembly structures and appoint premises for the combination of flexible segments with individual and variable cycle times and line segments with fixed uniform cycle times [18]. The control of the output sequence of the flexible segment in order to continue in a line production is the main challenge in hybrid structure and is investigated by using answer set programming [5].

The presented approaches illustrate the conflict between high flexibility for realizing a high product variance and efficient production. Reconfigurable and modular systems focus on positioning in this conflict. Often the original production principle is dissolved, and a flexibilization of the entire system is aimed, which can result in high investment costs. In the idea of hybrid production structures, only one subarea is designed to be transformable, and the original production principle is to be retained in the remaining subarea (usually flow production). On the one hand, this should produce large production quantities economically, but on the other hand, it should also be possible to map product variants and react to changing requirements. The challenge is to ensure the flow of materials between the subareas of different principles, to know the interactions and to produce economically overall. In particular, there is a research gap in investigating the interdependencies of different target variables to design the production system efficiently. Hybrid production principles will be presented in this paper as a cross-industry solution for the economic coverage of a high diversity of variants. In addition, the paper examines whether the hypothesis is confirmed that the integration of transformability through hybrid structures leads to a reduction in the production system's total life cycle costs.

3   Hybrid Production Structures

3.1   Combination of Production Principles

As the literature review shows, combining different production principles, especially the flexibilization of production lines, is not a novelty. These developments demonstrate that it is no longer possible to produce economically with the isolated implementation of one production structure and that the changing market requirements cannot be fulfilled in this way. A major challenge is implementing the many technical possibilities so that production is still economical. Therefore, hybrid structures represent a possible solution approach. They offer the opportunity to use the advantages of different production principles and to strategically position oneself in between them. However, the term hybrid production or assembly is used in various contexts; thus, a definition is given first:

"*A hybrid organizational structure generally describes a combination of at least two interlinked different organizational structures.*" By using this definition, there is no differentiation of manufacturing and assembly required, and it can be used for both and for production in general. When using the term 'hybrid assembly', it is important to differentiate that this term also describes semi-automated assembly as a combination of manual and automated assembly [19]. Although the degree of automation is a significant influencing factor when considering economic efficiency, it is not considered further in this paper. It will focus on the combination of flow and other principles like job shop and cell production to a hybrid production structure. Hybrid production structures can be designed in a variety of ways. The ratio between the flow and the job shop production can be determined, and the sequence and layout in different combinations is possible. Exemplary structures are shown in Fig. 1.

The challenge with hybrid organizational structures is that the different systems are interlinked and not decoupled by a warehouse as in the production stage concept [10]. On the one hand, interlinking offers the challenge of absorbing the spread in processing times; on the other hand, the logistical effects are not transparent due to complex inter-relationships and dependencies between structural and influencing parameters. These interdependencies have not yet been sufficiently investigated in previous research work. Also, guidelines for selecting the optimum configuration between several production structures do not exist. As an outlook, a research approach is presented to quantify the correlations and to determine optimal hybrid production structures.

The combination of production principles creates several subareas in the production system. If changes are required in the production system (e.g. the addition of new variants, changes in the production process or physical limitations), it is easier to adapt only one subarea or configure one subarea for higher changeability from the beginning. The effects on the life cycle costs and length of a production system are explained in the following section.

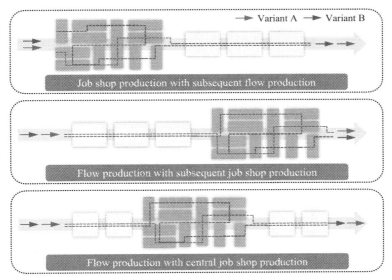

Fig. 1. Exemplary hybrid production structures

3.2 Solution for Changeable Production Systems for Longer Life Cycles

Hybrid production structures allow a certain flexibility to produce many variants with only one production system. A production system's flexibility means adapting to changing requirements within defined flexibility corridors without major investments in a fast time and without significant structure changes [7]. However, in its life cycle, a production system is confronted with long-term change drivers due to constantly increasing challenges, which cannot be managed by flexibility alone. This requires pre-planned potentials for structural adaptations, which can be activated as needed (transformability). A transformability potential integrated into the production system can extend the original function or shift the flexibility corridors by activating it [20]. In order to cover the uncertainties that can occur during a production system's life cycle, systems must be planned to be changeable. Flexibility and transformability as strategies of changeability [7] can be integrated into hybrid production structures.

By integrating transformable elements into a production system, occurring changes can be managed with marginally time and cost. However, the investment costs for the integration of transformable elements are higher. If this investment is not made from the beginning, the costs to change it afterwards are significantly higher [21]. In hybrid production structures, transformability potentials do not have to be integrated into all areas of the production system; instead, it is sufficient to make only subareas transformable. For example, when implementing a further product variant, only the subarea of the job shop principle in the hybrid production structure would have to be changed. This area concentrates the change potential in large flexibility ranges and high transformation level shifts, which can be used to adapt to changing requirements. By concentrating the changeability on only one subarea, both the cost of change and the duration of change are reduced. The part of the production system operating according to the flow principle can mostly remain unchanged. Dealing with an identical challenge for a mono-organized production system would be much more costly by extending it to the entire system.

Hybrid production structures with transformable subareas reduce the production system's overall life cycle costs. Mono-organized production systems require a higher capital investment to make them transformable, so the total life cycle costs are similarly high as if almost no transformability is integrated. The relationship between the degree of transformability and the life cycle costs of a production system is shown in Fig. 2. Life cycle costs distinguish between the costs of the initial investment as well as direct, and indirect transformation costs. Direct costs consider costs for adjustments and indirect costs mean costs due to production losses. [21] Hybrid production structures thus enable lower life cycle costs by integrating transformability potential in one subarea. In addition, the life cycle of a production system is extended since only subareas need to be adapted when change drivers occur, and it is not necessary to plan a new production system immediately.

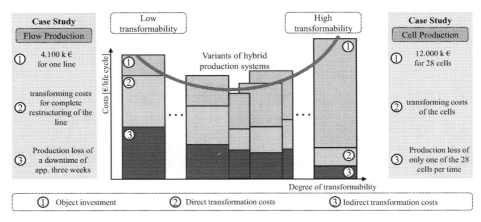

Fig. 2. Life cycle cost depends on transformability with case study; based on [21]

Case Study. For validation purposes, a case study was conducted at a manufacturer of electric motors. This company produces large quantities with short cycle times in a highly automated flow production (assembly line). Changing customer requirements are reducing the number of units, and the product portfolio is expanding, so alternative production concepts must be reviewed. An automated assembly cell was developed that produces the same product as the comparative line but, unlike the line, the cell can also produce other variants by changing only the software program. However, the assembly cell has a cycle time of 140 s, whereas the line can produce a significantly higher number of units with a cycle time of only 5 s. In order to achieve the same number of units per year, 28 assembly cells would be required to replace one line. The cell and the line represent respectively the two extreme scenarios of the illustrated life cycles costs in Fig. 2.

Scenario Flow Production (Line). The line describes the left column of the figure. The investment costs of the assembly line amount to approx. 4,100 k€. The line is designed for exactly one product, and the flexibility corridor is very small. Therefore, changes usually lead to a restructuring of the line, which has far-reaching effects. Within the stations of the line, the functioning of the tools must be adapted, a change in the sequence of the stations may be necessary, and the work plan and, thus, the programming of the line must be adapted. Such a change can take up to three weeks, so the indirect costs due to a complete production downtime are very high.

Scenario Cell Production. The production system of 28 assembly cells corresponds to the scenario of the right column in Fig. 2. The purchase of an assembly cell costs about 420 k€. Accordingly, for 28 cells is almost an investment of 12,000 k€ necessary if the production line is to be replaced only by cells. The object investment is very high, but the assembly cells achieve very high flexibility and transformability. Different variants can be produced simultaneously. If a major adjustment is required, this can be carried out sequentially to the assembly cells. Then, only one cell is out of operation during the transformation work, and the remaining 27 cells can continue to produce. Thus, the indirect transformation costs are very low.

Hybrid Production Structures. Both examples represent an extreme scenario. A hybrid production structure would be possible by splitting the work plan operations and producing one part in the line and the other in the assembly cells. As a result of the fact that part of the production process is already produced in the line, fewer operations are required in the assembly cell. This shortens the cycle time, and in total less investment in cells is required compared to pure cell assembly. Compared to the line, indirect transformation costs are reduced. In the case of a necessary adjustment in the cell area, only individual cells have to be down. Or, if an adjustment is necessary for the line area, the assembly cells can also cover the complete range of functions for the production of the product, so that there is no complete downtime. The line is directly interlinked with the assembly cells and distributes its semi-finished products to the assembly cells via conveyor belts.

Results and Limitations. This case study shows that implementing only one production principle can lead to high initial costs but low transformation costs or allow transformations only by high costs. A combination of production principles can reduce the production system's total life cycle costs and increase the production system's total life cycle by allowing changes. However, there are many factors to consider when designing a hybrid production structure. For example, challenges exist in balancing the line and cells, the possible utilization of the line, and the buffer stocks that may be required between the two systems. Also, it must be determined when a hybrid production structure will perform better than a conventional production principle. As an outlook on how to answer these questions, a research approach is presented below.

4   Research Approach to Investigate Hybrid Production Structures

While flow production can produce high volumes but can handle only limited variants, other production principles like job shop or cell production can handle many variants but cannot keep up with productivity [5]. Regarding the logistical performance, the flow principle convinces with high utilization and short throughput times. In contrast, the job shop principle is more flexible regarding the product, the operating sequence, and the process times, but the logistical performance is worse due to longer throughput times and high WIP. [8] Just the differences between the two production principles indicate that integrating one into the other is a challenge, and different objectives must be considered. However, the interdependencies between numerous influencing variables are very complex, and there is currently no method for quantifying the correlations as a basis for configuring hybrid production structures. Therefore, a research approach is presented in this paper. The first step is to collect the challenges in the design of production structures. This collection is carried out in close cooperation with industrial partners with different initial situations regarding production structures. After identifying challenges and influences, the cause-effect relationships will be analyzed qualitatively about the achievement of logistical targets. In order to quantify the identified impact relationships and the effects of different production structures, a simulation model is developed. Using the simulation model, the correlations can be investigated through simulation studies and analyzed using statistical methods. The correlation between the influence and target variables is to be simplified by using mathematical description and effect models to

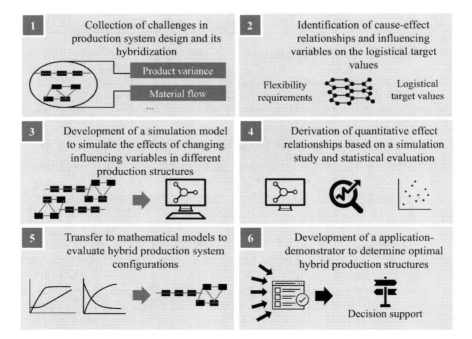

Fig. 3. Research approach to investigate hybrid production structures

determine the optimal configuration of a hybrid production system. By researching the quantitative correlations, guidelines for designing hybrid production structures can be derived and serve as a decision-making tool for industrial companies. The outcome of this research will be an application demonstrator to determine the optimal hybrid production structure in terms of logistical target achievement. By understanding the interrelationships between flexibility and logistical performance, companies can position themselves between these opposing objectives through hybrid production structures. The described steps of this research approach are illustrated in Fig. 3.

5 Discussion

Many companies are already interested in designing their production systems to be transformable, but therefore no systematic decision frameworks exist. So the presented research approach focuses on analyzing the interdependencies between economic and logistical targets and flexibility requirements to develop a decision framework for companies to use hybrid production structures and to allocate suitable production principles into the right areas. However, combining different production principles can cause problems, especially if a balanced line represents a subarea. When connecting the subareas, linkage must be considered; otherwise, production failures of individual stations will impact the rest of the chain. Thus, elastic linking using buffers should be investigated. Material supply also requires consideration in the design of hybrid production structures. For example, the production sequence of the first section in the case of a line in the second section must match the cycle to ensure just-in-time material supply. The effects on the production employees should be addressed too, because they must constantly attend to new tasks, especially in the transformable subarea, and may require higher qualifications.

Hybrid production structures are not the only option for companies to react to changes. Today, companies are globally interlinked and can, for example, cover higher production volumes and greater product variance through outsourcing or an extended workbench. If such possibilities can be considered, the corresponding relocation effects must be reflected when determining the life cycle costs of a production system. Therefore, the shown correlations between the degree of transformability and life cycle costs are only valid for a local unit of a production system and not for globally interconnected production chains.

6 Conclusion

This paper describes hybrid production structures as a solution for flexible and transformable production systems to handle increasing product variants and faster-changing conditions. It is described that hybrid production structures enable higher flexibility and reduce life cycle costs through transformable subareas. Subsequently, a research approach is presented for quantifying correlations to determine optimal hybrid production structures. The discussion highlighted the limitations of hybrid production structures while also providing an outlook on further research needs.

Acknowledgements. Funded by the Deutsche Forschungsgemeinschaft (DFG, German Research Foundation) – 471726131.

References

1. Koren, Y.: The global manufacturing revolution: Product-process-business integration and reconfigurable systems. Wiley, Hoboken, N.J (2010)
2. Fries, C., Fechter, M., Nick, G., Szaller, Á., Bauernhansl, T.: First results of a survey on manufacturing of the future. Procedia Comput. Sci. **180**(107 No. 3), 142–149 (2021)
3. Bley, H., Zenner C.: Variant-oriented assembly planning. CIRP Ann. **55**(1), 23–28 (2006)
4. Hu, S.J., Ko, J., Weyand, L., et al.: Assembly system design and operations for product variety. CIRP Ann. **60**(2), 715–733 (2011)
5. Kaiser, P., Thevapalan, A., Reining, C., Roidl, M., Kern-Isberner, G., Hompel, M.: Hybrid production: enabled by controlling the output sequence of a matrix production using answer set programming. Procedia CIRP **107**(3–4), 1305–1310 (2022)
6. Perwitz, J., Sobottka, T., Beicher, J.N., Gaal, A.: Simulation-based evaluation of performance benefits from flexibility in assembly systems and matrix production. Procedia CIRP **107**, 693–698 (2022)
7. Hingst, L., Park, Y.B., Nyhuis, P.: Life cycle oriented planning of changeability in factory planning under uncertainty. In: publishing, Hannover (2021)
8. Wiendahl, H.P., Reichardt, J., Nyhuis, P.: Handbook factory planning and design. Springer, Berlin Heidelberg, Heidelberg (2015)
9. Hüttemann, G., Buckhorst, A.F., Schmitt, R.H.: Modelling and assessing line-less mobile assembly systems. Procedia CIRP **81**(6), 724–729 (2019)
10. Grigutsch, M., Nywlt, J., Schmidt, M., Nyhuis, P.: Highly flexible final production stages - taking advantages of scale effects by reducing internal component variants. AMR **907**, 127–137 (2014)
11. Koren, Y., Gu, X., Guo, W.: Reconfigurable manufacturing systems: principles, design, and future trends. Front. Mech. Eng. **13**(2), 121–136 (2018)
12. Kern, W., Rusitschka, F., Bauernhansl, T.: Planning of workstations in a modular automotive assembly system. Procedia CIRP **57**(5), 327–332 (2016)
13. Greschke, P., Schönemann, M., Thiede, S., Herrmann, C.: Matrix structures for high volumes and flexibility in production systems. Procedia CIRP **17**, 160–165 (2014)
14. Hofmann, C., Brakemeier, Na., Krahe, C., Stricker, N., Lanza, G.: The impact of routing and operation flexibility on the performance of matrix production compared to a production line. In: Schmitt, R., Schuh, G. (eds.) WGP 2018, pp. 155–165. Springer, Cham (2019)
15. Kern, W., Rusitschka, F., Kopytynski, W., Keckl, S., Bauernhansl, T.: Alternatives to assembly line production in the automotive industry. In: 23rd international conference on production research. Manila (2015)
16. Küber, C., Westkämper, E., Keller, B., Jacobi, H.F.: Method for a cross-architecture assembly line planning in the automotive industry with focus on modularized, order flexible, economical and adaptable assembly processes. Procedia CIRP **57**, 339–344 (2016)
17. Göppert, A., Schukat, E., Burggräf, P., Schmitt, R.H.: Agile hybrid assembly systems: bridging the gap between line and matrix configurations. In: Weißgraeber, P, (ed.) Arena2036 Ser, advances in automotive production technology - theory and application: conference on automotive production, pp. 3–11. Stuttgart (2021)
18. Kampker, A., Kawollek, S., Marquardt, F., Krummhaar, M.: Potential of hybrid assembly structures in automotive industry. Indust. Manufac. Eng. (2020)

19. ElMaraghy, H.A. (ed.): Changeable and reconfigurable manufacturing systems. Springer, London, London (2009)
20. Zäh, M.F., Moeller, N., Vogl, W.: Symbiosis of changeable and virtual production–the emperor's new clothes or key factor for future success. In: Proceedings of the 1st conference on changeable, agile, reconfigurable and virtual production (CARV2005), München (2005)
21. Nyhuis, P., Heger C.: Adequate factory transformability at low costs. International conference on competitive manufacturing (COMA), Stellenbosch (2004)

A Methodology to Promote Circular Economy in Design by Additive Manufacturing

Simona Ianniello[1,2]($\boxtimes$), Giulia Bruno[2], Paolo Chiabert[2], Fabrice Mantelet[1],
and Frédéric Segonds[1]

[1] Laboratoire de Conception de Produits et Innovation LCPI – Arts et Métiers Institute of Technology – HESAM University, 151 Boulevard de l'Hôpital, 75013 Paris, France
simona.ianniello99@gmail.com
[2] Department of Management and Production Engineering - Corso Duca degli Abruzzi 24, Politecnico di Torino, 10129 Torino, Italy

Abstract. [Context] Within the framework of PLM; Circular Economy (CE) is an important concept that seeks to design out waste and pollution, keep products and materials in use and regenerate natural systems. Early Design Stages (EDS) are crucial because they set the foundation for the rest of the production process and can determine the product's overall functionality, usability, and manufacturing processes.

[Problem] As companies increasingly recognize the benefits of transitioning to a circular economy, there is a growing need for tools and methodologies to support the design of circular products and services.

[Proposal] This paper presents a CE card deck as a novel approach to facilitate the early stages of product development. The deck consists of 10 cards based on Morseletto's "Targets for a circular economy" work that represents mainstream circular economy principles and strategies and can be used by designers, engineers, and other stakeholders to generate ideas, evaluate options, and make informed decisions.

The results of a pilot study with design and engineering students (Master level) suggest that the card deck can support the exploration of CE concepts and facilitate the identification of circular solutions, in a Design by AM context (DbAM). The paper concludes with a discussion of the potential benefits and limitations of the card deck approach, and its integration into a PLM framework, and gives suggestions for future research.

Keywords: Circular Economy · Eco-design · Green Manufacturing · Environmental Impact · Life cycle engineering · Design by Additive Manufacturing

1 Introduction

In recent years, Circular Economy (CE) has gained significant traction as a solution to many environmental and economic challenges, by applying the reuse, recycling, and remanufacturing principles, with the aim of reducing waste and keeping products and

© IFIP International Federation for Information Processing 2024
Published by Springer Nature Switzerland AG 2024
C. Danjou et al. (Eds.): PLM 2023, IFIP AICT 702, pp. 300–312, 2024.
https://doi.org/10.1007/978-3-031-62582-4_27

materials in use for a longer period before they are discarded [1]. Product Lifecycle Management (PLM) should serve as a major help for companies looking to switch to a circular mode, by supporting them in the design and develop products more easily recyclable and repairable, thus with a longer life span [2]. Despite the crucial importance that CE will have in the coming years, there is a lack of tools to facilitate its application in the design phases of product development. In addition, researchers find promising the use of Additive Manufacturing (AM) in supporting CE strategies (e.g., providing recycled materials, repairing, remanufacturing, and recycling materials) [3]. For this reason, we have deeper analyzed the link between these principles as both, in the context of PLM, can help companies to stay competitive, adapt to the increasing demand for sustainable products, save costs, and improve environmental performance [4]. The Design by AM (DbAM) methodology prioritizes the integration of AM processes into Early Design Stages (EDS) to drive innovation. The concept generation phase provides an opportunity to fully utilize the distinctive capabilities of AM processes to create innovative solutions [5].

The research objective of this paper is to propose and validate in the context of DbAM a Circular Economy Card Deck as a tool that helps designers, engineers, and other stakeholders to generate ideas, evaluate options, and make informed decisions. Furthermore, a thorough examination of the strong connection between AM and CE has been conducted, including a detailed analysis of the opportunities provided by AM specific methods that assist designers in navigating the technical challenges of AM and how it aligns with CE principles [6].

The paper is structured as follows, Sect. 2 presents a literature review on the CE, the obstacles associated with implementing CE in product design, and the relationship between CE and AM. Section 3 gives an overview of the research design approach. Section 4 presents the results. Finally, Sect. 5 draws conclusions and presents future works.

2   State of the Art

2.1   Circular Economy vs Linear Economy

Circular Economy is a broad concept, and more than one hundred definitions, offering multiple, sometimes contradictory, ways of conceptualizing CE have been identified [7]. What emerges from the literature review is the urgent necessity to transition from a Linear Economy (LE) to a Circular one, moving away from a resource consumption and waste generation model, towards a more sustainable and efficient system of resource use [8].

Sauvé et al. put forward a definition of CE that differs from the traditional linear economy approach. They argued that while sustainable development in the linear model primarily concentrates on minimizing waste, recycling, and pollution, the CE model places a greater emphasis on resources and considers all inputs and outputs of the production process, with a specific focus on waste management [9].

To provide and develop a framework of strategies to guide designers and business strategists in the move from a linear to a CE, the terminology of slowing, closing, and narrowing resource loops must be introduced [1]. These concepts are useful in the transition to a CE reducing the demand for new resources and minimizing waste.

1. Slowing resource loops: the utilization period of products is extended and/or intensified, resulting in a slowdown of the flow of resources.
2. Closing resource loops refers to the practice of recycling and reusing materials, rather than discarding them.
3. Resource efficiency, also known as narrowing resource flows, refers to the practice of using fewer resources to produce a given product or service.

Furthermore, to switch from a linear to a CE, the R-strategies [10] are very significant as they define a system in which multiple options and targets can be applied to promote CE implementation. The 10 Rs [10] are a set of strategies that can be used to promote sustainability and reduce waste by encouraging resource conservation and reuse.

Potting et al. [11] developed a framework that categorizes strategies for achieving circularity in order of increasing power, with R9 being the most powerful and R0 being the least powerful. This hierarchy should not be considered a strict rule, as there may be exceptions and secondary effects that can affect the effectiveness of these strategies, the hierarchy must be considered with caution and used as a general guide when evaluating CE strategies. As shown in Fig. 1, R0, R1, and R2 strategies decrease the utilization of natural resources and materials applied in a product chain by fewer products being needed for delivering the same function. These three strategies are related to the Design Phase, strategies from R3 to R7 are related to consumption aspects of the products, and the latter two (Recycle and Recovery) are related to how to return the product after its life cycle came to an end.

CE	Smarter product use and manufacture	R0	Refuse	Make product redundant by abandoning its function or by offering the same function with a radically different product
		R1	Rethink	Make product use more intensive (e.g. through sharing products or by putting multi-functional products on market).
		R2	Reduce	Increase efficiency in product manufacture or use by consuming fewer natural resources
	Extend lifespan of product and its parts	R3	Reuse	Re-use by another consumer of discarded product which is still in good condition and fulfils its original function
		R4	Repair	Repair and maintenance of defective product so it can be used with its original function
		R5	Refurbish	Restore an old product and bring it up to date
		R6	Remanufacture	Use parts of discarded product in a new product with the same function
		R7	Repurpose	Use discarded products or its part in a new product with a different function
	Useful application of materials	R8	Recycle	Process materials to obtain the same (high grade) or lower (low grade) quality
LE		R9	Recovery	Incineration of material with energy recovery

Fig. 1. CE strategies adapted from Morseletto et al. [10]

2.2 Challenges for Promoting CE in the Early Design Stage

The EDS encompasses "project definition and planning," research and validation of the concept, and architectural design up to the preliminary layout creation [12]. In terms of production stages, one of the hurdles that CE faces is that it is usually more expensive to manufacture a durable long-lasting good than an equivalent quick and disposable version. This is a public good problem: the benefits of producing less or non-durable goods are private while the environmental cost is public [9].

Manufacturing companies need to prioritize sustainability to stay competitive in the market. One approach they can take is eco-design, which is the process of designing products with the goal of reducing their environmental impact throughout their entire lifecycle. This approach can be used as a strategy to improve the sustainability of a product during the early design phase [13]. Design for CE has recently come into focus as a new research area in the wider field of sustainable design [14]. Product life extension and complete recovery of products and materials form essential elements of this approach, design for a CE highlights the importance of high-value and high-quality material cycles [15]. An emergent field, in addition, is the potential contribution of AM to these circular strategies [16]. In fact, AM technologies are widely used in concept development [17] and could represent an important instrument for enhancing CE during the EDS of product development. Moreover, what emerges from the literature review is the "lack of environmental and lifecycle considerations in the curriculum for the Early Design Stages" [18]. Designers are facing a challenge in finding appropriate approaches for incorporating AM into the early design phases [19]. To assist designers in reducing the environmental impact of AM, two methods have been proposed. The first method is involving experts in AM during the EDS. The second method is providing designers with tools specifically designed for their work. These methods can also be applied to other areas beyond AM [20].

Within the context of DbAM, a useful supporting tool for inspiring the application of AM in product design is described by Lang et al. [6]. The authors defined 14 opportunities of the potential of AM (Topology optimization, material choices, multimaterials, monoblock...) and represented them in 14 inspirational objects, each associated with one opportunity, later shown. They proved that such methodology can help foster innovative ideas through associations between product sector-specific knowledge and the potential of AM [6]. The tool helps designers capture the design potential of AM to design creative solutions at the EDS by incorporating AM knowledge as early as possible during the ideation phase. The 14 opportunities of AM include shape complexity, hierarchical complexity, functional complexity, and material complexity, each with its own specific characteristics, such as freeform shapes, monoblock, material choices, and multi-materials. These 14 objects have been used in this work and have been implemented in the tool proposed for empowering CE, as better described in the next section.

2.3 Synthesis

In summary, we notice the necessity to transition from a linear economy to a CE in the field of product development. However, the implementation of CE principles during the design phase of product development is currently limited by a lack of tools and a lack of research. Additionally, there is a necessity to conduct further research on the intersection of CE and AM to fully understand the potential of these technologies to work together for sustainable production. These limitations highlight the need for further development of tools and deeper research in this area to fully realize the benefits of CE in product development. We can conclude there is a lack of methodology to support CE in the EDS/AM frameworks.

3 Research Design Approach

The methodology incorporates a card deck and 14 AM opportunities to assist designers in integrating CE principles and maximizing the opportunities for AM during the EDS. After that the experiment has been conducted, participants have been asked to evaluate the tool proposed.

3.1 Circular Economy Deck Tool

This study proposes a Card Deck tool, named "Circular Economy Deck", as cards are valuable for "sparking creativity, externalizing tacit concepts, constructing and organizing ideas, and working both playfully and collaboratively" [21]. Collaborative and open strategies for EDS have been demonstrated as crucial tools for the implementation of successful CE [22]. This tool can help designers to analyze, ideate, and develop the circularity potential in their projects during the EDS of production. The tool is based on the previous literature review of circular-oriented innovation principles and strategies to realize it. The principles are organized according to the intended circular strategy outcome that they pursue (narrow, slow, close, regenerate) [23], and one of the 10-R strategies [10] that each card is representing.

Figure 2 shows the Circular Economy Deck. Each of the 10 cards represents a CE strategy on which it is important to be focused. On the back side of the cards, the explanation of the strategy is presented with a user-friendly image, and on the front side a score and the strategic effect between slow, narrow, close, and regenerate is assigned.

Fig. 2. Circular Economy Deck is composed of 10 cards and 1 legend card front and back sides.

A legend card is provided with the 10 cards in the deck as well. In each card, each strategy's effects in terms of Life Cycle Assessment have been explained thanks to the symbols under the title. To decide the existing link between each strategy and each of the 4 symbols present or not on the card, an analysis with 3 experts on the CE framework has been conducted.

3.2 Experiment for Tool Exploitation

The experiment was conducted with 12 master's engineering students. The objectives of this experiment are to evaluate the effects of using the Circular Economy Deck during the EDS of a product and to deepen the relationship between CE and AM. To reach this goal, 5 phases have been conducted involving creativity, CE information, and AM knowledge.

Table 1. Phases of the experiment

		Phase 1: Introduction to Additive Manufacturing processes	Phase 2: Introduction to Circular Economy concepts	Phase 3: Presentation of the Card Deck for Circularity and of the AM Cubes	Phase 4: Creativity session "*The foldable helmet of the future*"	Phase 5: Questionnaires
🕐	Group using Card Deck				125'	10'
🕐	Group without Card Deck	5'	15'	15'	125'	10'

Table 1 describes the approach used for evaluating the Cards-based tool.

5 phases COMPOSED THIS APPROACH:

- 1st phase: Students are introduced to AM Processes through a presentation with PowerPoint.
- 2nd phase: Students are introduced to CE concepts through a presentation with PowerPoint.
- 3rd phase: Present and explain each card of the Circular Economy Deck to the students. Students have also used the AM opportunities [21].
- 4th phase: Creativity session: twelve students were separated into 2 groups. 6 students from the 1st group used the Circular Economy Deck for generating their idea sheet and 6 students, from group 2, did not use it. Both groups have used the 14 AM opportunities cubes. The teacher in charge of the creativity session presents the brief: "design the foldable helmet of the future".

- 5th phase: After having realized a creativity session using brainstorming, purge phase, and inversion phase to offer a maximum of Idea Sheets (IS), students have been asked to fill in different questionnaires.

During the 1st and 2nd phases, a short lesson about AM and CE has been provided to the students to briefly explain these concepts. In the 3rd phase, students have been introduced to the Circular Economy Deck and an explanation of how to use the cards is rendered. The 4th phase is shown in Fig. 3, representing how the Creativity session has been conducted.

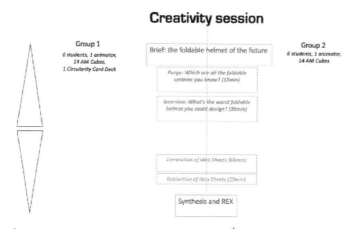

Fig. 3. Structure of the Creativity Session in the 4th phase has been conducted.

3.3 Evaluation of the Tool

During the 5th phase, different questionnaires were proposed to all participants to assess their feelings about this approach. Five questions, reported in Table 2, were asked to understand participants' interests, their perceived acquisition of CE knowledge and the application of these concepts for developing the idea, and their perceived acquisition of CE knowledge related to AM processes. In addition, students who used the Card Deck must declare which cards have been used in the development of their idea.

Students have been asked to complete a matrix shown in Table 3 for evaluating the relationship between each of the 14 AM cubes and the CE strategies presented in the cards. We asked the students to rank the top 3 CE strategies more powerful for each cube and more related to each opportunity the AM cubes want to refer to.

Table 2. List of questions to the participants

List of questions
1.   I feel more able to explain the main concepts of CE.
2.   I think that I have better understood the relationship between CE and AM.
3.   I feel capable of proposing ideas of the innovative object being more focused on the opportunities of the CE.
4.   I feel able to explain and exploit all the pursuable strategies for Circularity.
5.   I think I have mastered the CE strategies and their power.

Table 3. Matrix to fill in to rank the relationship between CE strategies and AM opportunities. Each cube is different from the others and represents one AM opportunity. This matrix has been used to evaluate how much each opportunity is impactful on CE strategies.

AM CUBES / CE STRATEGIES	1	2	3	4	5	6	7	8	9	10	11	12	13	14
REFUSE														
RETHINK														
REDUCE														
REUSE														
REPAIR														
REFURBISH														
REMANUFACTURE														
REPURPOSE														
RECYCLE														
RECOVERY														

1. Segmentation
2. Embedded components
3. Internal channels
4. Infilling
5. Auxetics structure
6. Material choices
7. Multi-materials
8. Freeform shapes
9. An object from 3D scans
10. Microstructure variation
11. Texture
12. Metablock
13. Topology optimization
14. Non-assembled mechanisms

4   Results

In total, 8 Idea Sheets (IS) were generated. 1st group, that used the card deck, realized 4 IS and 2nd group realized 4 Ideas Sheets as well. Figure 4 presents an example of one of them. It is a new concept of a foldable bike helmet as the Creativity Session theme was *"The foldable helmet of the future"*. In Fig. 4, the innovative helmet is thin and retractable, when it is closed it looks like a headband and it unfolds as painted protecting the head and spine. The materials involved in the production are entirely recycled.

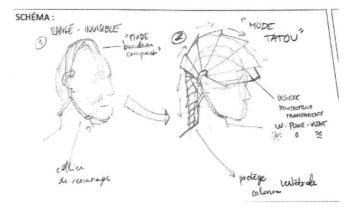

Fig. 4. Example of a spontaneous generation of an idea for the "Foldable helmet of the future".

4.1 Results on Circularity Level

Participants have been asked to declare which cards they used for developing the Idea Sheet and the results, shown in Table 4, reveal that the most used strategies are Reuse, Repair, and Recycle. These 3 cards have been used for each idea generated with the Card Deck. The average of cards used by the students is 4 cards for each Idea Sheet created, this result is a good clue to declare that the tool helps designers in being focused on circularity's aspects during the EDS.

Table 4. Table counting how many times each card has been used by participants with the Card Deck.

Card	Number of uses
REUSE	5
REPAIR	5
RECYCLE	5
REDUCE	4
REFURBISH	4
REMANUFACTURE	4
RETHINK	3
REPURPOSE	3
RECOVERY	2
REFUSE	1

4.2   Results on the Feeling of CE Performance

Students have been asked to answer the following questions, also presented in Table 2, choosing their answer on a scale of agreement. The answers permitted were "Totally Disagree", "Disagree", "Somewhat Disagree", "Somewhat Agree", "Agree", and "Totally agree" and they have been transposed to a scale from 1 to 6 (Fig. 5).

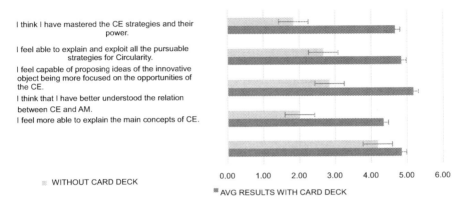

Fig. 5. Results of the satisfaction questionnaire (6 students per panel).

The results of this questionnaire indicate a better understanding of the CE strategies among the group of students who utilized the Circular Economy Deck. This is confirmed by the students themselves, who reported feeling more able to explain and apply the strategies related to CE. Additionally, those who used the Card Deck stated that they felt more capable of proposing ideas for innovative products, with a greater focus on CE opportunities – which aligns with the tool's intended purpose. Furthermore, the students reported that the relationship between CE and AM was clarified and better understood with the tool provided. To further confirm the effectiveness of the research approach, it has been asked to the students if they felt more able to explain the main concepts of CE. The results of this question show that the gap between the two groups is not as significant as for the other questions even if present, suggesting that Phase 2 of the research was useful and confirming the previous results. Overall, the students' positive feedback and a better understanding of CE strategies using Circular Economy Deck provide valuable insights into the field of PLM.

4.3   Results on the Link Between AM and CE

The results of this table confirm and empower the strong link already studied between AM and CE strategies. Participants were asked to rank the top 3 strategies most powerful for each cube and more related to each AM opportunity. Table 5 represents the correlation between each strategy and each cube because of the data collected after counting the times each strategy has been declared in correlation with the others from the data collected in Table 3. The red cells are those characterized by a low level of correlation between the strategies and the AM opportunities, this is also represented by values from 0 (lowest) to 10 (highest) grades of correlation.

The AM opportunities with the highest number of strategies associated are nonassembled mechanisms, material choices, segmentation, and objects from 3D scans. The strategies that resulted in being the most related to AM processes are Rethink, Reduce, Recovery, and Reuse. In contrast, the AM opportunities that are less related to CE strategies are multi-materials, embedded components, and infilling.

4.4 Synthesis of Results

The results of a survey on the usage of the Circular Economy Deck tool among students indicate a better understanding and application of CE strategies. The students reported feeling more capable of proposing innovative products with a focus on CE opportunities, and the relationship between CE and AM was clarified. The top three CE strategies identified by the students were Reuse, Repair, and Recycle. The AM opportunities with the highest correlation to CE strategies were non-assembled mechanisms, material choices, segmentation, and objects from 3D scans, while multi materials, embedded components, and infilling were less related. Overall, the results confirm the strong link between AM and CE strategies, the effectiveness of the research approach to empower circular aspects during the EDS of production and the students provided positive feedback on the tool.

Table 5. Table showing the relationship between AM opportunities and CE strategies.

	Non assembled mechanisms	Material Choices	Segmentation	Objects from 3D scans	Texture	Topology optimization	Internal Channels	Microstructure variation	Monoblock	Freeform shapes	Infilling	Auxetics structure	Embedded components	Multi-materials
RETHINK				4				6		2			1	1
REFURBISH	5	10		2	1		2		6				2	
REPURPOSE	6			2			1	2				2	4	2
RECOVERY		4	5			3			5	5	1		3	
REPAIR			6		2		4				6		4	
REDUCE	1	4						1	2	1		6		
REUSE	1	5		1	6			3	1	1				
RECYCLE			6				4		5		3			
REFUSE	5		1	2	1		1		6	4				
REMANUFACTURE	3	3		10	1			6		2		8		1

5 Conclusions and Future Work

The main research focus of this paper is discovering if the most important concepts related to CE can be learned and kept in consideration during the Early Design Stage of Product Development thanks to a Circular Economy Deck that aims at increasing awareness of CE concepts during the EDS of production, in a Design by AM context. The study involves 3 domains: creativity, CE, and AM. The results of a pilot study with master's design and engineering students suggest that the Circular Economy Deck can support the exploration of CE concepts and facilitate the ideation of eco-products. Moreover, by introducing concepts and tools related to AM during the testing phase, the study improved understanding and awareness of AM's potential applications in the CE field.

One of the limitations of this work is the number of students involved in it. Future research is vital to continuously update the tool with new strategies, and, more generally, to try to test the tool first with more groups of engineering students and then in companies. Later, it would be interesting to assess the cards by industrial experts in the domain. Furthermore, there is value for future research to develop sector-specific versions of this Circular Economy Deck such as for Industry 5.0, biotech firms, green buildings and constructions, sustainable agriculture and food systems, and similar companies.

References

1. EMF: Circularity indicators: an approach to measuring circularity. Ellen MacArthur Found. **12** (2015). https://doi.org/10.1016/j.giq.2006.04.004
2. Cholewa, M., Minh, et al.: PLM solutions in the process of supporting the implementation and maintenance of the circular economy concept in manufacturing companies. Sustainability **13**, 10589 (2021). https://doi.org/10.3390/su131910589
3. Kravchenko, M., Pigosso, D.C.A., McAloone, T.C.: Circular economy enabled by additive manufacturing: potential opportunities and key sustainability aspects, In: Balancing innovation and operation. presented at the proceedings of NordDesign 2020, The Design Society (2020). https://doi.org/10.35199/NORDDESIGN2020.4
4. Thompson, M., Moroni, G., Vaneker, T. et al.: Design for Additive Manufacturing: Trends, opportunities, considerations, and constraints. CIRP Ann. **65**(2), 737–760 (2016). ISSN 0007-8506. https://doi.org/10.1016/j.cirp.2016.05.004
5. Segonds, F.: Design By Additive Manufacturing: an application in aeronautics and defence. Virtual Phys. Prototyp. **13**, 237–245 (2018). https://doi.org/10.1080/17452759.2018.1498660
6. Lang, A., et al.: Augmented design with additive manufacturing methodology: tangible object-based method to enhance creativity in design for additive manufacturing. 3D Print. Add. Manufac. **8**, 281–292 (2021). https://doi.org/10.1089/3dp.2020.0286
7. Kirchherr, J., Reike, D., Hekkert, M.: Conceptualizing the circular economy: an analysis of 114 definitions. Resour. Conserv. Recycl. **127**, 221–232 (2017). https://doi.org/10.1016/j.resconrec.2017.09.005
8. Michelini, G., Moraes, R.N., Cunha, R.N., Costa, J.M.H., Ometto, A.R.: From linear to circular economy: PSS conducting the transition. Procedia CIRP **64**, 2–6 (2017). https://doi.org/10.1016/j.procir.2017.03.012
9. Sauvé, S., Bernard, S., Sloan, P.: Environmental sciences, sustainable development and circular economy: alternative concepts for trans-disciplinary research. Environ. Develop. **17**, 48–56 (2016). https://doi.org/10.1016/j.envdev.2015.09.002
10. Morseletto, P.: Targets for a circular economy. Resour. Conserv. Recycl. **153**, 104553 (2020). https://doi.org/10.1016/j.resconrec.2019.104553
11. Potting, J., Hekkert, M., Worrell E. and Hanemaaijer A., in January 2017. Circular Economy: measuring innovation in the product chain - Policy Report
12. Segonds, F., Cohen, G., Véron, P., Peyceré, J.: PLM and early stages collaboration in interactive design, a case study in the glass industry. Int. J. Interact. Des. Manuf. **10**, 95–104 (2016). https://doi.org/10.1007/s12008-014-0217-4
13. Folkestad, J.E.: Resolving the conflict between design and manufacturing: integrated rapid prototyping and rapid tooling (IRPRT). J. Indust. Technol. **17** (2001)
14. Ceschin, F., Gaziulusoy, I.: Evolution of design for sustainability: from product design to design for system innovations and transitions. Des. Stud. **47**, 118–163 (2016). https://doi.org/10.1016/j.destud.2016.09.002

15. Alamerew, Y.A., Brissaud, D.: Circular economy assessment tool for end of life product recovery strategies. J. Remanufac. **9**(3), 169–185 (2019). https://doi.org/10.1007/s13243018-0064-8.hal-01910562

16. Sauerwein, M., Doubrovski, E., Balkenende, R., Bakker, C.: Exploring the potential of additive manufacturing for product design in a circular economy. J. Clean. Prod. **226**, 1138–1149 (2019). https://doi.org/10.1016/j.jclepro.2019.04.108

17. Laverne, F., Bottacini, E., Segonds, F., Perry, N., D'Antonio, G., Chiabert, P.: TEAM: a tool for eco additive manufacturing to optimize environmental impact in early design stages. In: Chiabert, P., Bouras, A., Noël, F., Ríos, J. (eds.), Product lifecycle management to support industry 4.0, IFIP advances in information and communication technology, pp. 736–746. Springer, Cham (2018). https://doi.org/10.1007/978-303001614-2_67

18. Foteini, M., Frédéric, S., Maud, R., Nicolas, P.: A methodological proposal to link Design with Additive Manufacturing to environmental considerations in the Early Design Stages. Int. J. Interact. Des. Manufac. 1–14 (2017). https://doi.org/10.1007/s12008-017-0412-1

19. Valjak, F., Bojčetić, N.: Conception of design principles for additive manufacturing. Proc. Int. Conf. Eng. Des. **1**, 689–698 (2019). https://doi.org/10.1017/dsi.2019.73

20. Laverne, F., Marquardt, R., Segonds, F., Koutiri, I., Perry, N.: Improving resources consumption of additive manufacturing use during early design stages: a case study. Int. J. Sustain. Eng. **12**, 365–375 (2019). https://doi.org/10.1080/19397038.2019.1620897

21. Logler, N., Yoo, D., Friedman, B.: Metaphor cards. In: Proceedings of the 2018 on designing interactive systems conference 2018 - DIS 2018, pp. 1373–1386. ACM Press, NY (2018). https://doi.org/10.1113/JP275465

22. Panza, L., Faveto, A., Bruno, G., Lombardi, F.: Open product development to support circular economy through a material lifecycle management framework (n.d.)

23. Konietzko, J., Bocken, N., Hultink, E.J.: A tool to analyze, ideate and develop circular innovation ecosystems. Sustainability **12**, 417 (2020). https://doi.org/10.3390/su12010417

Design and Release Process for AM Parts

Daniel Schmid[✉]

ZHAW Zurich University of Applied Sciences, 8401 Winterthur, Switzerland
daniel.schmid@zhaw.ch

Abstract. The paper presents the results of a survey that analyses the additive manufacturing (AM) industry's application of design and release processes based on a proposed "ideal" release process, which was developed with the knowledge and experience of the Product Lifecycle Management and additive manufacturing (AM) groups at the University of Applied Sciences. This process reflects modern computer-aided design (CAD) tools and their capability, e.g. to add the form and position tolerancing by 3D annotation (product manufacturing information, PMI). As the complexity of additive manufacturing emphasizes the importance of release processes, the requirements of "design for manufacturing" (DfM), especially "design for additive manufacturing" (DfAM), are also considered. Furthermore, the present way of designing AM parts (raw and final part) is reflected.

A gap analysis is conducted on the survey results, whereby improvements to close the gap are discussed. This knowledge is used in a running scientific project with an industry partner to offer additive manufacturing services via an internet platform that calculates an immediate and quantity-dependent price offer.

Keywords: Release Process · Change Management · Additive Manufacturing (AM) · Product Lifecycle Management · Design for Manufacturing

1 Introduction

The purpose of this paper's survey is to analyse the additive manufacturing industry's common design practises and degree of digitalisation as well as its specific opinions on and its utilisation of predefined design and release processes. This not only reflects the current state of the additive manufacturing industry, but also provides insight into the specific shortcommings of the companys' digitalisation and design processes, how those can be optimised, and why it is beneficial to do so.

1.1 Additive Manufacturing

Additive manufacturing (AM) is one of the newest manufacturing technologies [1] and is considered a primary shaping method ([2], Chapter 39). As such, the design processes for additive manufacturing and casting deviate significantly from those of other manufacturing technologies, e.g. turning, where the designing engineer does not usually spend a considerable amount of time designing the raw part.

© IFIP International Federation for Information Processing 2024
Published by Springer Nature Switzerland AG 2024
C. Danjou et al. (Eds.): PLM 2023, IFIP AICT 702, pp. 313–322, 2024.
https://doi.org/10.1007/978-3-031-62582-4_28

From a process point of view, there are effectively no additive manufactured or casted parts that do not require postprocessing, e.g. by milling. Hence, it is ideal to design the raw part before the final part. However, often the design process is executed backwards, and the final part is designed first and then used as the basis from which the raw part's design is derived.

This leads to less than ideal designs, which is only exacerbated by another challenge in the design process of additive manufacturing worth mentioning, despite not being investigated in this study, which is the industry's little established knowledge of "design for manufacturing" (DfM) [3, 4] or, more specifically, "design for additive manufacturing" (DfAM) [5–7]. For example, the orientation of an additive manufacturing part in the machine [8, 9] is critical for the supporting structure's complexity, the part's deformation and tension, and the overall effectiveness and efficiency of the manufacturing process. Therefore, a close cooperation between the designer and the additive manufacturing expert is required to achieve an optimal result. In that respect also, additive manufacturing is similar to casting, as it requires a similar exchange of knowledge.

1.2 Release and Change Process

The above mentioned challenges in additive manufacturing parts' design process and the common problematic design methlogies employed in the industry make the utilsation of release and change processes in the context of advancing digitalisation in general even more important than in other manufacturing technologies.

Change and release processes are vital in the product lifecycle [10, Ch. 11.4] and can be seen as part of a digital thread [11] and the 3D Master [12] methodology. They ensure that manufacturing takes place on the right design [13, 14]. They are the starting point to ensuring traceability throughout the lifecycle of a product instance [15, 16]. As such, they are a cornerstone of Industry 4.0, related value chains [17], and are potentially even connected to technologies such as, e.g. the blockchain [18]. Traceability, also known as "Traceability 4.0" [19], is typically divided into tracking and tracing. The industry has incorporated this instrument to some extent and uses it, e.g. for quality assurance. Furthermore, the data of design (master data) and manufacturing codes are required to build specific digital twin frameworks [20] by using the digital shadow of production [21].

All the above is also valid for additive manufacturing (AM). The technology is even more complex than classic manufacturing technologies, e.g. turning and milling; therefore, the manufacturer should aim for good traceability. The additive manufactured item is the starting product for the subsequent machining and might contain a complex, topologically optimized structure [6, 22]. It is indicative of the technology's complexity that in a (German) textbook regarding the development methodology of AM [23], no word regarding a proper release process is mentioned.

This gap is more pronounced when considering change processes [24, 25]. For instance, if the raw part gets updated, the final machining must be reevaluated as well. An example would be the reduction of a hole's diameter (raw part); the final machining has to be adapted to avoid potentially damaging the tools. Figures 1 and 2 show the raw and final parts, where the centre bore, marked in red, is machined.

All the arguments above consolidate the motivation that should lead the establishment of a proper release and change process in an organization.

Fig. 1. Raw part   **Fig. 2.** Final machined part

2 Proposed Design and Release Process Description

As design and release processes and design practises can be an abstract topic that is prone to misunderstandings, an "ideal" design and release process (see Fig. 3) is proposed to act as a tool for both common understanding and for challenging the participant's design practises in their industrial environment. This "ideal" design and release process is developed through the collaboration and the exchange of knowledge and experience from the research groups for Product Lifecycle Management and Additive Manufacturing (AM) at the University of Applied Sciences. Hereinafter the "ideal" design and release process that is proposed to the survey's participants is explained.

Design for Manufacturing (DfM) and Design for Additive Manufacturing (DfAM) is key for modelling the raw part (1) and the final part (2). The process's prerequisite is that the engineers understand both additive manufacturing and classic manufacturing technologies. The raw part gets its unique part number for identification in the PLM, and the customers' unique article number for part identification is allocated (remains on the final part).

An associated copy (linked) of the raw part is used to define the final machining (2). This final part contains the related product manufacturing information (PMI) and is identifiable by a unique part number. When finalised, it gets released accordingly.

The preparation for manufacturing runs in parallel. On the one hand, the raw part gets exported (3), e.g. STL, and elements such as supporting structure and filler are defined in a respective slicing tool (4). On the other hand, the manufacturing of the final part is prepared in a CAM process (5). The material that needs to be milled is the difference between the raw and final parts.

Both manufacturing setups get released individually. The manufacturing and post-processing (6) of the raw part is conducted. That is the point in the process where the serial number is written on the part (e.g. engraved or lasered). Afterwards, the final part is manufactured (7). Note: With this process, it is also possible to design the finished model in the first step and the AM raw part afterwards (copy with link).

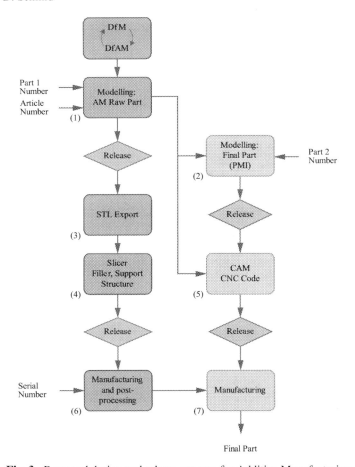

Fig. 3. Proposed design and release process for Additive Manufacturing

3   Survey and Results

3.1   Boundary Conditions

In total, 24 people who finalised the Certificate of Advanced Study (CAS) course Additive Manufacturing [26] have received the questionnaire. Three were from the same company, so only one of them answered the questionnaire.

Furthermore, two different networks have been approached: IBAM [27] and SAMG [28]. They have rejected approaching their members outside of their communication schedule. Therefore, the conclusions are based on the answers of 14 former CAS participants and two other contacts. In total, 16 answers constitute the sample. The response rate is 58%. Considering that three people from the CAS came from the same company, 14 out of 22 companies have answered the questionnaire (64% response rate).

3.2 Questionnaire and Answers

The following table lists the most relevant questions of the questionnaire, along with the respective possible answers and compiled results to grant a quick overview. The questions are aimed at evaluating and understanding the companys' data handling and utilisation as well as their understanding: {4} clarifies if design and manufacturing is within the same company, {7} to {10} explore the way of working to design AM parts, and {11} to {17} focus on the releas process.

3.3 Cross-Comparisons

Table 2 shows three relevant cross-comparisons between two questions each. The first two comparisions provide insight on which participants would like to apply the proposed design and release process, while the third comparison shows if those companies that go to the trouble of applying PMI in their company also use it for subsequent processes.

In Table 1, the relevant cross-comparisons are listed. The first considers only those who manufacture in-house, while the second and third consider those who apply PMI.

3.4 Bias

Bias, or survey error, is a common challenge in surveys and is discussed on related web tools [29] and internet pages [30], as well as in literature [31, 32]. The most common types of bias are [33]:

- Sampling bias
- Non-response bias
- Acquiescence bias
- Social desirability bias
- Question order bias
- Interviewer bias

This survey is primarily affected by the sampling bias and the non-response bias. The sampling bias occurs because only people with a certain degree of knowledge and experience in additive manufacturing or product lifecycle management have been approached (see Sect. 3.1). But it does not matter as the survey aims to conduct a gap analysis between release and change management theory and its application in the industry. If the conclusion is that there is a significant gap, it is even more pronounced when looking at the whole industry. If there is no gap, the conclusions can only be drawn for the pinnacle in the industry. Therefore, relevant conclusions can be drawn despite this sampling bias.

Much the same applies for the non-response bias. The assumption is that the approached people who have not answered the questionnaire would not increase the rate of participants who have applied the design and release process.

Table 1. Relevant questions, related answers, and results

{4} Is the AM part manufactured in-house or by a supplier?	
In-house	81 %
By a supplier	19 %

{7.1} Do you design a model for the raw part (AM part) and a separate one for the final part?

Yes	56 %
No	31 %
Different	13 %
Empty	0 %

{7.2} Are the models linked with each other (copy with link)? [Follow-up {7.1}]

Yes	31 %
No	19 %
Different	0 %
Empty	50 %

{8.1} Do you apply product manufacturing information (PMI, "3D annotation"), e.g. tolerances, in your 3D model?

Yes	19 %
No	81 %
Different	0 %
Empty	0 %

{8.2} If yes, are the PMI used digitally in the following processes as manufacturing and quality assurance? [Follow-up from {8.1}]

Yes	19 %
No	0 %
Different	0 %
Empty	81 %

{9} Which fundamental data get created within your design process?

Model of the raw part and model of the final part including product manufacturing information (PMI, "3D annotation")	13 %
Model for the raw part, model for the final part, and related drawing	38 %
Model of the final part and product manufacturing information (PMI, "3D annotation")	6 %
Model of the final part and related drawing	38 %
Other	6 %

{10} Which files are you managing in your product lifecycle management (PLM) system? Please select the related elements.

Model for the raw part (can include PMI)	25 %
Drawing of the raw part	56 %
Export of the raw part, e.g. as STL format	19 %
File of the slicing tool (e.g. Magics)	19 %
Export of the slicing tool (machine code)	19 %
Quality assurance data of the raw part as e.g. measurements	38 %
Model of the final part (can include PMI)	75 %

(*continued*)

Table 1. (*continued*)

Drawing of the final part	69 %
Machining code of the final part	6 %
Quality assurance data of the final part as e.g. measurements	13 %

{11} Which elements need to be released according to your release process?

Model for the raw part (can include PMI)	19 %
Drawing of the raw part	44 %
Export of the raw part, e.g. as STL format	6 %
File of the slicing tool (e.g. Magics)	6 %
Export of the slicing tool (machine code)	6 %
Model of the final part (can include PMI)	44 %
Drawing of the final part	69 %
Machining code of the final part	6 %

{12} Are parts manufactured, e.g. for prototyping, without being released according to your release process?

Yes	63 %
No	38 %

{14} Is there a differentiation in the release process between AM and conventionally manufactured parts (e.g. milling or turning)?

Yes	19 %
No	81 %

{17} Would you like to apply this proposed release process, not considering the effort for its introduction?

Yes	38 %
No	63 %

Table 2. Cross-comparisons

{4} & {17} Liking to apply the proposed release process while production takes place today in-house

Yes	46 %
No	54 %

{4} & {8.1} Liking to apply the proposed release process while using today product manufacturing information

Yes	100 %
No	0 %

{8.1} & {8.2} Applying PMI ("3D annotation") and using those further

Yes	100 %
No	0 %

4 Conclusions and Gap Analysis

Dissemination of additive manufacturing among participant companies:

- The people who answered work for large, medium, and small-sized companies. What the companies have in common is that only a few people are deeply involved in additive manufacturing technology, working in teams of 1 to 11 people per company. Only one answer from a large-sized company stated a number above this range: 20 people. Therefore, the conclusion is that additive manufacturing is still a niche technology.
- The average share of AM parts among all types (not quantity) of parts manufactured in a company is 3.5%. This figure does not include entities such as schools or service providers focusing on this manufacturing technology. For example, one such excluded answer came from a company that exclusively offers the manufacturing of AM parts as a service. Thus, they claim to have a 100% share of additive manufactured part types.

Dissemination of "design for additive manufacturing" and utilisation of optimised release processes and PMI among participant companies:

- The "design for additive manufacturing" has not yet been established throughout the industry to a particularly high degree. In many cases, the final part is designed, and the AM part is derived from it. Optimisation requires loops and potentially several releases and changes within the PLM. This is particularly true, if it is the production's task to figure out the raw part's design.
- Few companies/entities are using the full potential of modern design tools and related platforms. Typically, because the alternative ways of working seem more effortless. As a result, the traceability regarding the part identification (type) and the digital shadow of production [21] is not optimal.
- The classic setup that includes having a model and drawing(s) for a part is still the most common. Product manufacturing information (PMI) in the form of 3D annotations is only utilised in around 20% of cases. Those who do use it are also using the information in subsequent production steps. The remaining 80% are not exploiting this potential regarding digitalisation and Industry 4.0.
- Around half of the companies/entities see the proposed release process as beneficial. This ratio is slightly more pronounced by those who manufacture the AM parts in-house. It achieves 100% among those that apply PMI in their workflow. Therefore, the conclusion is that the more digitalisation and higher integration are driven, the more applicable the release process is.

In conclusion, there is still a significant gap between the application and the potential of digitalisation and modern development tools in the still niche additive manufacturing industry. However, the responses from the companies that already exploit said potential show the benefits of a high degree of digitalisation as well and well implemented release processes. Those responses also confirm the benefits of the proposed design and release process.

As we aim to achieve a high level of digitalisation in the industry, the teaching and integration of the proposed release process will continue. The path taken with this PLM approach should be maintained, as it is vital for maintaining traceability, building the base for a digital shadow in production, and developing additive manufacturing and Industry 4.0.

References

1. Ritter, S.: AM Field Gide Compact. Frankfurt am Main. Mesago Messe Frankfurt GmbH (2022). https://formnext.mesago.com/frankfurt/de/themen-events/am-field-guide/downloads.html
2. Bender, B., Göhlich, D. (eds.): Dubbel Taschenbuch für den Maschinenbau 2: Anwendungen. Springer, Heidelberg (2020). https://doi.org/10.1007/978-3-662-59713-2
3. Introduction to Mechanical Engineering. https://link.springer.com/book/https://doi.org/10.1007/978-3-319-78488-5. Accessed 18 Jan 2023
4. Advances on Mechanics, Design Engineering and Manufacturing. https://link.springer.com/book/https://doi.org/10.1007/978-3-319-45781-9. Accessed 18 Jan 2023
5. Lachmayer, R., Lippert, R.B., Kaierle, S.: Konstruktion für die Additive Fertigung 2018. Springer, Berlin/Heidelberg (2019). https://doi.org/10.1007/978-3-662-59058-4
6. Thompson, M.K., et al.: Design for Additive Manufacturing: trends, opportunities, considerations, and constraints. CIRP Ann. **65**(2), 737–760 (2016). https://doi.org/10.1016/j.cirp.2016.05.004
7. Xiong, Y., Tang, Y., Zhou, Q., Ma, Y., Rosen, D.W.: Intelligent additive manufacturing and design: state of the art and future perspectives. Addit. Manuf. **59**, 103139 (2022). https://doi.org/10.1016/j.addma.2022.103139
8. Das, P., Chandran, R., Samant, R., Anand, S.: Optimum part build orientation in additive manufacturing for minimizing part errors and support structures. Procedia Manuf. **1**, 343–354 (2015). https://doi.org/10.1016/j.promfg.2015.09.041
9. Goguelin, S., Dhokia, V., Flynn, J.M.: Bayesian optimisation of part orientation in additive manufacturing. Int. J. Comput. Integr. Manuf. **34**(12), 1263–1284 (2021). https://doi.org/10.1080/0951192X.2021.1972466
10. Meier, H., Uhlmann, E. (eds.): Industrielle Produkt-Service Systeme. Springer Berlin Heidelberg, Berlin, Heidelberg (2017). https://doi.org/10.1007/978-3-662-48018-2
11. Daase, C., et al.: Following the digital thread – a cloud-based observation. Procedia Comput. Sci. **217**, 1867–1876 (2023). https://doi.org/10.1016/j.procs.2022.12.387
12. Kitsios, V., Haslauer, R.: 3D-Master: Zeichnungslose Produktbeschreibung mit CATIA V5. Springer Fachmedien Wiesbaden, Wiesbaden (2014). https://doi.org/10.1007/978-3-658-05845-6
13. Robin, V., Rose, B., Girard, P., Lombard, M.: Management of engineering design process in collaborative situation. In: ElMaraghy, H.A., ElMaraghy, W.H. (eds.) Advances in design, pp. 257–267. Springer-Verlag, London (2006). https://doi.org/10.1007/1-84628-210-1_21
14. Pfalzgraf, P., Bopoungo, A.P., Trautmann, T.: Cross enterprise change and release processes based on 3D PDF. In: Stjepandić, J., Rock, G., Bil, C. (eds.) Concurrent engineering approaches for sustainable product development in a multi-disciplinary environment: proceedings of the 19th ispe international conference on concurrent engineering, pp. 753–763. Springer London, London (2013). https://doi.org/10.1007/978-1-4471-4426-7_64
15. Schmid, D., Nyffenegger, F.: Metro Map Illustrating the Digitalisation in Industry, PLM Transit. Times Place Human Transformation Technology Organ 19th IFIP WG 51 International Conference PLM 2022 July 2022 Grenoble Fr. (2022). https://doi.org/10.1007/978-3-031-25182-5_29
16. Bougdira, A., Akharraz, I., Ahaitouf, A.: A traceability proposal for industry 4.0. J. Ambient. Intell. Humaniz. Comput. **11**(8), 3355–3369 (2020). https://doi.org/10.1007/s12652-019-01532-7
17. Ustundag, A.: Industry 4.0: managing the digital transformation, 1st ed. 2018. In: Springer series in advanced manufacturing. Springer, Cham (2018). https://doi.org/10.1007/978-3-319-57870-5

18. Voigt, K.-I., Müller, J.: Digital business models in industrial ecosystems: lessons learned from Industry 4.0 across Europe. In: Future of business and finance. Springer, Cham, Switzerland (2021)

19. Aosel, A., Yamakawa, K.: Traceability 4.0: The fundamental element of global manufacturing (2020). https://www.precicon.com.sg/wp-content/uploads/2021/04/A-Better-Way-to-Understand-Traceability-in-Smart-Manufacturing.pdf#page=1&zoom=auto,-205,798

20. Rasheed, A., San, O., Kvamsdal, T.: Digital Twin: Values, Challenges and Enablers (2019)

21. Bauernhansl, T., Hartleif, S., Felix, T.: The Digital Shadow of production – a concept for the effective and efficient information supply in dynamic industrial environments. Procedia CIRP 72, 69–74 (2018). https://doi.org/10.1016/j.procir.2018.03.188

22. Lachmayer, R., Lippert, R.B., Fahlbusch, T. (eds.): 3D-Druck beleuchtet: Additive Manufacturing auf dem Weg in die Anwendung. Springer, Heidelberg (2016). https://doi.org/10.1007/978-3-662-49056-3

23. Lachmayer, R., Ehlers, T., Lippert, R.B: Entwicklungsmethodik für die Additive Fertigung. Springer, Berlin, Heidelberg (2022). https://doi.org/10.1007/978-3-662-65924-3

24. Stekolschik, A.: Engineering change management method framework in mechanical engineering. IOP Conf. Ser. Mater. Sci. Eng. 157(1), 012008 (2016). https://doi.org/10.1088/1757-899X/157/1/012008

25. Shakirov, E.: Integrated Engineering and Manufacturing Change Management in the Additive Manufacturing Context, p. 224

26. CAS Additive Fertigung. ZHAW School of Engineering. https://www.zhaw.ch/de/engineering/weiterbildung/detail/kurs/cas-additive-fertigung/. Accessed 24 Jan 2023

27. Innovation Booster Additive Manufacturing – The network to support innovation. https://ibam.swiss/. Accessed 24 Jan 2023

28. Swiss Additive Manufacturing Group. Swissmem. https://www.swissmem.ch/de/produkte-dienstleistungen/netzwerke/industriesektoren/swiss-additive-manufacturing-group.html. Accessed 24 Jan 2023

29. Common types of survey bias and how to avoid them. SurveyMonkey. https://www.surveymonkey.com/mp/how-to-avoid-common-types-survey-bias/. Accessed 23 Jan 2023

30. Survey Bias: Common Types of Bias and How to Avoid Them. Qualtrics. https://www.qualtrics.com/uk/experience-management/research/survey-bias/. Accessed 23 Jan 2023

31. Tourangeau, R.: The science of web surveys. Oxford University Press, New York (2013)

32. Total Survey Error in Practice, 1st ed. Wiley (2017). https://doi.org/10.1002/9781119041702

33. quantilope, '6 Types of Survey Biases and How To Avoid Them'. https://www.quantilope.com/resources/glossary-six-types-of-survey-biases-and-how-to-avoid. Accessed 23 Jan 2023

Investigation on Additive Manufacturing Processes Performed by Collaborative Robot

Khurshid Aliev[1,2] (ID), Mansur Asranov[1,2] (ID), Tianhao Liu[3] (ID),
and Paolo Chiabert[1,2(✉)] (ID)

[1] Politecnico di Torino, Corso Duca degli Abruzzi 24, 10129 Torino, Italy
{khurshid.aliev,mansur.asranov,paolo.chiabert}@polito.it
[2] Turin Polytechnic University in Tashkent, Kichik Halqa Yuli 17, Tashkent, Uzbekistan 100095
[3] Simpro S.P.A., via Torino 454, 10032 Brandizzo (TO), Italy
tianhao.liu@simpronet.com

Abstract. The additive manufacturing (AM) applications using collaborative robots (cobot) are rapidly increasing in the manufacturing field. The integration of AM with a cobot abilities can help prototyping and manufacturing custom-made parts in a more efficient way. This paper relies on manufacturing cell that combines a fused deposition modeling (FDM) extruder with a 6-axis cobot controlled by IoT edge computing devices. The production processes are designed in a robot simulation software, where digital twin (DT) of the manufacturing cell is available. Direct and reverse communication between the simulation software and the physical manufacturing cell allows for implementing the real industrial cases. The manufacturing cell has been tested to demonstrate the viability of replacing traditional 3D printers in the industrial sector while taking advantage of working in a complex and dynamic environment. According to this approach this paper promotes the enlargement of the set of robot-abilities by adding additive manufacturing capabilities.

Keywords: Additive manufacturing · Cobot 3D printing · Industry 4.0/5.0 · Smart manufacturing

1 Introduction

In recent years Additive Manufacturing (AM) technology has been applied in different fields, including automotive, aerospace, food, bioengineering, architecture and manufacturing [1].

AM is defined as the process of joining materials to make parts from three-dimensional (3D) model data, whereas 3D printing is a technique that builds 3D objects layer by layer from a 3D digital model (either by computer-aided design or by scanning the object) [2]. A print head, extruder, nozzle, or other printer technologies are used in the 3D printing process to build components by deposing material.

Fused Deposition Modeling (FDM) is the most widely used 3D printing technique. FDM uses a heating chamber to liquefy polymer that is fed into the system as a filament.

C. Danjou et al. (Eds.): PLM 2023, IFIP AICT 702, pp. 323–332, 2024.
https://doi.org/10.1007/978-3-031-62582-4_29

The filament is pushed into the chamber by a tractor wheel that generates the extrusion pressure [3]. Typical FDM equipment has three degrees of freedom to describe the shape of the object by moving deposition head.

This approach is easy to implement due to the declining prices of FDM machines, which have become affordable for individuals, but it has some drawbacks in introducing manufacturing constraints, such as production speed, material options, material density and accuracy. Additionally, it can result in a stairwell effect on the surface. The FDM manufacturing process deposits material layer by layer resulting in significant product anisotropy. Due to layering, the sloping surface of printed parts will suffer from the staircase effect, which affects surface quality and leads to stress concentration. These issues weaken the FDM product's strength and limit its application scenarios, prompting the researchers to conduct extensive exploratory work.

Multi-axis robot-manipulated manufacturing methods, which are widely used for assembly, welding and handling or pick and place tasks, provide high quality and consistency, maximum productivity, safety and accuracy for repetitive task, and low labor cost [4, 5]. Cobots' adaptable and flexible functionalities fulfill the dynamic demands of manufacturing. The use of multi-axis robot systems in combination with additive manufacturing technologies enables multi-axis additive manufacturing and the fabrication of complex geometries in a variety of industrial environment.

Nevertheless, the path planning of multi-axis 3D printing end-effector is more complex and less developed task [6]. The majority of multi-axis printing researches aim to reduce or eliminate support. There are few research studies on 3D toolpath planning and fabrication strategy for printing parts with complex geometry and mechanical properties. The deposition direction of multi-directional printing differs from traditional 3D printing, which deposits material in a series of parallel planes, therefore the mechanical properties of the multi-axis printed parts requires deeper analyzes. Moreover, there are no general control languages available for multi-axis printing. Finally, different platforms typically use different scripting languages to control the hardware, making software development more difficult.

This paper proposes a platform integration for multi-axis cobot assisted additive manufacturing with FDM. It describes the manufacturing process where a CAD model, embedded into Cyber-Physical System, drives 3D multi-axis printer, based on open source architecture, which produces the physical object. The rest of the paper reviews related works of existing multi-axis 3D printing systems and demonstrates various case studies of 3D printed objects using robotic-assisted additive manufacturing before presenting and discussing results.

2   Related Works

The use of robots to perform 3D printing is a research question currently under development.

Robotic manipulators have been used to print complex 3D geometries without the use of supporting materials, and many research efforts are being directed toward the development of a robotic arm assisted 3D printing platform [7].

Yoa et al. propose a framework for 6-DOF robot arm based 3D printing and continuous toolpath planning method to improve the strength and surface quality [8]. The continuous toolpath planning method enables full use of robot arm-based additive manufacturing, achieving smooth printing on surfaces with high curvature while avoiding the staircase effect and collision in the process. 3D printing path planning has been investigated to improve the structural rigidity of manufactured parts [9].

Other authors proposed a 3D printing simulator based on an off-line robot programming system. This method usually requires data transfer to the real robotic system as well as adjustments to various parameters related to both robot operation and the AM process itself [10].

Ščetinec et al. recommended on-line layer height control and in-process toolpath replanning to improve geometric accuracy for tall shell parts [11].

Another research proposes an integrated framework for collaborative robot-assisted AM using fused deposition modeling (FDM) as an AM sub-process. This is a generic platform with position and orientation control that can be easily integrated (hardware and software) in any robot or multi-axis machine [12].

The integration of collaborative robot and AM systems present some challenges: in terms of technology integration, the definition of interconnected robot and AM parameters needs improvement, especially regarding availability of paths that are collision-free and reduce the stair scale effect on the complex surfaces. Like CNC machines, the ordinary 3D printer follows G-code instructions for positioning, but G-code does not specify how the robot should move to avoid collisions. Simulation are necessary to address this issue. Although there are a few route generation software options for multi-axis robot AM available in the market, they cannot fully unleash their potential. Additionally, the real time control of the collaborative robots introduces relevant difficulties in managing the whole system.

3   Development of the Additive Manufacturing Robotic Cell

The AM robotic cell was developed in the Mind4Lab laboratory of Politecnico di Torino to demonstrate the innovative robot-abilities in additive manufacturing.

The proposed additive manufacturing workstation includes a collaborative robot UR10e from Universal Robots, an On-Robot RG6 flexible dual finger robot gripper with a wide stroke (160 mm) and a gripping force range from 25 N to 120 N, a NEMA 17 stepper motor to drive the filament extruder, an Arduino board (Uno R3) to control and communicate with the robot controller, a heated nozzle, a heated bed to perform printing and RoboDK software to design the Digital Twin (DT) of the AM workstation.

The UR10e, a 6-axis cobot with force accuracy of 5.5 N in the axes (x,y,z) of the tool flange and position repeatability of ±0.05 mm, has a maximum payload capacity of 12.5 kg and a spherical workspace. The cobot working distance is 1300 mm radius and 300 mm radius around the robot base (Fig. 1) with 0.05 mm pose repeatability, making it more suitable for complex AM applications while maintaining machining accuracy.

The 3D printing device is composed of a NEMA 17 stepper motor that drives a filament extruder and a heated nozzle with diameter 0.4 mm taken from a commercial 3D printer, that performs a fused deposition modelling (FDM) process.

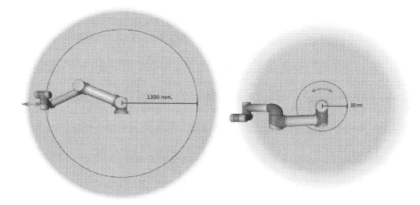

Fig. 1. UR10e working area.

All the components are fastened to the bracket and the On-Robot RG6 dual finger gripper fixes the structure to the cobot flange through its quick exchange connector resulting in a payload of 1.850 kg for the end-effector prototype. Figure 2 depicts the 3D printing end-effector prototype integrated on the UR10e cobot.

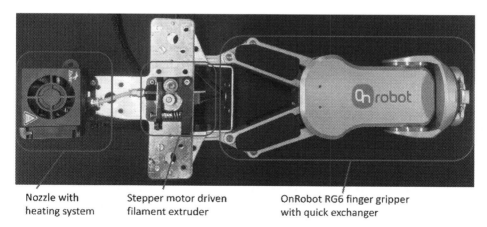

Nozzle with   Stepper motor driven   OnRobot RG6 finger gripper
heating system   filament extruder   with quick exchanger

Fig. 2. A prototype of the end effector.

Digital Twin, IoT communication and hardware control system of the AM cell is logically divided into several functions as shown in Fig. 3. The teach pendant and UR10e control box are used to receive the printing program and control the movements of the cobot, while sending ON/OFF signals to the Arduino board, synchronizing the extrusion process with the cobot motion. The nozzle temperature control is a separate system operated by the motherboard from the commercial 3D printer.

In the RoboDK software, a virtual AM cell, including the cobot, its accessories and the cell structure, is created by arranging all necessary Computer Aided Design (CAD) models of the components. Then, the CAD model of the ready-to-produce object is imported and placed in the desired position.

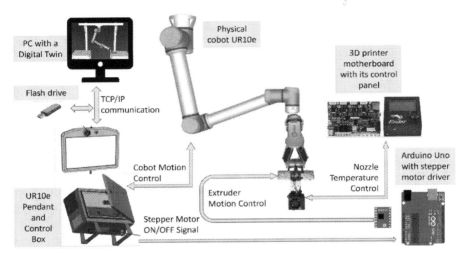

Fig. 3. Robot based AM work cell, IoT communication and control system.

Figure 4 summarizes the complex activities for AM production. RoboDK software allows for creating and simulating the tasks that contain the manufacturing path.

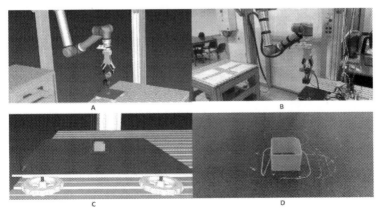

Fig. 4. A) CAD model of the AM cobot cell; B) Physical model of the cell; C) 3D printing of the simulated environment; D) 3D printing of the workpiece.

The RoboDK tasks are converted to robot programs through a dedicated postprocessor and, finally, the robot programs are executed on the integrated HW platform.

Figure 4A and 4C show virtual environment and simulation of AM cell, while Fig. 4B with 4D illustrate physical environment of the cell and 3D printing of the workpiece.

4 Case Studies

Three different objects have been designed in CAD software and printed in robot additive manufacturing cell. The Slic3r application available in RoboDK software converts 3D models into printing instructions. The objects are sliced into multiple layers 0.35 mm thick parallel to the predefined reference coordinate system. The tool path is defined and optimized by RoboDK to prevent collisions during printing. To minimize excess vibration, the cobot printing speed is set to 10 mm/s.

Case A: The task evaluates the multi-planar printing capabilities of the AM work cell. A tube-shaped object is divided into four parts, each with a unique reference coordinate system. The first part is created on the hotbed using conventional methods. The subsequent parts are then created on the surface of the previous part, with the printing reference system being elevated 15 mm and inclined −20° along the Y-axis for each part. The printing process is shown in Fig. 5, where CAD model of the workpiece, 3D printing simulation and CAD model of the workstation are demonstrated.

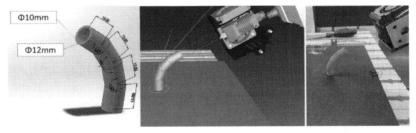

Fig. 5. Case A: On the left, CAD model of the workpiece and its dimensions, in the middle CAD model of the workstation, on the right side the printed tube.

Case B: A box with a hole is created to evaluate the performance and accuracy of the system when producing a simple and basic shape. The dimension of the box is 20 × 15 × 20 mm, and the diameter of the through hole is 10 mm. Figure 6 shows the original CAD model, the CAD model in the virtual environment and the physical environment during printing.

Case C: A thin wall shape object is used to analyze the adhesion of each single layer and the vertical printing quality. The inspection of the object allows for analyzing the material flow and evaluating its uniformity. Figure 7 shows the thin wall shape printed object: the CAD model of the object, and its workstation in digital and physical realization.

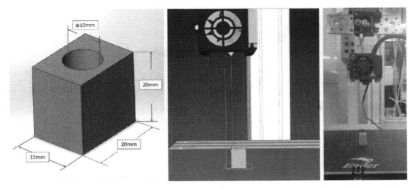

Fig. 6. Case B: CAD model of the workpiece with dimensions (left), CAD model of the workstation (middle), and printed box with a hole (right).

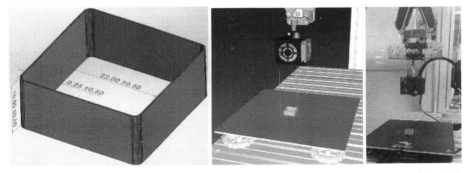

Fig. 7. Case C: CAD model of the workpiece with dimensions (left), CAD model of the workstation (middle), and real system of printed object and devices (right).

5   Results and Discussions

The printed objects in Cases A, B and C, were tested on their simulation environment and measured using caliper. The simulation environment allows to simulate a workstation and to evaluate the system in real-time through online programming. The setup of the simulation environment of the workstation supports the online and offline programming.

Figure 8 shows the simulation results and real printed objects from the three case studies.

Figure 9 summarizes the measurement results for each of the three scenarios. The measured data highlight an observable difference between nominal and physical dimensions.

According to the results, the performances of the multi-axis 3D printing system are comparable with performances of an ordinary 3D printing machine using FDM technology.

The proposed system has several limitations that require investigation to improve important characteristics of the system itself.

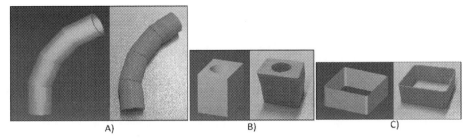

Fig. 8. A) Tube 3D model (left) and Cobot printed Tube (right); B) Holed Cube 3D model (left) Cobot printed cube (right); C) Squared thin wall 3D (left) and Cobot printed workpiece.

COMPARISON OF THE CAD AND PRINTED MODELS

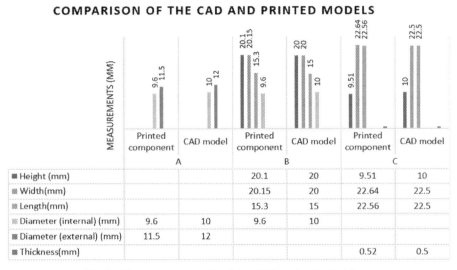

	Printed component	CAD model	Printed component	CAD model	Printed component	CAD model
	A		B		C	
▦ Height (mm)			20.1	20	9.51	10
▦ Width(mm)			20.15	20	22.64	22.5
▦ Length(mm)			15.3	15	22.56	22.5
▦ Diameter (internal) (mm)	9.6	10	9.6	10		
▦ Diameter (external) (mm)	11.5	12				
▦ Thickness(mm)					0.52	0.5

Fig. 9. Measurement results of the CAD and printed objects.

Robot Speed: The printing speed is set to 10 mm/s to achieve high positioning accuracy, but this results in a slower printing process, about 4–6 times longer than conventional 3D printing process whose maximum speed is 60 mm/s when PLA (Polylactic Acid Resin) is used.

Quality of the Printed Objects: Despite the low printing speed, we observed noticeable oscillation and layer shifting in the finished products. This is believed to be caused by the length of the end effector, which causes the nozzle to shake when the cobot arm comes to a sudden stop.

Additionally, the weight of the extrusion system may contribute to layer shifting, especially when the extrusion system is working in an inclined position. Redesign of bracket supporting the extruding system could increase the structural robustness and reduce the nozzle vibration.

Complex Geometry Structures: Multi-planar printing has the potential to print more complex objects than conventional methods, but the complexity of the generated structures is currently limited by the building material and tool path generation software.

The AM work cell has the potential to build structures with a continuously varying coordinate system, but further research is needed to fully realize this potential and make it easier to use.

Cost of the System: Replacing a conventional AM system with a cobot arm is economically viable only if the quality or complexity of the generated objects reaches a level that cannot be achieved by existing solutions on the market.

Despite these limitations, the proposed system demonstrated its potential for AM in increasing its flexibility within the other processes of manufacturing. The proposed system is capable of generating products on any plane, whether parallel to the ground or fixed.

Additionally, with the use of replaceable tools, AM can be integrated with other processing techniques and applied to a wider range of scenarios, resulting in the production of more complex structures with improved quality.

Overall, the system showcases that AM can be no longer considered as a fixed processing step but as a flexible technology able to cooperate with other ones.

6 Conclusions

This paper demonstrates the use of cobots to manage AM processes and proposes a physical tool with a digital setup to create a cyber physical system of printing process. Cobot's end-effector for additive manufacturing demonstrated its effectiveness in producing quality parts of various dimensions and geometries.

A cobot-based AM workstation integrates both software and hardware applications and assists the manufacturing process from the CAD model to the final part production. Design and simulation of AM properties in the CAD model of the workstation allows for controlling the cobot end-effector orientation, thus improving the production of higher quality parts with no stair-like structure on their finished surface.

Furthermore, the system improves the appearance, mechanical properties, and load support in specific directions.

Finally, the proposed solution replaces a 3D printing machine with a printing robot and opens up the possibility of new solutions that take advantage of the flexibility of the robot. On the other side the proposed solution requires further investigations in order to improve its reliability, manageability and productivity.

References

1. Ngo, T.D., Kashani, A., Imbalzano, G., Nguyen, K.T.Q., Hui, D.: Additive manufacturing (3D printing): A review of materials, methods, applications and challenges. Composit. Part B: Eng. **143**, 172–196 (2018)
2. International Organization for Standardization. ISO/ASTM 52900:2021. Additive manufacturing-general principles-fundamentals and vocabulary

3. Gibson, I., Rosen, D., Stucker, B.: Extrusion-based systems. In: Gibson, I., Rosen, D., Stucker, B. (eds.) Additive manufacturing technologies: 3d printing, rapid prototyping, and direct digital manufacturing, pp. 147–173. Springer New York, New York, NY (2015). https://doi.org/10.1007/978-1-4939-2113-3_6

4. Chiabert, P., Aliev, K.: Management of laser-cut sheet-metal part using collaborative robots. In: Junior, O.C., Noël, F., Rivest, L., Bouras, A. (eds.) Product lifecycle management. green and blue technologies to support smart and sustainable organizations: 18th IFIP WG 5.1 international conference, PLM 2021, Curitiba, Brazil, July 11–14, 2021, Revised Selected Papers, Part I, pp. 59–73. Springer International Publishing, Cham (2022). https://doi.org/10.1007/978-3-030-94335-6_5

5. Chiabert, P., Aliev, K.: Analyses and study of human operator monotonous tasks in small enterprises in the era of industry 4.0. In: Nyffenegger, F., Ríos, J., Rivest, L., Bouras, A. (eds.) Product lifecycle management enabling smart X: 17th IFIP WG 5.1 international conference, PLM 2020, Rapperswil, Switzerland, July 5–8, 2020, Revised Selected Papers, pp. 83–97. Springer International Publishing, Cham (2020). https://doi.org/10.1007/978-3-030-628 07-9_8

6. Wu, L., Yu, M., Gao, Y., Yan, D.M., Liu, L.: Multi-DOF 3D printing with visual surveillance. In SIGGRAPH Asia 2017 Posters, pp. 1–2 (2017)

7. Bhatt, P.M., Malhan, R.K., Shembekar, A.V., Yoon, Y.J., Gupta, S.K.: Expanding capabilities of additive manufacturing through use of robotics technologies: a survey. Addit. Manuf. **31**, 100933 (2020)

8. Yao, Y., Zhang, Y., Aburaia, M., Lackner, M.: 3D printing of objects with continuous spatial paths by a multi-axis robotic FFF platform. Appl. Sci. **11**(11), 4825 (2021)

9. Eder, S., Kwok, T.-H., Chen, Y.: Function-aware slicing using principal stress line for toolpath planning in additive manufacturing. J. Manufac. Process. **64**, 1420–1433 (2021)

10. Zhang, G.Q., Spaak, A., Martinez, C., Lasko, D.T., Zhang, B., Fuhlbrigge, T.A.: Robotic additive manufacturing process simulation-towards design and analysis with building parameter in consideration. In: 2016 IEEE international conference on automation science and engineering (CASE), pp. 609–613. IEEE (2016)

11. Ščetinec, A., Klobčar, D., Bračun, D.: In-process path replanning and online layer height control through deposition arc current for gas metal arc based additive manufacturing. J. Manuf. Process. **64**, 1169–1179 (2021)

12. Mohammad, S., Bearee, R., Neto, P.:An integrated framework for collaborative robot-assisted additive manufacturing. J. Manufac. Process. **81**, 406–413 (2022)

Optimization Framework for Assembly Line Design Problem with Ergonomics Consideration in Fuzzy Environment

Elham Ghorbani[✉], Samira Keivanpour, Firdaous Sekkay, and Daniel Imbeau

Department of Mathematics and Industrial Engineering, Polytechnique Montreal, Montreal, Canada
elham.ghorbani-totkaleh@polymtl.ca

Abstract. This paper presents a framework for solving assembly line design problems by considering ergonomics aspects. Although ergonomics factors have been ignored in conventional optimization problems in this area, in the long term, ergonomics risks and work-related injuries can impose considerable expenses on production systems. Moreover, in the design stage, different types of uncertainty in operational and ergonomics aspects can affect the optimization model. Therefore, the optimization framework in this study includes the results of ergonomics assessment tools and employs fuzzy logic to tackle imprecise factors. In the context of our problem, the sources of imprecision are twofold: environmental uncertainty and system uncertainty. Environmental uncertainty is related to demand uncertainty derived from market variations and customers' behavior. System uncertainty includes the uncertainties within the production process that partially relate to human aspects, such as uncertainty in task execution time and the physical capacity of the operators.

Keywords: Assembly Line Balancing Problem · Ergonomics · Fuzzy Set Theory

1 Introduction

The main goal of planning assembly lines (ALs) is to increase productivity and efficiency. This is achieved through optimization problems called assembly line balancing problems (ALBP) [1]. However, market fluctuations and changing customer needs introduce uncertainty and require flexible production systems. Manual tasks in ALs contribute to system flexibility but also pose risks to operators' ergonomics and line efficiency. Therefore, Ergo-ALBP considers both operational and ergonomics factors in optimizing the AL system. Manual tasks also introduce variability due to workers' physical characteristics, gender, age, experience, and skills, impacting the optimization model [2–4].

There are two critical research gaps in Ergo-ALBP literature. Firstly, traditional optimization models have overlooked human factors and ergonomics (HFE) in favor of operational factors like cost and time. Additionally, most studies have considered

© IFIP International Federation for Information Processing 2024
Published by Springer Nature Switzerland AG 2024
C. Danjou et al. (Eds.): PLM 2023, IFIP AICT 702, pp. 333–343, 2024.
https://doi.org/10.1007/978-3-031-62582-4_30

ergonomics aspects in existing ALs rather than in the design stage. However, considering ergonomics parameters in the design stage (Ergo-ALDP) is strategic planning that can prevent future costs for redesigning and taking corrective actions to solve ergonomics problems [5]. Secondly, vague and imprecise factors, such as market uncertainty and varying worker characteristics, need to be included in the optimization problem. This includes accounting for variable takt time, inconsistent ergonomics risk levels, and task times that depend on individual workers. Notably, no previous Ergo-ALBP studies have addressed both uncertain conditions and HFE aspects during the design phase of ALs.

This study contributes by proposing a framework to optimize Ergo-ALDPs in uncertain conditions. Fuzzy logic is used to handle imprecise parameters and variables. The framework is applicable to any type of AL in the design step and beneficial for engineers and ergonomics practitioners in production systems.

This manuscript is organized as follows: In Sect. 2, a brief introduction to Ergo-ALBP is provided. Section 3 and Sect. 4 present the optimization model and solution approach, respectively. In Sect. 5, the practical perspective of the proposed framework is discussed. Finally, Sect. 6 presents the concluding remarks.

2 Background

In the early 1900s, Ford's car manufacturing plants were a good example of assembly lines used in the context of mass production. Since then, this crucial element of mass and lean production systems has significantly evolved and transformed into a more agile system. Assembly lines are the final stage of most production systems and are the closest part to customers. Thus, optimizing them involves balancing them and eliminating any issues that prevent them from working smoothly. In the following subsections, the main aspects of these optimization problems are explained.

2.1 Assembly Line Balancing Problem (ALBP)

Optimization models ALs aim to eliminate unbalanced points, such as bottlenecks, that decrease efficiency and worsen Key Performance Indicators (KPIs). Balancing ALs involves optimizing them with respect to productivity and efficiency goals [1]. The formulation of ALBPs as linear programming (LP) models dates back to 1955 [6]. While a solution approach was introduced in 1961 [7], trial-and-error techniques have been the primary solving method for several decades. ALBPs are NP-hard combinatorial optimization problems (COPs) that require finding an optimal solution from a finite set of feasible solutions (FSs). ALBPs are categorized into simple (SALBP) and general (GALBP) problems [8]. SALBPs consider one-sided straight ALs with deterministic operation times, optimizing one or two objectives. They are classified into four types: Type 1 minimizes the number of workstations based on a given cycle time, Type 2 minimizes the cycle time based on a fixed number of workstations, Type F checks the feasibility of the problem with a fixed number of workstations and cycle time, and Type E minimizes both the cycle time and the number of workstations. GALBPs encompass more complex conditions, such as multiple product types, multiple sides, or non-straight assembly lines.

While SALBPs have been extensively studied, there is a need for more research on GALBPs to tackle sophisticated real-world problems [9]. The past decade has seen a positive trend towards considering these general problems to address complexity.

2.2   Assembly Line Balancing Problem with Ergonomics Aspect (Ergo-ALBP)

Assembly tasks pose ergonomics risks and work-related musculoskeletal disorders (WMSDs) due to their repetitive and prolonged nature. Considering ergonomics and operational factors together is crucial for preventing injuries, and Gunther et al. were the first to consider ergonomics risks in the ALBPs in 1983 [10]. Since then, there have been few contributions in this domain until 2011 when Otto and Scholl included an ergonomics objective in the optimization model [11]. Their study motivated other scholars to focus on Ergo-ALBPs.

Although many research studies have examined the balancing of different types of ALs, limited research has been conducted in Ergo-ALBPs. It is proved that neglecting ergonomics factors in the design stage can lead to health-related issues for workers in the long run, which may require corrective measures that cost 9.2 times more than preventive actions taken during the design phase [5]. However, in the case of Ergo-ALBPs, few studies have focused on design problems (Ergo-ALDPs). Baykasoğlu et al. [2] addressed a SALBP in the design phase and developed a heuristic solution to solve it. Finco et al. [12] modeled an optimization problem for designing a semi-automatic AL. They attempted to minimize the design cost and ergonomics risks by analyzing the vibration of automatic hand-held tools. In recent years, collaborative human-robot assembly line design problems (CALBPs) or (RALBPs) that integrate ergonomics aspects with assigning exoskeletons and robots have become more common. For instance, Abdous et al. [13] examined a CALDP and developed an optimization model to reduce the overall equipment cost (design cost of AL) and minimize the ergonomics risk level.

Based on the definition by the International Ergonomics Association (IEA), human factors and ergonomics (HFE) is a scientific discipline that examines interactions between humans and system components, aiming to enhance operator safety and system performance. Numerous ergonomics assessment tools (EATs), such as OCRA, REBA, RULA, OWAS, and NIOSH's RNLE, have been developed to evaluate ergonomics risk factors in workspaces. Some of these methods serve as foundations for national and international ergonomic standards, including EN1005-2, EN1005-5, ISO11228-1, and ISO11228-3. While no method is universally superior [14], EATs categorize ergonomics risk levels into ranges from low to high [3].

2.3   Uncertainties

Limited research has incorporated ergonomics aspects into optimization models for ALBPs, particularly in the design stage (Ergo-ALDP). Most studies have focused on deterministic problems, overlooking the impact of uncertainty. However, in the design phase, uncertainty affects the assembly design in some way. In general, there are two types of uncertainty: environmental and system uncertainty [15]. In the context of the problem under study, environmental uncertainty includes uncertainties in demand variations resulting from market fluctuations. Moreover, system uncertainty is related to any imprecision in the manufacturing process. It partially consists of human aspects, such as uncertainty in system reliability, task time, and the physical capacity of the workers. In addition, Golabchi et al. [16] found that the inputs of EATs are often imprecise, which could significantly impact the results. To model these uncertainties, researchers employ stochastic programming models when historical data is available to identify the probability distribution of imprecise factors. Otherwise, fuzzy programming is helpful.

To the best of the authors' knowledge, only Tiacci and Mimmi [17] have included uncertainty in their Ergo-ALBP model. They incorporated stochastic task times and introduced penalties for cases where ergonomics constraints and/or predicted cycle times were not addressed. To evaluate ergonomics parameters, they utilized OCRA, and for cost minimization, they employed a genetic algorithm (GA).

2.4 Fuzzy Approaches in Balancing Problems

Ergonomics aspects and operational parameters in optimization problems of ALs represent conflicting objectives. To overcome the vagueness in multi-objective models with ergonomics and operational functions, some articles such as Cheshmehgaz et al. [18] and Ozdemir et al. [4] employed fuzzy goal programming. Although considerable research studies in ALBPs have employed fuzzy set theory (FST) to address uncertain and imprecise conditions in ALs, only Mutlu and Özgörmüş [19] considered ergonomics risks as fuzzy numbers among Ergo-ALBPs literature. They applied Bellman and Zadeh's approach [20] to minimize the number of workstations and perceived workload.

3 Problem Context

As mentioned in previous sections, in the design stage of Ergo-ALBP, two types of uncertainty must be addressed in the optimization model. Since takt time, derived from the demand rate, is not deterministic in the design phase, cycle time is imprecise. Task execution times also vary based on the worker's skill and experience level. Furthermore, ergonomics risk factors are vague because, in the planning step, the works situations are not precisely determined nor who will perform the task in each workstation. Various characteristics of workstations (e.g., the force required to use tools or lift parts to be assembled, types of tools used, physical dimensions of workstation components, repetition and frequency of sub-tasks, thermal environment) and of operators (e.g., gender, age, skill level, experience, and physical and work capacity, prior training), can impact the ergonomics risk level of each task. Therefore, this section defines the optimization problem by taking some steps and developing the model from an initial mathematical problem, SALBP-Type1, to the final state by identifying fuzzy time and ergonomics parameters.

3.1 Initial Model

In the design stage, the optimum number of workstations should be identified, thus the mathematical problem is the same as SALBP-Type1. The notations of the initial optimization model are as follows:

Sets & Indexes:
N Set of tasks (i,j = Indices for tasks: i,j $\in \{1, ..., n\}$)
S Set of workstations (s = Index for workstations: s $\in \{1, ..., m_{max}\}$
P_i Set of immediate predecessors of task i
Parameters:
n = number of tasks m = number of workstations
t_i = execution time of task CT = cycle time
Decision variables:
x_{si} = Binary variable, if task i is assigned to station s, it will be equal to 1 otherwise 0

y_s = Binary variable, if at least one task is assigned to station s, it will be equal to 1 otherwise 0

Then the initial mathematical model for the proposed ALDP is the same as SALBP-Type1:

$$\text{Min} \sum_{s=1}^{m_{max}} y_s \tag{1}$$

$$\text{Subject to :} \sum_{s \in S} x_{si} = 1, \forall i \in N \tag{2}$$

$$\sum_{s \in S} s.x_{si} \leq \sum_{s \in S} s.x_{sj}, \forall i, j \in N | i \in P_j \tag{3}$$

$$\sum_{i \in N} x_{si}.t_i \leq CT, \forall s \in S \tag{4}$$

$$\sum_{i \in N} x_{si} \leq M.y_s, \forall s \in S \& \sum_{i \in N} t_i < M \tag{5}$$

$$x_{si}, y_s \in \{0, 1\}, \forall s \in S; \forall i \in N \tag{6}$$

In Eq. (1), the objective function minimizes the number of workstations. The constraint in Eq. (2) guarantees that every task i is assigned to a single workstation. Equation (3) checks that the assigned stations satisfy precedence relations between tasks i and j. Equation (4) ensures that the total stations' time cannot exceed the cycle time. Equation (5) guarantees the utilization of a workstation when any task is assigned to it. The last equation indicates that decision variables x_{si} and y_s are binary variables.

3.2 Fuzzy Ergo-ALDP

As mentioned before, in the design step, time-related parameters are imprecise. Cycle time depends on takt time (available time divided by demand) and the execution time of tasks. Thus, according to inconstant demand and variable task times, cycle time is imprecise and varies to some extent (α). As a result, fuzzy logic can help us to define this parameter's variability. Figure 1(a) illustrates the membership function of cycle time (CT) by considering α as an acceptable increase in CT which imposes some overtime in the production system, and it should be minimized.

Based on the output of most EATs and considering the imprecise input of these tools which brings uncertainty to our model, fuzzy sets can define the results in the best way. Figure 1(b) depicts the membership function for a typical EAT. In this function, if the result of EAT is between ER_L and ER_U, it is interpreted as a moderate risk. While the results lower than ER_L show a low-risk level, the outputs upper than ER_U entail a high ergonomics risk. Taking advantage of the research done by Cheshmehgaz et al. [18], ergonomics risks can be evaluated as accumulated factors. As a result, various EATs can be applied to assess desired ergonomics parameters in the final optimization model. Some studies employed fuzzy set theory in literature to tackle the uncertainty in assessing ergonomics parameters. For instance, Ghasemi and Mahdavi [21] developed a new REBA scoring system based on fuzzy sets and several fuzzy membership sets. Furthermore, Wang et al. [22] integrated a 3D automated posture-based ergonomics risk assessment with a specialized rule-based fuzzy inference algorithm to solve the issue derived from the imprecise nature of inputs.

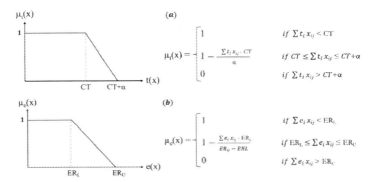

Fig. 1. The membership functions: (a) For cycle time (b) For ergonomics risk factor

4 Proposed Optimization Framework

In Ergo-ALBP literature, Mutlu and Özgörmüş [19] were the only ones to consider fuzzy ergonomics risks in their research and solve their model using the Bellman-Zadeh method. However, in their approach [20], the constraints and objectives are treated together, even though they convey different meanings. Therefore, the proposed solution procedure in this paper employs a bipolar view [23]. In this perspective, negative preferences play the role of constraints and restrict the number of FSs. In contrast, positive preferences act as the objective function(s) and evaluate FSs to find the best one. Figure 2 shows the procedure of the proposed heuristic method for solving the fuzzy Ergo-ALDP. This heuristic approach combines the COMSOAL (Computer Method of Sequencing Operations for Assembly Lines), Fuzzy goal programming, and a fuzzy inference system.

The first step involves identifying FSs based on time and precedence constraints, using the mathematical model developed in the previous section. The task assignment

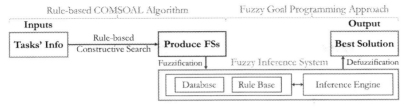

Fig. 2. Schematic of the proposed heuristic solution approach

rules for this step are consistent with those proposed by Baykasoglu et al. [2], and the pseudo-code for the first part is presented in Fig. 3. It is worth noting that FSs can be generated using different CTs by varying the value of α. Moreover, tasks' execution times vary in a range of $[t_i, t'_i]$, t_i is the average time of executing the task and t'_i is the maximum duration for doing it by the lowest skill operator.

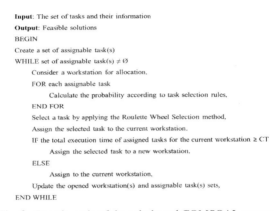

Fig. 3. Pseudo code of the rule-based COMSOAL approach

The solution method's second part involves applying a fuzzy inference system, as illustrated in Fig. 2. In this stage, ergonomics considerations are expressed as rules, and FSs are evaluated based on these rules, including the membership function defined in Sect. 3.2.

5 Application Perspective

The fuzzy optimization framework is advantageous for real-world scenarios due to its ability to handle complex and uncertain information that is challenging to quantify precisely. It is particularly useful in ergonomics risk prediction during the design phase, where data may be incomplete or uncertain. The proposed framework can incorporate multiple objectives and constraints, enabling a more balanced approach to decision-making and comprehensive analysis of the optimized system. This framework is suitable for various industries, including the automotive sector, which is prominent in Ergo-ALBP literature. Incorporating realistic uncertain conditions in planning and designing production systems is a challenge. Thus, addressing uncertainty in optimization problems is expected to become more prevalent in the future to identify robust solutions.

As mentioned before, the proposed optimization approach consists of two steps. In the first step, FSs are identified based on technical parameters and constraints, making it useful for different configurations of assembly lines (e.g., 2-sided, U-shape lines). The second step involves considering ergonomics aspects by developing fuzzy rules to evaluate FSs and determine the best among them. Various ergonomics fuzzy rules can be applied, such as the assessment methodology developed by [21], which uses fuzzy sets and REBA, or the fuzzy REBA and RULA risk rating proposed by [22].

A numerical example is provided to exhibit the relevance of the suggested mathematical model and the efficiency of the proposed solution approach. The example is generated randomly and consists of 10 tasks with the precedence diagram that is shown in Fig. 4.

Fig. 4. Precedence diagram of the example

By applying the proposed heuristic algorithm, four FSs are found with the minimum number of workstations which is four stations. For each workstation, ergonomics risk factors are assessed by the model developed in the Gallagher and Heberger study [24]. They evaluated MSD risk factors by examining the interaction of force and repetition of tasks. Table 1 indicates the results of their model in the form of fuzzy rules. By applying four fuzzy rules, the ergonomics risk level of each workstation is calculated as shown in Fig. 5. In the next step, the FSs should be evaluated to find the best solution. For this final step, we can consider the following three approaches to detect the optimum solution:

1) Highly Conservative Approach: No red area task assignment is permitted.
2) Conservative Approach: Limiting the number of moderate-risk task assignments (minimize orange area).
3) Less Conservative Approach: Limiting the number of minor-risk task assignments (minimize yellow area).

Table 1. Ergonomics assessment fuzzy rules derived from [24]

Rule No.	Rule Statement
1	**IF** Repetition <u>AND</u> Force are low, **THEN** the Risk Level is Acceptable
2	**IF** Repetition is high <u>AND</u> Force is low, **THEN** the Risk Level is minor
3	**IF** Repetition is low <u>AND</u> Force is high, **THEN** the Risk Level is moderate
4	**IF** Repetition <u>AND</u> Force are high, **THEN** the Risk Level is high

Based on the first approach, the first FS is eliminated in finding the optimum solution. The second approach removes the third FS. Finally, the optimum solution, in this case, will be the fourth FS which does not have any high-risk level and the number of its moderate-risk and minor-risk workstations is minimum in comparison with other FSs.

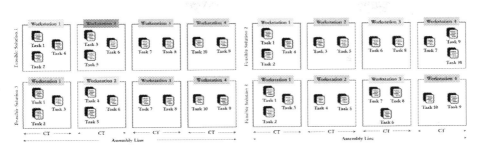

Fig. 5. Ergonomics assessment of workstations in each FS

This numerical example was developed just for explanation of our proposed optimization framework. However, it is expected that implementation of this algorithm on real case studies with proper fuzzy ergonomics rules and optimum detection approaches can find an effective solution in Ergo-ALDPs. The developed optimization framework is a versatile tool that can be customized to solve a wide range of problems under uncertain conditions, making it applicable to various domains.

6 Conclusions

Due to the importance of considering HFE in the design of manual assembly processes, as well as the vital role of ALs in manufacturing systems, this paper presents a practical procedure for optimizing the Ergo-ALDP. This study proposes a framework that integrates ergonomics factors with operational parameters to optimize the ALDP in uncertain conditions. The objective is to determine the optimal task assignments for a minimum number of workstations while adhering to time restrictions and minimizing the ergonomics risk level. To achieve this objective, the study adopts a bipolar view, where operational aspects are considered negative preferences for producing FSs, while ergonomics aspects are positive preferences for evaluating FSs and identifying the best one. The fuzzy set theory is employed through several membership functions and a fuzzy inference system that conveys various rules based on different ergonomics assessment techniques (EATs).

Based on the importance of considering inconsistent conditions in the design phase of ALs, future study directions can include stochastic optimization models for general industries with typical tasks and historical data. Furthermore, more sophisticated ALs can be considered, and more complicated optimization models can be developed to probe more realistic problems. This proposed fuzzy framework could be applied to some case studies to be verified and validated.

Acknowledgements. We would like to acknowledge our industrial partner (Dassault Systèmes) and Mitacs for funding under Mitacs Accelerate program IT28031.

References

1. Hazır, Ö., Delorme, X., Dolgui, A.: A review of cost and profit oriented line design and balancing problems and solution approaches. Annu. Rev. Control. **40**, 14–24 (2015)
2. Baykasoglu, A., Tasan, S.O., Tasan, A.S., Akyol, S.D.: Modeling and solving assembly line design problems by considering human factors with a real-life application. Hum. Factors Ergon. Manuf. Serv. Ind. **27**(2), 96–115 (2017)
3. Vig, C.: The use of ergonomic and production KPIs to evaluate workstation performance-How human variation affects the performance and how Industry 4.0 technology can improve it (Master's thesis, NTNU) (2020)
4. Ozdemir, R., et al.: Fuzzy multi-objective model for assembly line balancing with ergonomic risks consideration. Int. J. Prod. Econ. **239**, 108188 (2021)
5. Falck, A.C., Rosenqvist, M.: A model for calculation of the costs of poor assembly ergonomics (part 1). Int. J. Ind. Ergon. **44**(1), 140–147 (2014)
6. Salveson, M.E.: The assembly-line balancing problem. Trans. Am. Soc. Mech. Eng. **77**(6), 939–947 (1955)
7. Halgeson, W.B., Birnie, D.P.: Assembly line balancing using the ranked positional weighting technique. J. Ind. Eng. **12**(6), 394–398 (1961)
8. Baybars, I.: A survey of exact algorithms for the simple assembly line balancing problem. Manage. Sci. **32**(8), 909–932 (1986)
9. Becker, C., Scholl, A.: A survey on problems and methods in generalized assembly line balancing. Eur. J. Oper. Res. **168**(3), 694–715 (2006)
10. Gunther, R.E., Johnson, G.D., Peterson, R.S.: Currently practiced formulations for the assembly line balance problem. J. Oper. Manag. **3**(4), 209–221 (1983)
11. Otto, A., Scholl, A.: Incorporating ergonomic risks into assembly line balancing. Eur. J. Oper. Res. **212**(2), 277–286 (2011)
12. Finco, S., Abdous, M.A., Battini, D., Calzavara, M., Delorme, X.: Assembly line design with tools vibration. IFAC-PapersOnLine **52**(13), 247–252 (2019)
13. Abdous, M.A., Delorme, X., Battini, D.: Cobotic assembly line design problem with ergonomics. In: Boosting Collaborative Networks 4.0: 21st IFIP WG 5.5 Working Conference on Virtual Enterprises, PRO-VE 2020, Valencia, Spain, 23–25, November, 2020, Proceedings 21, pp. 573–582. Springer International Publishing, Cham (2020).https://doi.org/10.1007/978-3-030-62412-5
14. Takala, E.P., et al.: Systematic evaluation of observational methods assessing biomechanical exposures at work. Scand. J. Work Environ. Health **36**(1), 3–24 (2010)
15. Ho, C.J.: Evaluating the impact of operating environments on MRP system nervousness. Int. J. Prod. Res. **27**(7), 1115–1135 (1989)
16. Golabchi, A., Han, S., Fayek, A.R.: A fuzzy logic approach to posture-based ergonomic analysis for field observation and assessment of construction manual operations. Can. J. Civ. Eng. **43**(4), 294–303 (2016)
17. Tiacci, L., Mimmi, M.: Integrating ergonomic risks evaluation through OCRA index and balancing/sequencing decisions for mixed model stochastic asynchronous assembly lines. Omega **78**, 112–138 (2018)
18. Cheshmehgaz, H.R., Haron, H., Kazemipour, F., Desa, M.I.: Accumulated risk of body postures in assembly line balancing problem and modeling through a multi-criteria fuzzy-genetic algorithm. Comput. Ind. Eng. **63**(2), 503–512 (2012)
19. Mutlu, Ö., Özgörmüş, E.: A fuzzy assembly line balancing problem with physical workload constraints. Int. J. Prod. Res. **50**(18), 5281–5291 (2012)
20. Bellman, R.E., Zadeh, L.A.: Decision-making in a fuzzy environment. Manage. Sci. **17**(4), B-141 (1970)

21. Ghasemi, F., Mahdavi, N.: A new scoring system for the Rapid Entire Body Assessment (REBA) based on fuzzy sets and Bayesian networks. Int. J. Ind. Ergon. **80**, 103058 (2020)

22. Wang, J., Han, S., Li, X.: 3D fuzzy ergonomic analysis for rapid workplace design and modification in construction. Autom. Constr. **123**, 103521 (2021)

23. Dubey, D., Mehra, A.: Fuzzy multiobjective linear programming: a bipolar view. In: Advances in Computational Intelligence: 14th International Conference on Information Processing and Management of Uncertainty in Knowledge-Based Systems, IPMU 2012, Catania, Italy, 9–13, July, 2012, Proceedings, Part IV 14, pp. 458–468. Springer, Heidelberg (2012). https://doi.org/10.1007/978-3-642-31724-8_48

24. Gallagher, S., Heberger, J.R.: Examining the interaction of force and repetition on musculoskeletal disorder risk: a systematic literature review. Hum. Factors **55**(1), 108–124 (2013)

Optimization of the Operation Management Process of a Company in the Electronic Manufacturing Sector

Marcelo Carneiro Gonçalves[1]([✉]) [ORCID], Katuzi Hamasaki[1,2] [ORCID],
Izamara Cristina Palheta Dias[1] [ORCID], and Elpidio Oscar Benitez Nara[1] [ORCID]

[1] Pontifical Catholic University of Paraná (PUCPR) - Industrial and Systems Engineering
Graduate Program (PPGEPS), Curitiba, Paraná CEP: 80215-901, Brazil
{carneiro.marcelo,elpidio.nara}@pucpr.br,
izamara.dias@pucpr.edu.br
[2] Ernst and Young Business Advice (EY) - Technology Solutions Delivery, Curitiba, Paraná,
CEP 80410-201, Brazil

Abstract. The use of mathematical programming models for production planning has been proposed since the 1950s, being a widely applied tool, since it can provide optimal solutions for production planning problems. For manufacturing companies, it is a great challenge to plan in uncertain environments when there are large variations in planning parameters. Thus, the greatest difficulty in dealing with Mathematical Programming models in production planning is that, in general, with the intention of simulating reality through these models, it is necessary to estimate values for the planning parameters, which may not always be possible accurately, and consequently, the model's optimal solution may not represent the best solution to the problem. In this context, the classic approach to deal with a dynamic economic scenario is the use of robust optimization models, which propose a suboptimal solution in relation to the deterministic model. Therefore, the objective of this paper is to apply a robust optimization model in the operations management process of an electronic components manufacturing company. First, a content analysis was performed, then company data was collected, the model was proposed, and the results were analyzed. Results suggested more than 80% of the production should be done in anticipation. The optimal solution, at the lowest cost, was obtained from the minimal scenario. The worst and robust solution, bringing the highest cost, came from the intermediate scenario, proving that the production plan could be performed even with adversities on sight.

Keywords: Linear Programming · Robust Optimization · Operations Management

Published by Springer Nature Switzerland AG 2024
C. Danjou et al. (Eds.): PLM 2023, IFIP AICT 702, pp. 344–354, 2024.
https://doi.org/10.1007/978-3-031-62582-4_31

1   Introduction

Carrying out production planning in manufacturing companies has progressively become a complex and expensive task. In addition, the increase in competition and market competitiveness has forced companies to increasingly seek production systems that are simultaneously effective and efficient, so that they can achieve their organizational objectives using the available resources in the best possible way. Or to put it another way, the organizational objectives of manufacturing companies incorporate at least one new challenge: Obtaining a good production decision-making system, which has the property of having a minimum cost.

Since the 1970s, sophisticated decision-making support systems have been implemented in a large number of medium and large organizations, such as Master Production Schedule - MPS, Material Requirements Planning - MRP, and Enterprise Resource Planning – ERP's. However, such models only provide viable solutions to the production planning problem, in the sense of not considering the fulfillment of some optimality criteria, consequently the quality of the solution found does not provide an adequate analysis both in relation to cost and in relation to sensitivity. In addition, the use of the MRP technique for large problems becomes unfeasible in relation to the amount of effort used to find a viable solution.

Given this scenario, Linear Programming (LP) models have been proposed and widely used to solve production planning problems. However, carrying out production planning efficiently is a great challenge, especially when, in the day-to-day experience of the factory floor, there is great variability in the parameters used in the models.

To deal with the fact that the parameters of the problem are subject to variations along the planning horizon, it is proposed the use of a robust approach, which is a worst-case technique, which seeks to reach viable solutions for a problem, considering the worst scenario of realization of uncertainties. In other words, it seeks to solve the mathematical model by minimizing the maximum deviation of the random variables chosen to be analyzed in the model [1].

Therefore, the modeling of the production planning problem through a robust approach arises from the need to consider the action of uncertainties in the model parameters, and with its use, a suboptimal solution is considered in relation to the value estimated by the deterministic problem, where, in general, the parameters are estimated using means, without association to a specific standard deviation, as in the robust approach [2].

This paper seeks to model the production planning problem considering the concepts of robust optimization in a company in the electronic equipment manufacturing segment, in order to provide the planner with greater security in decision-making in the face of the variability of the economic scenario in the environment where the company operates. To do so, a content analysis was first performed in the literature, to identify related works and possible gaps in the literature in relation to robust optimization, then a mathematical model was proposed, data collected in the company and the main results analyzed with the help of the Lingo software.

2 Robust Optimization

Real-life optimization problems often contain uncertain data, e.g., demand variability, cycle times, setup times, productive capacity, etc. The reasons for these uncertainties in the data could come from measurement/estimation errors that come from lack of exact knowledge of the parameters of the mathematical model or because of the business' dynamics.

There are two distinct approaches for dealing with data uncertainty in optimization: robust and stochastic optimization. Stochastic optimization assumes an important premise, which is that the true probability distribution of uncertain data must be known or estimated. Robust optimization, on the other hand, does not assume that the probability distributions are known, but assumes that the uncertain data resides in a set of uncertainties [3].

Robust optimization is a relatively young research field and has been mainly developed in the last 15 years. Especially in the last 5 years, there have been many publications that show the value of robust optimization in applications in many fields including finance, management science, supply chain, healthcare and engineering ([4–14]).

In general, two types of uncertainty sources are considered in robust optimization problems: uncertainty in constraints and in the objective function. In the first case, the variation of the model's parameters can cause an infeasible solution, while, in the second, the variations in the objective function parameters can lead to solutions considered optimal to be very far from the best solution.

The classic optimization problem consists of minimizing (or maximizing) an objective function, subject to a set of constraints:

$$
\begin{aligned}
& f(x) \\
\text{Subject to :}\quad & g(x) \leq 0 \\
& h(x) = 0
\end{aligned}
\tag{1}
$$

where: $f(x) : R^n \to R, f(x) \in c^1(R^n)$, $g(x) : R^n \to R^p$, $g(x) \in c^1(R^n)$ and $h(x) : R^n \to R^m$, $h(x) \in c^1(R^n)$.

The uncertainties can be either in the objective function, $f(x)$, or in the inequality constraints, $g(x)$, or equality constraints, $h(x)$.

A possible treatment for the problem is to analyze the worst case, that is, to determine the solution that minimizes the maximum possible objective function when considering all possible instances of the problem. That is, the robust optimization problem is fundamentally a minimax problem.

Thus, to robustly minimize the model (1), with uncertainties that may be present both in the objective function, which will be called: $f(x, \alpha_f)$, as in the restrictions of inequality and equality, which will be called as $g(x, \alpha_g)$, $h(x, \alpha_h)$, respectively, where these uncertainties can oscillate over a continuous vector set, $\Omega_f \in \left[\begin{array}{c} \alpha \\ -f \end{array} \pm \varepsilon_f \right]$, $\Omega_g \in \left[\begin{array}{c} \alpha \\ -g \end{array} \pm \varepsilon_g \right] e \Omega_h \in \left[\begin{array}{c} \alpha \\ -h \end{array} \pm \varepsilon_h \right]$, respectively, there is the following problem (2).

$$
\begin{aligned}
&F(x, \alpha_f) \\
s.t. \quad &F(x, \alpha_f) = \{f(x, \alpha_f) | g(x, \alpha_g) \le 0; h(x, \alpha_h) = 0\}
\end{aligned}
\tag{2}
$$

Thus, a robust optimization model can be developed by applying the minimax concept.

The minimax concept allows the planner to obtain a robust production planning model capable of incorporating variations in the planning parameters. Since these variations occurred under a given set of analyzed uncertainty, the formulated planning will still remain feasible, thus, there is no need for a production re-planning.

Depending on the adopted range of uncertainty level, that is, the level of confidence that the manager decides to consider, the objective function may deteriorate if the adopted range is considerably wide. However, it ensures that the model remains feasible in the range of variation of the considered uncertain parameters, bringing a solution suboptimal. This fact is called the price of robustness [1].

[2] robust model was the pioneer in robust optimization, and it is extremely conservative, in the sense that the value of the objective function deteriorates too much to guarantee the robustness, in terms of feasibility, of the solution.

[2] used the term "uncertainty box" to refer to the space of realization of uncertain parameters in the model, that is, the vector space has as its center an average vector that can vary symmetrically over a given range (deviation) along the "box". The advantage of this approach is the simplicity of its application. The disadvantage is the high level of conservatism.

3   Content Analysis

A content analysis was performed on the Scopus database, no year restriction, with the terms "Production Planning Problem", "Mathematical Model" and "Robust". Only 9 scientific articles were obtained that deal with production planning via robust optimization. 4 of them analyzed "demand" as an uncertain parameter, or robust parameter, 3 of them analyzed the "production level", 1 analyzed the "budget available for production" and 1 the "assembly time". With that, we verified a gap in the literature regarding robust optimization analysis in parameters associated with production capacity (Table 1).

Table 1. Index descriptions and symbols.

Title	Authors	Year	Source	Robust Parameter
Energy and carbon-constrained production planning with parametric uncertainties [15]	Chaturvedi, N.D., Kumawat, P.K., Keshari, A.K	2021	IFAC-PapersOnLine	Budget
Robust optimization approach to production system with failure in rework and breakdown under uncertainty: Evolutionary methods [16]	Rabbani, M., Manavizadeh, N., Aghozi, N.S.H	2015	Assembly Automation	Demand
A minimax p-robust optimization approach for planning under uncertainty [17]	Seo, K.-K., Kim, J., Chung, B.D	2015	Journal of Advanced Mechanical Design, Systems and Manufacturing	Demand
A robust optimization model for multi-product two-stage capacitated production planning under uncertainty [18]	Rahmani, D., Ramezanian, R., Fattahi, P., Heydari, M	2013	Applied Mathematical Modelling	Production Cost and Demand
Semiconductor production planning using robust optimization [19]	Ng, T.S., Fowler, J	2007	IEEM 2007: 2007 IEEE International Conference on Industrial Engineering and Engineering Management	Production
A robust optimization model for multi-site production planning problem in an uncertain environment [20]	Leung, S.C.H., Tsang, S.O.S., Ng, W.L., Wu, Y	2007	European Journal of Operational Research	Production loading plan and workforce level

(continued)

Table 1. (*continued*)

Title	Authors	Year	Source	Robust Parameter
A robust optimization model for production planning of perishable products [21]	Leung, S.C.H., Lai, K.K., Ng, W.-L., Wu, Y	2007	Journal of the Operational Research Society	Production loading plan
A robust dynamic planning strategy for lot-sizing problems with stochastic demands [22]	Raa, B., Aghezzaf, E.H	2005	Journal of Intelligent Manufacturing	Demand
Robust production planning for a two-stage production system [23]	Morikawa, Katsumi, Nakamura, Nobuto	1996	Proceedings of the Japan/USA Symposium on Flexible Automation	Assembly time

4 Application of Robust Optimization Concept

4.1 Company Description

Company M, founded in 2012, operates in the electronics manufacturing sector, specializing in the assembly of Surface Mount Device (SMD) components, integration of electronic products and formation of Plated-through Holes (PTH) components. Its headquarters is in southern Brazil. Among a highly complex product portfolio, the products selected were WAFFER 371000490 (Product 1) and WAFFER 371000493 (Product 2), for being both heavily demanded items, with a monthly output, from 1 to 2 batches of 1200 pieces.

4.2 Mathematical Model

[24–28] presented models that were used as a basis for modeling the production problem developed in this work.

$$Z = \sum_{i=1}^{U} \sum_{j=0}^{M} \sum_{k=1}^{Q} \left(T_{ik}X_{ijk}\right) * V_{ik} + H_{jk}W_k + A_{jk}Y_k + (X_{ijQ} - D_{ij})Z_i + \left(G_{ik}X_{ijk}\right)$$
$$* \alpha_{ik} \qquad \text{(O.F}$$

$$\sum_{j=0}^{M} X_{ijki} = \sum_{j=0}^{M} D_{ij} \; \forall i, k = Q \qquad \forall i, k = Q \qquad (3)$$

$$X_{ijk} \geq D_{ij} \qquad \forall i, j = 0, k = Q \qquad (4)$$

$$\sum_{j=1}^{j} X_{ijk} \geq D_{ij} \qquad\qquad \forall i, j = 1, 2, \ldots, M, k = Q \qquad (5)$$

$$X_{ijk} = X_{ijQ} \qquad\qquad \forall i, \forall j, k = 1, 2, \ldots, Q - 1 \qquad (6)$$

$$H_{jk} \leq 7200 \qquad\qquad \forall j, \forall k \qquad (7)$$

$$\sum_{i=1}^{U} \sum_{k=1}^{Q} (T_{ik} + G_{ik}) X_{ijk} \geq \sum_{k=1}^{Q} O_k \underline{N_k} \qquad \forall j, O_k = 0 \qquad (8)$$

$$\sum_{i=1}^{U} (T_{ik} + G_{ik}) X_{ijk} \leq \underline{N_k} + H_{jk} + A_{jk} \qquad \forall j, \forall k \qquad (9)$$

$$\sum_{k=1}^{Q} (T_{ik} + G_{ik}) X_{ijk} \leq \underline{N_Q} + H_{jQ} + A_{jQ} \qquad \forall i, \forall j \qquad (10)$$

$$T_{11}X_{1j1} + T_{12}X_{1j2} + T_{21}X_{2j1} + T_{22}X_{2j2} \leq \underline{N_1} + H_{j1} + A_{j1} \qquad \forall j \qquad (11)$$

$$X_{ijk}, H_{jk}, A_{jk} \geq 0 \qquad (12)$$

The constraints and objective function presented below represent the result of modeling the scenario found in the production problem of company M, after numerical experiments and validations.

Where : $R^n \rightarrow R, T_{ik} \in R^n, X_{ijk} \in R^n, H_{jk} \in R^n, W_k \in R^n, A_{jk} \in R^n, Y_k \in R^n, D_{ij} \in R^m, Z_i \in R^n, \alpha_{ik} \in R, G_{ik} \in R^n, O_k \in R^n, N_k \in R^n$.

Above, you can see the cost-minimizing objective function (O.F.), (3) General demand Constraint, (4) Demand constraint per period, (5) Demand constraint considering stock, (6) Bottleneck-guided Production and Work-In-Process Inventory Block constraint, (7) Maximum working hours constraint, (8) Productive Capacity Minimal Occupation, (9) Pure Capacity Constraint, (10) Per Product Capacity Constraint, (11) Single Resource Constraint, (12) non-Negativity.

Input data to what is described below, were given by the Production Planner, who already had most of the information, such as nominal capacity, minimum occupation, processing time and product setup. Altogether, 47 data entries were received and all of them were used. Due to internal reasons, information about product and processes were not available, therefore, cost data used in this model were all arbitrarily chosen. Index descriptions are: i, j and k indicate, respectively, the product, period and process.

4.3 Results

Results were obtained using Microsoft Excel data export functionality of Lingo software, performing a direct extraction to the desired tables, but only objective function (cost-minimization) results will be presented in this study, due to page limitation.

Table 2. Variables descriptions and symbols (Tables 2 and 3).

Variable Name	Variable Type	Description	Symbol
Outsourced hours	Continuous	Additional production capacity of j product in k process, given in seconds, by purchased outsourced work	A_{jk}
Production	Continuous	Amount of i products, that passed through k process, during j period, given in products units	X_{ijk}
Overworked hours	Continuous	Additional production capacity of j product in k process, given in seconds, by overworked hours	H_{jk}

Table 3. Parameters description and symbols.

Parameter	Description	Symbol
Nominal Capacity	Nominal process k production capacity, given in hours. Robustness is further applied in this value, so it is noted with a line on its top	N_k –
Minimum Occupation	Minimum desired occupation ratio, for each process k	$\mathbf{O_k}$
Processing Time	Passing-thru time spent by product i in process k, given in hours	$\mathbf{T_{ik}}$
Nominal Process Cost	Processing cost of a single product i unit, on process k, given in reais (Brazilian currency)	$\mathbf{V_{ik}}$
Overtime Process Cost	Process k additional cost, during overtime working hours, given in reais	$\mathbf{W_k}$
Outsourced Process Cost	Process k additional cost, during outsourced hired hours, given in reais	$\mathbf{Y_k}$
Inventory Cost	Stocking cost for a single product i unit, over a period	$\mathbf{Z_i}$
Product Setup Time	Time spent setting the process k for product i production	$\mathbf{G_{ik}}$
Setup Time Cost	Cost of time spent setting process k in order to produce product i	$\mathbf{\alpha_{ik}}$

After disturbing model's robust parameter, which was nominal capacity, $\underline{N_k}$, intermediate capacity scenario presented the robust solution (Table 4).

Under another circumstances, for instance, model could have chosen to produce the whole 1200 pieces in only one day or break it into daily production. This is possible due to the gap between order placement by the customer and delivery date, which is 11 days as informed by the company and represented here as 11 periods of production (0 to 10).

The productive capacity, in the 3 possible scenarios, was affected by the decision to hire overtime in the exact same amount, and in the same periods.

Table 4. Scenario comparison

Scenario	Objective Function value	Ranking
Intermediate	$2.665.193,30	Robust
Possible	$2.665.053,77	Sub-optimal
Minimum	$2.662.980,37	Optimal

There are the values of 7200 s, or two additional hours, totaling 62 contracted over-time, over periods 5, 6, 7, 8, 9 and 10 for all machines, and for machine 2 in the zero period.

5   Final Considerations

It was possible to analyze and extract a large amount of data and information from company's current scenario, enough to generate a feasible model. Even if not trustful to the daily reality of the company. Results that met the proposed demand for the planning periods were obtained, even with a sub optimal solution.

Results shows that Robust Linear Programming is an option, when Materials Requirement Planning, Manufacturing Resource Planning and Enterprise Resource Planning costs and flexibility does not match companies demand.

Also, Robust Linear Planning is not limited to deliver a production plan that only complies with all the plant's capabilities, such as models previously said. Through Robust Linear Programming, it is possible to obtain optimized solutions for minimizing pro-duction time, or reducing costs as seen in this work. All of this, with the use of an eco-nomically viable tool, especially when compared to the Materials Requirement Planning solutions offered in the market.

Small-scale industries, which may struggle with less resources and lower budget, could improve results by using practices that build resilient infrastructure, promote inclu-sive and sustainable industrialization and foster innovation, becoming more competitive when lowering the costs of production planning and mitigating production planning infeasibility risks, making it possible to small-scale industries have a better portion of total industry value added.

Implications for practices are related to United Nation Sustainable Development Goals (UN SDG) no. 9 and 13. Indicator 9.3.1 verifies small-scale industries participation in total industry value added, and indicator 9.4.1 checks on the CO_2 emission per unit of value added. Indicator 13.2.2 look towards total greenhouse emissions per year [29, 30].

Lowering general CO_2 emissions or emissions per unit of value added could be incorporated in the mathematical model, defining which level of emission is desired or helping analyze the impacts on production when minimizing emission.

Furthermore, this study offers a glimpse of what could be another meaningful appli-cation of Operations Research, by showing practical results, validating what previous authors already proposed and mathematically modeling a production process.

Future studies are suggested, applying it to a greater mix of products, with more extensive bills of materials and in other sectors of the industry, which operate with pushed production. The development of solutions to the same problem, however based on methods of decision-making processes, appear as possible proposals for improvement.

References

1. Bertsimas, D., Sim, M.: The price of robustness. Oper. Res. **52**, 35–53 (2004)
2. Soyster, A.L.: Convex programming with set-inclusive constraints and applications to inexact linear programming. Oper. Res. **14**, 1154–1151 (1973)
3. Alem, D., Morabito, R.: Programação estocástica e otimização robusta no planejamento da produção de empresas moveleiras. Prod. J. **25**(3), 657–677 (2015)
4. Souza, B.F.: Aplicação de algoritmos de morfogênese adaptativa modificado e de otimização robusta em problemas de placas finas enrijecidas. Thesis, University of Brasilia UNB, Brazil (2021)
5. Barbosa Filho, A.C.B.: Desenvolvimento de um framework para otimização robusta e aplicação para o planejamento tático de uma cadeia de suprimentos de argônio. Thesis Federal University of Uberlândia, Brazil (2020)
6. Virtudes, P.T.B.C.: Robust optimization in finance. Thesis, Coimbra University, Portugal (2019)
7. Martinez, J.J.L.V.: Otimização robusta e programação estocástica para o problema de roteamento de veículos com múltiplos entregadores: formulações e métodos exatos. Ph. Dissertation, Federal University of São Carlos UFSCAR, Brazil (2019)
8. Tavares, C.S.: Seleção de fornecedores sob incertezas via otimização robusta. Thesis, Federal University of São Carlos UFSCAR, Brazil (2019)
9. Silva, M.A., Cavalini, A.A., Steffen Junior, V., lobato, F.S.: Otimização do Processo de Fermentação Batelada-Alimentada usando Evolução Diferencial e Otimização Robusta. In: Proceeding Series of the Brazilian Society of Applied and Computational Mathematics, vol. 6, no. 1 (2018)
10. Centeno, D.L.R.: Otimização robusta multiobjetivo por análise de intervalo não probabilística: uma aplicação em conforto e segurança veicular sob dinâmica lateral e vertical acoplada. Ph. Dissertation, Federal University of Rio Grande do Sul UFRGS, Brazil (2017)
11. Alves, J.: Modelo de otimização robusta orientado por dados aplicado na alocação de renda fixa. Thesis, Pontifical Catholic University of Rio de Janeiro PUC-Rio, Brazil (2017)
12. Nora, L.D.D., Siluk, J.C.M., Júnior, A.L.N., Nara, E.O.B., Furtado, J.C.: The performance measurement of innovation and competitiveness in the telecommunications services sector. Int. J. Bus. Excell. **9**(2), 210–224 (2016)
13. Baierle, I.C., Schaefer, J.L., Sellitto, M.A., Furtado, J.C., Nara, E.O.B.: Moona software for survey classification and evaluation of criteria to support decision-making for properties portfolio. Int. J. Strateg. Property Manage. **24**(2), 226–236 (2020)
14. Schaefer, J.L., Baierle, I.C., Sellitto, M.A., Furtado, J.C., Nara, E.O.B.: Competitiveness scale as a basis for Brazilian small and medium-sized enterprises. EMJ – Eng. Manage. J. **33**(4), 255–271 (2021)
15. Chaturvedi, N.D., Kumawat, P.K., Keshari, A.K.: Energy and carbon-constrained production planning with parametric uncertainties. IFAC-PapersOnLine **54**(3), 560–565 (2021)
16. Rabbani, M., Manavizadeh, N., Aghozi, N.S.H.: Robust optimization approach to production system with failure in rework and breakdown under uncertainty: evolutionary methods. Assem. Autom. **35**(1), 81–93 (2015)

17. Seo, K.-K., Kim, J., Chung, B.D.: A minimax p-robust optimization approach for planning under uncertainty. J. Adv. Mech. Des. Syst. Manuf. **9**(5), 1–10 (2015)
18. Rahmani, D., Ramezanian, R., Fattahi, P., Heydari, M.: A robust optimization model for multi-product two-stage capacitated production planning under uncertainty. Appl. Math. Model. **37**(20–21), 8957–8971 (2013)
19. Ng, T.S., Fowler, J.: Semiconductor production planning using robust optimization. In: IEEM 2007: 2007 IEEE International Conference on Industrial Engineering and Engineering Management, vol. 4419357, pp. 1073–1077 (2007)
20. Leung, S.C.H., Tsang, S.O.S., Ng, W.L., Wu, Y.: A robust optimization model for multi-site production planning problem in an uncertain environment. J. Oper. Res. Soc. **181**(1), 224–238 (2007)
21. Leung, S.C.H., Lai, K.K., Ng, W.-L., Wu, Y.: A robust optimization model for production planning of perishable products. J. Oper. Res. Soc. **58**(4), 413–422 (2007)
22. Raa, B., Aghezzaf, E.H.: A robust dynamic planning strategy for lot-sizing problems with stochastic demands. J. Intell. Manuf. **16**(2), 207–213 (2005)
23. Katsumi, M., Nobuto, N.: Robust production planning for a two-stage production system. In: Proceedings of the Japan/USA Symposium on Flexible Automation, vol. 2, pp. 1279–1282 (1996)
24. de Sampaio, R., Wollmann, R., Vieira, P.: A flexible production planning for rolling-horizons. Int. J. Prod. Econ. **190**(8), 31–36 (2017)
25. Gonçalves, M.C, Wollmann, R.R.G., Sampaio, R.J.B.: Proposal of a numerical approximation theory to solve the robust convex problem of production planning. Int. J. Oper. Res. (2022)
26. Hamasaki, K., Gonçalves, M.C., Junior, O.C., Nara, E.O.B., Wollmann, R.R.G.: Robust linear programming application for the production planning problem. In: Fernando Deschamps, Edson Pinheiro de Lima, Sérgio E. Gouvêa da Costa, Marcelo G. Trentin, (ed.) Proceedings of the 11th International Conference on Production Research – Americas: ICPR Americas 2022, pp. 647–654. Springer Nature Switzerland, Cham (2023). https://doi.org/10.1007/978-3-031-36121-0_82
27. Gonçalves, M.C., Nara, E.O.B., Dos Santos, I.M., Mateus, I.B., Do Amaral, L.M.B.: 11th International Conference on Production Research –Americas 2022 (ICPR AMERICAS 2022). Springer Proceedings in Mathematics and Statistics, (2022). https://doi.org/10.1007/978-3-031-36121-0
28. Serra, F.N.T., Nara, E.O.B., Gonçalves, M.C., da Costa, S.E.G., Bortoluzzi, S.C.: Proceedings of the 11th International Conference on Production Research –Americas 2022 (ICPR AMERICAS 2022). Springer Proceedings in Mathematics and Statistics (2022). https://doi.org/10.1007/978-3-031-36121-0
29. UN department of economic and social affairs, sustainable development goal number 9 webpage. https://sdgs.un.org/goals/goal9. Accessed 21 Jan 2023
30. UN department of economic and social affairs, sustainable development goal number 13 webpage. https://sdgs.un.org/goals/goal13. Accessed 21 Jan 2023

Towards Zero-Defect Manufacturing in the Silicon Wafer Production Through Calibration Measurement Process: An Italian Case

Federica Acerbi[1]([envelope]) [iD], Andrea Pranzo[2], Cristina Sanna[2], Marco Spaltini[1] [iD], and Marco Taisch[1] [iD]

[1] Department of Management, Economics and Industrial Engineering, Politecnico di Milano, Via Lambruschini 4/B, 20156 Milan, Italy
{federica.acerbi,marco.spaltini,marco.taisch}@polimi.it
[2] MEMC Electronic Materials S.P.A, Viale Gherzi, 31, 28100 Novara, NO, Italy
{apranzo,csanna}@gw-semi.com

Abstract. In electronic devices, the number of transistors and components per unit of area is still significantly increasing. In this context, the quality of the substrate of these devices is consequently increasing, making stricter and more pervasive the quality controls that a silicon wafer undergoes during its production process. This issue is extremely important to be addressed to reduce costs in quality controls in production moving towards zero-defect manufacturing. The purpose of this paper is to reduce time and costs spent in calibration procedures of the instruments needed to measure the mechanical parameters of silicon wafers, by revising and standardizing the already adopted procedures. To address this goal, the extant literature, patents, and standard about procedures employed for measuring the mechanical parameters of silicon wafers are studied. The results are elaborated and applied to an industrial case study, the Italian branch of a Taiwanese manufacturing company. In particular, the focus of the case is on the Bow/Warp machine's calibration which needs to be performed periodically to guarantee a correct measurement accuracy. Such calibration has strong implications for production efficiency and flow. The results are reported and discussed to highlight the key practical and theoretical implications.

Keywords: Silicon Wafer · Mechanical Parameters · Bow/Warp Calibration · Zero Defect Manufacturing · Waste Management

1 Introduction

In 50 years, technology improvements in electronics drove innovation in the whole society becoming a cornerstone element in daily life. A common element for most of the electrical devices lays in the reliance on silicon, a semiconductor material used to manufacture most of the components such as transistors [1]. Transistors are key to determine

© IFIP International Federation for Information Processing 2024
Published by Springer Nature Switzerland AG 2024
C. Danjou et al. (Eds.): PLM 2023, IFIP AICT 702, pp. 355–364, 2024.
https://doi.org/10.1007/978-3-031-62582-4_32

the performances of the final product since device's power, efficiency and capability depends strongly on circuit density, that can be expressed as the number of transistors per chip. Over the decades, size of transistors has decreased from 20 μm in 1970 to few nanometres in recent years [2]. However, the technology pace in the Semiconductor industry keeps increasing and, in such a fast-changing environment, companies must have the capabilities to quickly adapt their processes to new technological requirements. From Silicon wafer production to the final assembly of circuit, the number of quality controls on the product increases with the tightening of product's specifications. Quality control is not a value adding activity but, especially in this high value context, it can save the additional costs of transforming material that will be further processed before being scrapped [3]. Moreover, to ensure the elimination of any defect, the several quality controls needed in the industry are extremely expensive and require a huge amount of sample products wasted. To reduce avoidable quality controls limiting waste creation and costs increment, semiconductor Industry is pursuing a Zero-Defect Manufacturing (ZDM) strategy [4] which aims at producing without delivering non-conforming products to the next production step. To achieve ZDM, [5] underlined two main approaches: i) Product-oriented ZDM identifies and studies the defects on the actual parts and ii) Process-oriented ZDM studies the defects of the manufacturing equipment and based on that can evaluate whether the manufactured products are good or not. Nevertheless, at the best of authors knowledge, few contributions explored the potential benefits of introducing at process level, more specifically in the calibration phase, the ZDM principles. Therefore, in this paper the focus will be put on the application of Process-oriented ZDM approach to the measurement process of shape parameters of silicon wafers which, should be perfectly round and flat disks, through the definition of reusable test wafers. To address this goal covering the envisioned gap in terms of limited attention over the introduction of ZDM principles in calibration procedures, a case study is proposed.

Entering more in detail, the shape of the silicon wafers may deviate due to deformations and thickness variations. Indeed, one of the most important parameter to be measured is the wafer flatness, that is defined as the variation of thickness with respect to a reference plane. The reference plane can be chosen in two ways: i) Three-point method or ii) Best fit method (the one adopted in this study) [6]. In brief, the main parameters that define the variations of wafer's shape and require machine calibration are [6]: (i) Total thickness variation (TTV): the maximum variation in wafer thickness; (ii) Bow (B): the distance between the reference plane and the central point of the median surface; (iii) Warp (W) or warpage: the difference between the maximum and the minimum distance of the median surface from the reference plane. Therefore, in this contribution, the calibration process of the measurement machines located in different departments of a company is investigated considering the calibration as one of the processes where start implementing a preliminary ZDM approach. More in detail, one of the ZDM pillars is the standardisation [7], hence, to address the goal of this contribution the authors aim to evaluate the possibility to standardize the calibration procedure of the different departments, thus identifying a unique procedure putting the basis for ZDM while reducing costs and wastes.

The remainder of the contribution is the following: Sect. 2 describes the research objective and the methodologies employed to address it, Sect. 3 provides a background about the Bow and Warp (B/W) formation and calibration theory, Sect. 4 describes the adoption of the approach to standardise the calibration procedure inside the company and discusses the results obtained through experiments and Sect. 5 concludes the research highlighting the key contributions of the paper, limitations and future outlooks.

2   Research Objective and Methodology

2.1   Research Objective and Research Questions

The goal of the research is to introduce the ZDM principles into the calibration measurement process considering that ZDM is usually applied to product while the application at process level, especially regarding the calibration procedures, is still limited. Indeed, from a practical point of view, the introduction of ZDM could cope with the challenges faced in the industrial environment during the calibration procedure of silicon wafer defining a standard procedure including the nominal wafer characteristics to be used. In this regard, the research aims to introduce ZDM by understanding how to standardize the B/W calibration procedures of a group of machines, which are currently located in different departments. In this way, it is expected to obtain better operational performances by reducing waste and putting the basis to embrace a ZDM approach at plant level while ensuring high measurement accuracy. To pursue the declared research objective, the following research questions (RQ)s had been formulated and addressed. RQ1): *"What are the factors and core elements affecting the quality of measurement corrections applied through B/W calibration?"* RQ2): *"How to standardize B/W calibration methodologies and practices to maintain a high measurement accuracy and improve operational efficiency?"*.

2.2   Research Methodology

To answer to RQ1, both scientific literature (using Scopus) and grey literature (such as patents and standards) were reviewed based also on the suggestions from experts in the field (i.e., the person who developed one of the patents). The string of keywords used for the review was focused on specific technical topics, namely: TITLE-ABSTRACT-KEYWORD (("Silicon" or "Silicon Wafer") AND "Mechanical properties" AND "Young's Modulus" AND ("Warp" OR "Warpage") AND "Bow" AND "Manufacturing"), as also suggested by the experts. The results were scanned and selected according to their relevance to the industrial problem faced: the calibration procedure and its standardization. The selected papers, registered patents, standards, and machine's builder instructions were reviewed (RQ1). The results were discussed and applied to an industrial case, the Italian plant of a semiconductor company, to define the standard procedure to be followed by the whole plant addressing RQ2. The application case analysed was the Italian plant of MEMC Electronic Materials, controlled by GlobalWafers, a Taiwanese company producing and supplying silicon wafers for the semiconductor industry. The application case was conducted through semi-structured interviews with company's managers to investigate the AS-IS and by a 2-factors Design of Experiment (DOE) approach defining the TO-BE scenario [8].

3   Background on Bow and Warp

3.1   Mechanical Parameters Measurement: Bow and Warp Definition

In 200 mm wafers, flatness is typically measured using non-contact capacitance metrology [9]. This can be performed in different ways although one of the most widespread is prescribed by the SEMI MF1390-0218 [10]. According to it, the wafer is supported by a holding device (chuck) in its centre. The chuck has a diameter of about 3.6 cm for 200 mm wafers. The measurement system is composed by two probes. The distance between the probes must be accurately known and it is calibrated periodically. Wafers are rotated and scanned through the capacitance probe system to scan the selected area. The probes are capable of independent measurement of the distances between the probe itself and the nearest surface of the wafer. The result of the measurement is a set of numbers describing the thickness of the wafer in each scanned point (that result to be more than 8.000 for a 200 mm wafer). Each parameter related to flatness and shape (i.e., B/W) is then computed through different algorithms from this raw data. Although the process appears to be relatively simple, it presents some issues due to the intrinsic complexity given by the low accepted tolerances. Among all, there is the gravity force effect exerted on the wafer. Since the wafer is held by a chuck whose diameter is considerably smaller than the one of the wafers themselves, the gravity force tends to pull down its edges. To not transfer this gravitational effect on the final values of the parameters, a gravitational correction algorithm must be applied as reported by the SEMI MF1390-0218 [10]. This correction is performed using data collected during the "Bow/Warp calibration" procedure, in which a representative wafer is scanned, flipped, and scanned again (see sub-Sect. 3.2). In Fig. 1 the effect of gravity is schematically illustrated.

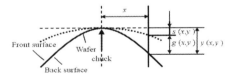

Fig. 1. Effects of gravity on a wafer under measurement

According to SEMI MF1390-0218 [10], the factors that affect the gravity induced deflection are mainly the Young's modulus (or Elastic modulus quantifying the elastic behaviour of the material), that can be affected by the crystallographic orientation (that is the orientation of the structure described by "*hkl*" Miller indices, that are the reciprocals of the coordinates of the intercepts on the XYZ axes, multiplied by the lowest common denominator) [11, 12], the nominal thickness and the nominal diameter. Moreover, also the backside conditions may influence the calibration procedure results.

This measurement method is the most adopted for measuring mechanical parameters of wafers which diameter is lower or equal than 200 mm [9]. For higher diameters, different methodologies have been developed in recent years [13]. In particular, such methodologies comprise optical methods based on spectral-domain interferometer. In this measurement set-up, the wafer is unclamped and supported at the edges. For this reason, the gravitational effects do not cause deflection on the measured wafers and the measurement embodiment does not need to correct for such deflection [14].

3.2 Bow/Warp Calibration Procedure

B/W calibration, also defined as "Representative wafer inversion calibration" is a measurement machine set-up that needs to be performed periodically. Details about the measurements and computation behind B/W calibration can be found in [15]. The measurement apparatus is composed by a two-probe system and a chuck, on which the wafer to be measured relies. The coordinate system works as follows: the wafer's surface is identified by the angle θ and the distance from the centre X. At each measured point, identified by a couple (x, θ), corresponds a vertical distance z from the plane passing by middle of the two probes. As already mentioned, the B/W calibration is necessary to apply a correction to the measured wafer: this correction becomes necessary due to the gravitational force that is applied to the wafer (especially at the edges), when it relies on the chuck at its centre. In addition to that, as claimed in *US patent 4750141*[15], the calibration is also necessary to correct for the errors induced by the measurement apparatus itself. Overall, the B/W calibration can be subdivided into two steps: 1) X-Calibration: A first measurement M1 of the representative wafer is performed. This measure can be split into one desired wafer related component Mw and an undesired fixture and gravitational related component Mc (where M1, Mc, Mw are matrices)); 2) θ-Calibration: after the measurement is completed, the wafer is released by the chuck and the chuck alone is rotated by a predefined angle $\theta 1$, defining a second home position for the chuck. The flipped wafer is then chucked and measured again.

4 Defining the Nominal Standard Wafer: Application Case

In this section, the objective is to define the nominal wafer characteristics that the testing wafer should have. To achieve such goal the authors analysed the factors affecting the physical and metrological aspects that determine B/W calibration adjustments (emerged in Sect. 3). This is done through an industrial use case to propose a standard procedure to be adopted homogeneously at plant level. As reported in Table 1, at its current state the company under analysis is adopting different calibration procedures in the different areas of the company. Moreover, a nominal wafer is often used for the test and then it is discarded. Through the analysis suggested by the findings from the extant literature and patents (see chapter 3) it is aimed to evaluate the characteristics a nominal wafer should have to be used in the different areas to reduce scraps and waste while preserving quality putting the basis to embrace ZDM.

Table 1. Summary of B/W calibration procedures in the different areas

Area	Calibration frequency	Wafer used for calibration	Control Plan
A	Every lot; with a wafer taken from the production	Wafer from the lot, non-destructive process	Sampling based on position in the machine
B	Every shift	Nominal wafer, destructive process	Random sampling
C	Every change in measurement mode and every change in lot backside condition	Nominal wafer, destructive process	Random sampling
D	Every change in measurement mode and every change in lot backside condition	Wafer from the lot, destructive process	100% of wafers

As previously anticipated, the main factors affecting the gravitational deflection are the Young's modulus, the thickness and the diameter. Considering keeping stable the wafer diameter (around 200 μm), the other aspects were investigated by relying on a DOE analysis to identify a standard procedure to be applied in the whole plant.

4.1 Young's Modulus and Crystallographic Orientation

The first experiment performed is about the Young's modulus and the crystallographic orientations since, as previously mentioned, the Young's modulus of a thinned silicon wafer differs according to wafer crystallographic orientation as reported in Table 2.

Table 2. Young's modulus in different crystallographic orientations [16]

Crystallographic orientation	<100>	<110>	<111>
Young's modulus [MPa]	130	169	168.9

First, an experiment was executed to test the actual significance of crystallographic orientation on the goodness of gravity correction applied by the measurement instrument. Indeed, this experiment was set up to quantify the effects of crystallographic orientation of the nominal wafer on warp measurements. For this purpose, two entire lots, with different crystallographic orientation, were measured twice, each time calibrating with a sample wafer taken from the lots themselves (see Table 3).

Table 3. Experimental design on effects of calibration wafer's crystallographic orientation

Measured Lot	Lot's orientation	Calibration Wafer	Calibration wafer's orientation
A	<100>	A	<100>
A	<100>	B	<111>
B	<111>	A	<100>
B	<111>	B	<111>

Once the data were acquired for each lot, the measurements obtained were compared wafer by wafer. To verify if the change in calibration procedure significantly affects the W measurements, the distributions of paired differences in the Warp Best Fit (WBF) numerical values are computed as $WarpBF_{WFRB} - WarpBF_{WFRA}$ and plotted in Fig. 2.

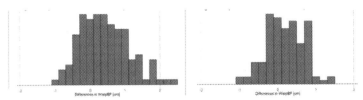

Fig. 2. Distribution of WBF differences LOT A (left) vs LOT B (right)

In the analysis performed, the most appropriate W value is the one measured when the orientation of the calibration wafer and the measured lot coincide. Nevertheless, the results show lower variability for LOT B distributions (the paired differences are always below the arbitrary ± 2 μm threshold and they appear to be symmetric around 0). Indeed, when measuring a <111> oriented wafer, less differences are highlighted when changing the properties of calibration wafer. In conclusion it is acceptable to measure a <111> wafer even calibrating for B/W with a <100> oriented wafer while the opposite situation introduces more variability in the results. For these reasons, and assuming that for the Bow Best Fit (BBF) would work accordingly, a crystal with <100> orientation can be used to create standard wafers to be used across the plant for B/W calibrations.

4.2 Backside Conditions

The second experiment performed is about the backside condition of the standard nominal silicon wafer. Silicon wafers can be produced with different backside conditions which means that different materials can be deposited on wafer's backside, potentially causing a difference in its stiffness and elastic properties. Different materials that can be deposited on the backside are: Polycrystalline silicon, Silicon oxide and Epitaxial monocrystalline silicon. Also, different combinations of the above-mentioned materials can be deposited. To test for the significance of the backside condition on the gravitational induced deflection, 238 wafers presenting very different backside conditions were measured twice. The two measurements of the same wafer differ from each other for the representative wafer used for B/W calibration and the measurement machine used. Firstly, the B/W calibration was performed with a wafer from the lot or with a wafer with the same backside condition of the measured ones. Then, the B/W calibration was performed with one of the nominal wafers produced. Both the WBF and the BBF paired differences are centred around zero as showed in Fig. 3 and Fig. 4. Indeed, given the variety of backside conditions tested, it can be stated that there is no evidence that the wafer's backside condition affects significantly the gravitational induced deflection of a 200 mm wafer supported by a 3 cm diameter chuck at its centre.

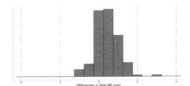 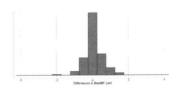

Fig. 3. Histogram of WBF paired differences   **Fig. 4.** Histogram of BBF paired differences

4.3 Nominal Thickness

Last, the thickness of the standard nominal wafer is studied. Thanks to the empirical formula to determine the deflection induced by gravity at the edge of a wafer reported in the SEMI MF1390-0218, it is possible to estimate the gravity-induced deflection at the edge for different nominal thicknesses. Considering that the highest request by the customers in terms of thickness is 725 μm, the standard nominal wafer thickness to be used for calibration will be 725 μm.

4.4 Empirical Results from the Application

In light of the research conducted about articles, registered patents and standards and the experiments performed, it was possible to update the B/W calibration procedure uniformizing it in all the different areas of the plant. A set of wafers from the same ingot have been produced to act as nominal samples to be used for B/W calibration. Considering that the previous procedure required to use wafers from production for the calibration (with consequent scrap due to contamination), this led to a decrease in yield losses leading the company to get closer to the ZDM approach. Indeed, thanks to this evidence it is possible to reduce the number of B/W calibrations performed across the plant as well as the number of wafers scrapped for this practice (e.g., area C passed from 35 calibrations daily to 17,2 on average every day). The first benefit is the reduction of the number of machine set-ups performed every day by increasing the capacity of the measurement gates. The second benefit is the reduction of scraps that directly impacts the plant overall output volumes leading towards ZDM. A detailed description of the results achieved is summarized in Table 4.

Table 4. TO-BE control plan proposal

Area	TO-BE
A	the number of calibrations is reduced, from a required frequency of one calibration per lot, to one calibration per day, unless the nominal thickness of the measured lot is significantly different from the nominal one (±25 μm). Actually, most of the products requires a standard nominal thickness, thus the expected number of calibrations due to higher or lower thickness is low
B	the number of calibrations is slightly increased: this is due to the new calibration needed when the thickness is outside the $\pm$ 25μm range from the nominal one; the proposed procedure considers the effects of thickness on the gravity induced deflection, improving measurement accuracy
C	the amount of B/W calibrations performed is halved, increasing measurement machine capacity by more than 13.000 wafers/month
D	the amount of B/W calibrations performed is reduced, and the wafers used for calibrating the machines are not extracted from production wafers, but, as in all other areas, are nominal wafers purposefully created. Thus, yield losses are reduced since no wafer from production is scrapped to perform B/W calibration

5 Conclusions and Future Research Opportunities

The present contribution aims at putting the basis to embrace ZDM approach in semi-conductor industry by acting on the calibration process of silicon wafer through the introduction of a standard procedure. A review was conducted on the existing literature, patents, and standards pertaining to calibration methods. The purpose was to identify the critical factors influencing the calibration procedure and to establish a standardized approach that would lead to ZDM. The identified approach was applied to an industrial case to test its validity and to identify the characteristics needed for a nominal wafer in this specific case. Based on this result, it was possible to reduce both waste and costs ensuring high quality too leading towards a preliminary embracement of ZDM. The present research has both practical and theoretical implications. Regarding the practical implications, this research supported the company in reducing the yield losses and related costs by introducing a standard calibration procedure. This standardised procedure represents for the company the initial step towards ZDM approach which is nowadays required to be competitive on the market. In addition, the selected sector may have positive impacts on the whole society considering how diffused these components are. Also, theoretical implications are worth being mentioned, since the ZDM has been explored in several sectors, but not yet investigated in the calibration procedures area within the metrology field of research. Indeed, most of the previous ZDM-related research were focused either on product or manufacturing process and never considered the opportunities in specific stage of the asset lifecycle management, like the calibration of industrial assets. Last, in addition to the opening of this research field, other future research opportunities might be mentioned based on some research limitations. A comparison with similar industrial entities might be performed to evaluate the best calibration procedure according to their past experience. This can be complemented by an extensive literature review to evaluate whether additional tests might be performed.

References

1. Venema, L.: Silicon electronics and beyond. Nature **479**, 309 (2011)
2. Waldrop, M.M.: The chips are down for Moore's law. Nature **530**, 145–146 (2016)
3. Elshennawy, A.K.: Quality in the new age and the body of knowledge for quality engineers. Total Qual. Manag. Bus. Excell. **15**(5–6), 603–614 (2004). https://doi.org/10.1080/147833 60410001680099
4. Psarommatis, F., May, G., Dreyfus, P.A., Kiritsis, D.: Zero defect manufacturing: state-of-the-art review, shortcomings and future directions in research. Int. J. Prod. Res. **58**(1), 1–17 (2020). https://doi.org/10.1080/00207543.2019.1605228
5. Powell, D., Magnanini, M.C., Colledani, M., Myklebust, O.: Advancing zero defect manufacturing: a state-of-the-art perspective and future research directions. Comput. Ind. **136**, 103596 (2022). https://doi.org/10.1016/j.compind.2021.103596
6. Lindroos, V., Tilli, M., Lehto, A., Motooka, T.: Handbook of Silicon Based MEMS Materials and Technologies. Elsevier (2010).https://doi.org/10.1016/C2009-0-19030-X
7. Psarommatis, F., Sousa, J., Mendonça, J.P., Kiritsis, D.: Zero-defect manufacturing the approach for higher manufacturing sustainability in the era of industry 4.0: a position paper. Int. J. Prod. Res. **60**(1), 73–91 (2022). https://doi.org/10.1080/00207543.2021.1987551
8. Ranga, S., Jaimini, M., Sharma, S.K., Chauhan, B.S., Kumar, A.: A review on Design OF Experiments (DOE). Int. J. Pharm. Chem. Sci. **3** (2014). www.ijpcsonline.com
9. Valley, J.F.: The transition to optical wafer flatness metrology. In: AIP Conference Proceedings, pp. 413–420 Oct. 2003. https://doi.org/10.1063/1.1622504
10. SEMI MF1390-0218: Test method for measuring bow and warp on silicon wafers by automated noncontact scanning (2003)
11. Hopcroft, M.A., Nix, W.D., Kenny, T.W.: What is the young's modulus of silicon? J. Microelectromech. Syst. **19**(2), 229–238 (2010). https://doi.org/10.1109/JMEMS.2009.203 9697
12. Lee, S., Kim, J.H., Kim, Y.S., Ohba, T., Kim, T.S.: Effects of thickness and crystallographic orientation on tensile properties of thinned silicon wafers. IEEE Trans. Compon. Packag. Manuf. Technol. **10**(2), 296–303 (2020). https://doi.org/10.1109/TCPMT.2019.2931640
13. Ito, Y., Kunieda, M.: Warp measurement for large-diameter silicon wafer using four-point-support inverting method. Int. J. Autom. Technol. **11**(5), 721–727 (2017). https://doi.org/10.20965/ijat.2017.p0721
14. Park, J., Bae, J., Jang, Y.-S., Jin, J.: A novel method for simultaneous measurement of thickness, refractive index, bow, and warp of a large silicon wafer using a spectral-domain interferometer. Metrologia **57**(6), 064001 (2020). https://doi.org/10.1088/1681-7575/aba16b
15. Judell, N.H., Noel, N.J.: US patent 4750141- Method and apparatus for separating fixture-induced error from measured object characteristics and for compensating the measured object characteristic with the error (1988)
16. Pang, X.F., Chua, T.T., Li, H.Y., Liao, E.B., Lee, W.S., Che, F.X.: Characterization and management of wafer stress for various pattern densities in 3D integration technology. In: Proceedings - Electronic Components and Technology Conference, pp. 1866–1869 (2010). https://doi.org/10.1109/ECTC.2010.5490707

Author Index

A

Acerbi, Federica II-75, II-96, II-355
Agard, Bruno II-197
Ahmed, Shourav I-161
Aliev, Khurshid I-14, II-323
Altun, Osman I-116
Aoussat, Améziane II-28
Arista, Rebeca II-173
Asranov, Mansur I-14, II-85, II-323
Auer, Sören I-116
Awouda, Ahmed II-85

B

Bandinelli, Romeo II-130
Barienti, Khemais I-116
Bartoloni, Niccolò II-130
Beducci, Elena II-96
Belhi, Abdelhak II-140
Belkebir, Hadrien I-287
Benitez, Guilherme Brittes I-171, I-191
Berwanger, Sthefan I-243
Bindi, Bianca II-130
Blampain, Félix I-253
Bleckmann, Marco II-289
Borzi, Giovanni Paolo I-36
Boton, Conrad I-263
Bouras, Abdelaziz II-140, II-210
Bricogne, Céline I-253
Bricogne, Matthieu I-253, I-298
Brovar, Yana I-215, II-244
Bruno, Giulia II-119, II-300

C

Callupe, Maira II-254
Camara, Jean René I-309
Canciglieri Junior, Osiris I-181
Canciglieri, Matheus Beltrame II-151
Cardoso, Ana Maria Kaiser I-191
Casata, Johanna II-61
Cherifi, Chantal I-151
Cheutet, Vincent I-151

C

Chevrier, Pierre I-235
Chiabert, Paolo I-14, II-85, II-300, II-323
Choi, Hye Kyung II-221
Coutinho, Cristiano I-243

D

da Rocha Loures Robell, Pedro I-191
da Silva Neto, Osmar Moreira I-70
Dadouchi, Camélia II-197
Dautaj, Marco II-254
De Carolis, Anna II-75, II-96
de Castro e Silva, Andreia II-107
De Marco, Giuseppe I-36
de Moura Leite, Athon Francisco Staben II-151
de Oliveira, Michele Marcos I-181
Deguilhem, Benjamin I-309
Demoly, Frédéric I-287
Dias, Izamara Cristina Palheta II-107, II-344
Dibari, Camilla II-130
do Prado, Márcio Leandro I-171
Domingos, Enzo II-50
dos Santos Domingos, Gabrielly II-107
Douass, Mohamed I-151
Dupuis, Ambre II-197

E

El Anbri, Ghita II-14
Eynard, Benoît I-253

F

Fani, Virginia II-130
Farshad, Sabah II-244
Favre, Justin I-151
Fortin, Clement I-215, I-225, II-244
Foufou, Sebti I-287

G

Gardoni, Mickaël I-105, I-127
Gerhard, Detlef I-59, I-80, II-187

Ghorbani, Elham   II-333
Göbel, Jens C.   II-233
Gomes, Samuel   I-287
Gonçalves, Marcelo Carneiro   II-107, II-344
Gonzalez Lorenzo, Aldo   I-25
Grafinger, Manfred   II-233

H

Hamasaki, Katuzi   II-344
Hehenberger, Peter   II-61
Helbling, Samuel   I-47
Hinterthaner, Marc   I-116
Hiraoka, Hiroyuki   II-269
Hoffmann, Juliana   II-50
Hori, Arata   II-269

I

Ianniello, Simona   II-300
Imbeau, Daniel   II-333
Inoyatkhodjaev, Jamshid   I-14

J

Jung, Philipp   II-279
Junior, Osiris Canciglieri   I-277, II-151
Juresa, Yannick   II-233

K

Kärkkäinen, Hannu   I-136
Katzmayr, David   II-61
Kazanskii, Arkadii   I-215
Keivanpour, Samira   II-14, II-38, II-333
Kneidinger, Christian   II-61
Koepler, Oliver   I-116
Kuismanen, Olli   I-136

L

Lachmayer, Roland   I-116
Lamghari, Amina   II-38
Lee, Whan   II-221
Leite, Luciana Rosa   II-50
Lessa, Lucas Sydorak   I-181
Levrat, Eric   II-3
Liu, Tianhao   II-323
Lobo, José Roberto Alcântara   I-277
Locquet, Hugo   I-298
Lombardi, Franco   II-119
Lou, Gaopeng   I-3
Luu, Duc-Nam   II-28

M

Maier, Claus-Jürgen   II-28
Mallet, Antoine   I-309
Mantelet, Fabrice   II-300
Marange, Pascale   II-3
Maranzana, Nicolas   II-28
Mari, Jean-Luc   I-25
Marongiu, Giovanni   II-119
Mas, Fernando   II-163, II-173
Massot, Simon   II-28
Menon, Karan   I-136
Menshenin, Yaroslav   I-235
Merschak, Simon   II-61
Millescamps, Maxence   I-151
Miroite, Pierre   I-127
Miyangaskary, Mina Kazemi   II-38
Morales-Palma, Domingo   II-173
Moriondo, Marco   II-130
Mossa, Giorgio   I-36
Mozgova, Iryna   I-116, II-279
Musa, Tahani Hussein Abu   II-210

N

Nara, Elpidio Oscar Benitez   II-107, II-344
Neges, Matthias   I-80
Nogueira, João Cláudio   I-91
Noh, Sang Do   II-221
Nürnberger, Florian   I-116
Nyffenegger, Felix   I-47
Nyhuis, Peter   II-289

O

Oliva, Manuel   II-173
Orukele, Oghenemarho   I-25

P

Padovan, Gloria   II-130
Pereira, Carla Roberta   II-50
Pernot, Jean-Philippe   I-25
Pinon, Sébastien   I-253
Pinquié, Romain   I-235
Piovesan, Camila Vitoria   II-107
Polette, Arnaud   I-25
Pourzarei, Hamidreza   I-263
Pranzo, Andrea   II-355
Prod'hon, Romaric   I-287
Proulx, Mélick   I-105

Q
Quadrini, Walter   I-36

R
Rehman, Mubeen Ur   I-225
Reynoso-Meza, Gilberto   I-171
Rivest, Louis   I-263, I-298
Rolf, Julian   II-187
Rossi, Monica   I-203, II-254
Rudek, Marcelo   I-70, I-91

S
Sajadieh, Seyed Mohammad Mehdi   II-221
Saliger, Alexandra   II-233
Sanna, Cristina   II-355
Sassanelli, Claudio   I-36
Schmid, Daniel   II-313
Schumann, Dorit   II-289
Seelent, João Felipe Capioto   I-171
Segonds, Frédéric   I-309, II-300
Sekhari, Aicha   I-151
Sekkay, Firdaous   II-333
Serio, Francesco   II-85
Shendi, Milad Attari   I-3
Siewert, Jan Luca   I-80
Silva, Henrique Diogo   I-243
Sim, Seung Bum   II-221
Skrzek, Murillo   II-163
Soares, António Lucas   I-243
Son, Hyun Sik   II-221

Spaltini, Marco   II-75, II-96, II-355
Sullivan, Brendan   I-203
Szejka, Anderson Luis   I-277, II-163, II-173

T
Taisch, Marco   II-75, II-96, II-355
Tapia, Betania   I-215
Terzi, Sergio   I-36, II-254
Thomson, Vince   I-161
Thomson, Vincent   I-3
Trombi, Giacomo   II-130
Tuschen, Klaas   II-279

V
Valandro, Richard   I-91
Vanson, Gautier   II-3
Véron, Philippe   I-309
Vogt, Oliver   I-59

W
Wang, Haoqi   I-3
Wolf, Mario   I-59, II-187

Y
Yahia, Esma   I-309
Yamada Macicieski, Ana Paula Louise   II-50

Z
Zagatta, Kristin   II-279